Mixed Surfactant Systems

ACS SYMPOSIUM SERIES **501**

Mixed Surfactant Systems

Paul M. Holland, EDITOR

General Research Corporation

Donn N. Rubingh, EDITOR

The Procter & Gamble Company

Developed from a symposium sponsored
by the ACS Division of Colloid and Surface Chemistry
at the 65th Colloid and Surface Science Symposium,
Norman, Oklahoma,
June 17–19, 1991

American Chemical Society, Washington, DC 1992

TP994
M 59
1992

Library of Congress Cataloging-in-Publication Data

Mixed surfactant systems: developed from a symposium / sponsored by
  the ACS Division of Colloid and Surface Chemistry at the 65th
  Colloid and Surface Science Symposium, Norman, Oklahoma, June
  17–19, 1991; Paul M. Holland, editor, Donn N. Rubingh, editor.

  p.   cm.—(ACS Symposium Series, ISSN 0097–6156; 501).

  Includes bibliographical references and index.

  ISBN 0–8412–2468–4

  1. Surface active agents—Congresses.

  I. Holland, Paul M., 1947–   . II. Rubingh, Donn N., 1945–   .
III. American Chemical Society. Division of Colloid and Surface
Chemistry. IV. Colloid and Surface Science Symposium (65th: 1991:
Norman, Okla.) V. Series.

TP994.M59   1992
668′.1—dc20   92–26167
                                     CIP

The paper used in this publication meets the minimum requirements of American National
Standard for Information Sciences—Permanence of Paper for Printed Library Materials, ANSI
Z39.48–1984.   ∞

Foreword

THE ACS SYMPOSIUM SERIES was first published in 1974 to provide a mechanism for publishing symposia quickly in book form. The purpose of this series is to publish comprehensive books developed from symposia, which are usually "snapshots in time" of the current research being done on a topic, plus some review material on the topic. For this reason, it is necessary that the papers be published as quickly as possible.

Before a symposium-based book is put under contract, the proposed table of contents is reviewed for appropriateness to the topic and for comprehensiveness of the collection. Some papers are excluded at this point, and others are added to round out the scope of the volume. In addition, a draft of each paper is peer-reviewed prior to final acceptance or rejection. This anonymous review process is supervised by the organizer(s) of the symposium, who become the editor(s) of the book. The authors then revise their papers according the the recommendations of both the reviewers and the editors, prepare camera-ready copy, and submit the final papers to the editors, who check that all necessary revisions have been made.

As a rule, only original research papers and original review papers are included in the volumes. Verbatim reproductions of previously published papers are not accepted.

M. Joan Comstock
Series Editor

Contents

ADSORPTION OF MIXED SURFACTANTS
AT INTERFACES

PHASE BOUNDARIES AND SOLUBILIZATION
IN MIXED SURFACTANT SYSTEMS

Preface

MIXED SURFACTANT SYSTEMS ARE ENCOUNTERED in nearly all practical applications of surfactants. This situation is due to the inherent difficulty of preparing chemically pure surfactants and the performance advantage or synergism that often results from deliberately mixing different surfactant types, and it has led to considerable theoretical and experimental work to understand the properties and behavior of these complex systems. Six years ago, volume 311 in this series, *Phenomena in Mixed Surfactant Systems,* was the first to address this topic in book format. Now, expanded interest and additional work in this field lead us to reexamine this topic.

The present volume is divided into six sections dealing with various aspects of mixed surfactant systems. First, the overview section introduces the topic with a review of mixed surfactant systems, a basic introduction to modeling, and comments on the terminology used to describe nonideal mixing. The second section examines different approaches to modeling mixed surfactant aggregates based on thermodynamics, geometry, and molecular interactions. This section includes two new molecular modeling approaches, as well as a pseudophase approach for modeling polydisperse mixtures. Because mixed micelle formation exerts a controlling influence on the behavior of mixed surfactant systems, the third section addresses mixed micellar solutions and associated phenomena. Included are thermodynamic studies of mixed micellization, electron spin resonance (ESR) and NMR studies of mixed micelles, partitioning from micellar solution into an oil phase, and the effect of mixed micelles on chemical reaction rates. The fourth section covers mixtures containing unusual surfactant types, such as fluorocarbon, siloxane, bolaform, and bile salt surfactants. These surfactants often exhibit marked differences in behavior compared with standard hydrocarbon surfactants. Most practical benefits of surfactants arise from their effects on interfacial properties, and the fifth section deals with the adsorption of mixed surfactants at interfaces. Topics include synergism at various interfaces, modeling of contact angles and Langmuir monolayers, and experimental studies of adsorption at solid–liquid interfaces. The final section of the book deals with phase boundaries and solubilization in mixed surfactant systems. These topics include both precipitation and cloud point phenomena, as well as the solubilization of insoluble surfactants into mixed surfactant aggregates.

This volume is the result of the collective effort and expertise of many leading specialists in the area of mixed surfactant systems. We hope that by summarizing progress in this challenging area the book will benefit those developing new surfactant technology, students wishing to learn about the effects of surfactant mixtures, and other researchers in colloid and surface science seeking to advance understanding of these complex systems.

We wish to express our appreciation to the many authors who participated in this effort and without whom this volume could not have been completed. We would also like to acknowledge the organizers of the 65th Colloid and Surface Science Symposium, and we thank John Scamehorn for both encouraging the sessions on which this book is based and leading the way with his previous volume.

PAUL M. HOLLAND
Advanced Technologies Division
General Research Corporation
Santa Barbara, CA 93111

DONN N. RUBINGH
Corporate Research Division
The Procter & Gamble Company
Cincinnati, OH 45239

June 2, 1992

OVERVIEW

Chapter 1

Mixed Surfactant Systems

An Overview

Paul M. Holland[1] and Donn N. Rubingh[2]

[1]General Research Corporation, Santa Barbara, CA 93111
[2]Corporate Research Division, The Procter & Gamble Company, Cincinnati, OH 45239

The properties and behavior of mixed surfactant systems are discussed in the context of experimental techniques and modeling. The overview begins with a general description of mixed surfactant solutions, followed by a more detailed examination of mixed micelles, surfactant mixtures at interfaces, and the phase behavior of mixed systems. Topics include experimental measurements, approaches to modeling, nonideality, unusual surfactant types, micellar demixing, adsorption at various interfaces, chemical reactions in micelles, precipitation, cloud point phenomena and perspectives on the direction of future research on mixed surfactant systems.

Mixed surfactant systems are encountered in nearly all practical applications of surfactants. These mixtures arise from several sources. First is the natural polydispersity of commercial surfactants which results from impurities in starting materials and variability in reaction products during their manufacture. These are less expensive to produce than isomerically pure surfactants and often provide better performance. Second is the deliberate formulation of mixtures of different surfactant types to exploit synergistic behavior in mixed systems or to provide qualitatively different types of performance in a single formulation (e.g. cleaning plus fabric softening). Finally, practical formulations often require the addition of surfactant additives to help control the physical properties of the product or improve its stability.

Because of the growing theoretical interest and practical importance of mixed surfactant systems, this topic has seen much activity in recent years. Previous comprehensive overviews (1,2) and a book devoted to mixed surfactant systems (3) are now several years old, and a more recent book chapter is focused primarily on mixed systems which contain cationic surfactants (4). This overview is intended to provide both an overall introduction to the many areas discussed in detail by the authors of the chapters which follow, and an up-to-date general survey of work on the topic of mixed surfactant systems.

0097–6156/92/0501–0002$08.25/0

General Description of Mixed Surfactant Solutions

Aqueous solutions of surfactants, whether mixtures or single surfactants, exhibit a variety of interesting and useful phenomena. Even at low concentrations of surfactant in water, significant effects can be observed at interfaces with the solution. The most pronounced effect is a lowering of interfacial tension due to preferential adsorption of surfactant molecules at solution interfaces. With increasing surfactant concentration, the experimentally measured decrease in surface tension becomes linear with the logarithm of the surfactant concentration and surfactant monolayers or bilayers are formed at solution interfaces. Finally, a concentration is reached where this linear decrease in surface tension suddenly stops and striking changes occur in the light scattering in solution. At this critical micelle concentration (CMC), surfactant molecules undergo cooperative self-association to form large surfactant aggregates (micelles) with the hydrophobic chains of the molecules residing in the interior of the aggregates and the hydrophilic head groups at the surface in contact with aqueous solution. In this process both ideal and nonideal mixing contributions may occur. Since the hydrophobic effect (5) which drives the overall process is not specific to surfactant "head" group, the formation of randomly mixed surfactant aggregates will tend to be favored. This can be viewed as leading to the "ideal" component of mixing in the aggregate. In the case of mixtures with different surfactant types, electrostatic interactions between "head" groups can provide the basis for the "nonideal" component of mixing in the aggregate.

A mixed micellar solution is schematically illustrated in Figure 1 where a mixture of two surfactant types is indicated by open and filled circles representing the head groups. Shown are representations of a mixed micelle, mixed monolayer at the air/solution interface, and mixed bilayer aggregate at the solid solution interface. Rapid equilibrium between monomers and micelles establishes the concentration of each of the surfactant species in solution and thereby controls their chemical potential. The chemical potentials in turn represent the thermodynamic driving force for processes such as interfacial tension lowering and contact angle changes, and at equilibrium establish a thermodynamic link between surfactant molecules at interfaces and the mixed micellar solution.

Mixtures of different surfactant types often exhibit synergism in their effects on the properties of the system (6-14). The observed synergism can be attributed to nonideal mixing effects in the aggregates, and results in substantially lower CMCs and interfacial tensions than would be expected based on the properties of the unmixed surfactants alone. This situation has lead to both theoretical and practical interest in developing a quantitative understanding of the behavior of mixed surfactant systems, and can be exploited in applications such as detergency (14-16), enhanced oil recovery (17) and mineral flotation (18). We begin with the topic of mixed micelles, since monomer-micelle equilibrium exerts a key controlling influence on the behavior of the overall mixed surfactant system.

Mixed Micellar Solutions

Mixed micellar systems involving a wide range of surfactant types have been studied. These include all combinations of typical nonionic, anionic, and cationic surfactant

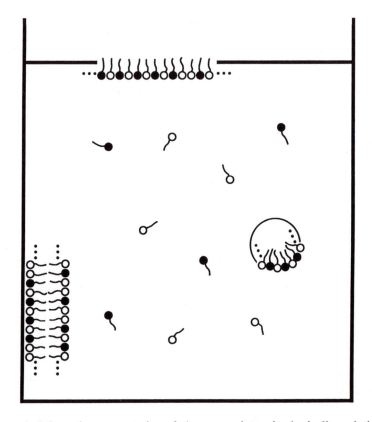

Figure 1. Schematic representation of phenomena in a mixed micellar solution of two surfactant types, illustrated by open and filled head groups. Shown are representations of surfactant monomers, a mixed micelle, mixed monolayer at the air/solution interface, and a mixed bilayer aggregate at the (hydrophilic) solid-solution interface (see text).

types, as well as binary mixtures of each of these with zwitterionic surfactant types. Some examples of these types of mixtures (9-12,19-26) are listed and discussed in Chapter 2 of this volume (27). Mixtures containing various surfactants with unusual hydrophobic groups and molecular structures have also been studied, and are covered in a separate section below.

Experimental Mixed CMC Measurements. Measurements show that the formation of mixed micelles in aqueous solution leads to striking changes in solution properties such as surface tension and light scattering. Among the experimental techniques routinely used for measuring mixed CMCs are surface tension measurements, conductivity, dye solubilization and light scattering. Other methods have also been demonstrated including calorimetry and oil-solution interfacial tension measurements.

The most widely used technique for the determination of mixed CMCs employs surface tension measurements. Here, the surface tension of a solution is measured at a number of concentrations both above and below the CMC and results plotted against the logarithm of surfactant concentration. The CMC is then taken as the "break point" or intersection of straight lines drawn through data representing the limiting surface tension above the CMC and the linear slope below the CMC (22). Measurement of surface tensions for this purpose have been performed using tensiometers with du Nouy ring (21,28-30) or Wilhelmy plate (6,22,31,32) and the sessile-drop method (33,34). A closely related technique has been demonstrated using interfacial tension at the oil-solution interface and the spinning-drop method (35,36).

Conductivity measurements (28,36-38) have also been employed in measurements of mixed CMCs but are necessarily limited to mixtures which contain ionic surfactants. Here, a break in the plot of conductivity versus concentration is taken to be the CMC. A related method which involves the use of specific ion electrodes to monitor the activity of counterions has also been used (29,39,40).

The dye solubilization method is a traditional technique that was used in much of the early work for determining mixed CMCs (41,42). It is currently not widely used for measuring mixed CMCs however, due in part to the potential problem of solubilized dye interacting with surfactant in the micelles and affecting the value of the CMC.

Light scattering measurements can also be applied for the determination of mixed CMCs (32). The most common technique is to measure the intensity of scattered light perpendicular to the incident beam at different surfactant concentrations. Since the intensity of scattered light increases significantly upon micellization, the CMC can be determined from the intersection of the linear regions above and below this discontinuity. An added benefit of using the light scattering technique is that it is possible to determine the micelle molecular weight provided other optical constants of the solution are available. This application is covered in a separate section below.

A related light scattering technique, known as quasi-elastic light scattering (QELS), measures the spectral distribution of the incident single frequency light upon scattering. The spectral half-width is proportional to the diffusion coefficient of the particle doing the scattering and hence also changes abruptly at the CMC when micelles are formed. The application of QELS to mixed micelles is discussed elsewhere in this volume (43).

Titration calorimetry has also been demonstrated for the measurement of mixed

CMCs (20,44), but has seen little use to date.   Here, an isoperibol titration is employed to monitor the heat of demicellization of a concentrated mixed micellar solution as it is diluted.   When the CMC is reached, a sharp break is observed.   Good results for mixed CMCs can be obtained as long as sufficient heats of micellization are present.

Monomer Concentration and Micelle Composition.   Fewer experimental measurements of monomer concentrations and micelle composition have been carried out in mixed micellar systems.   If either of these is known at a given surfactant concentration, the other can be calculated using the mixed CMC.   One of the primary techniques for determining monomer concentrations and micelle composition is ultrafiltration using membranes with molecular weight cutoffs which allow surfactant monomers to pass the filter but prevent micelles from passing.   This has been applied to various anionic-nonionic (21,45,46) and zwitterionic (20) hydrocarbon surfactant mixtures, as well as fluorocarbon-hydrocarbon mixtures (47).   A similar approach using gel filtration has been applied to fluorocarbon-hydrocarbon surfactant mixtures (48).

A very promising new development is the use of surfactant selective electrodes to study mixed surfactant systems.   This experimental approach has been applied in a detailed thermodynamic study of cationic-sulfobetaine mixtures reported in this volume by Hall et al. (49).

Micelle compositions have also been probed using NMR self-diffusion measurements (50), and surfactant activities in "mixed" prontonated/unprotonated amine-oxide micelles determined from titrations (51).   A new technique based on the decay of hydrated electrons has been developed to measure monomer concentrations of fluorocarbon surfactants in mixed anionic fluorocarbon-hydrocarbon surfactant systems.   This is discussed in this volume (52).   Structural details of various mixed anionic-nonionic, cationic-nonionic and anionic-cationic micelles have been probed using electron spin echo modulation and electron spin resonance techniques and this is also discussed in this volume (53).

Micelle Size and Aggregation Numbers.   Various techniques have been applied to experimentally determining mixed micelle sizes, with those based on the scattering of photons (or other particles) most widely used.   Binary mixed micelles of hydrocarbon surfactants studied by light scattering include nonionic-nonionic (54), anionic-nonionic (54-56), cationic-nonionic (54) and zwitterionic-anionic (32) mixtures.   Mixed fluorocarbon-hydrocarbon surfactant systems have been studied by light scattering (56), and small angle neutron scattering studies have been made on both mixed micelles of anionic-nonionic (57) and cationic-nonionic (58) hydrocarbon surfactants and anionic-anionic fluorocarbon-hydrocarbon surfactant mixtures (59).   A fluorescence probing technique has also been used to investigate micelle sizes in both anionic-nonionic hydrocarbon surfactant mixtures (60), and fluorocarbon-hydrocarbon anionic-nonionic (60,61), nonionic-anionic (61), nonionic-nonionic (61) and anionic-anionic (60) surfactant systems.

Mixtures with Unusual Surfactant Types.   Mixed micellar solutions containing unusual surfactant types present an interesting area for study.   These mixtures include surfactants with unusual hydrophobic groups such as fluorocarbon and siloxane chains, as well as surfactants with unusual structures such as the "two-headed" bolaform or α,ω type surfactants, and di-chain surfactants.

Mixtures of fluorocarbon and hydrocarbon surfactants have been extensively studied (47,48,50,56,59-73) and are especially interesting because of the potential immisibility of the two types of hydrophobic groups in mixed micelles and the possibility of demixing. The different types of fluorocarbon-hydrocarbon mixed surfactant systems studied have included anionic-anionic (47,48,50,59,60,62,64-69), cationic-cationic (63), anionic-cationic (66,69), cationic-anionic (66), and anionic-nonionic (56,60), and nonionic-nonionic (64,69) surfactant mixtures.

For mixed systems which have like charge (i.e. anionic-anionic, cationic-cationic or nonionic-nonionic) results often show evidence which suggests that demixing to fluorocarbon-rich and hydrocarbon-rich micelles is occuring. Evidence for this includes significant observed positive deviations from ideal mixing in the behavior of mixed CMCs determined from surface tension (62-65,67-69), ultra-filtration experiments (47,48), and the results of NMR studies of self-diffusion (50). However, the results of neutron scattering experiments on at least one anionic-anionic system indicate that mixed micelles do form (59). The problem of mixed systems of fluorocarbon and hydrocarbon surfactants and demixing is addressed in several other chapters in this volume (43,52,74,75).

In the case of other types of fluorocarbon-hydrocarbon surfactant mixtures where electrostatic interactions between surfactant head groups can occur, quite different results are observed. Here, significant negative deviations from ideal mixing in the behavior of mixed CMCs can be seen in both anionic-nonionic mixtures and those of opposite charge (i.e. various mixtures of anionic and cationic surfactants). The significant negative deviations from ideality observed for mixed CMCs in each of these cases (69) indicate that head group interactions are capable of overcoming the mutual phobicity of the hydrocarbon and fluorocarbon chains in mixed micelles.

Fluorocarbon-fluorocarbon mixed micellar systems have also been studied including anionic-anionic surfactant mixtures (68). As in the case of anionic-anionic hydrocarbon surfactant mixtures (42), these form mixed micelles and exhibit ideal mixing behavior.

Bolaform or α,ω type surfactants have two head groups, one on each end of the hydrophobic chain, and represent a particularly interesting new structural class of surfactants. Binary surfactant mixtures of these compounds such as alkane-α,ω-bis trimethylammonium bromides with alkyltrimethylammonium bromide cationic surfactants show the formation of mixed micelles to be highly dependent on chain length (76). This is especially apparent at low surfactant concentrations near the CMC of the pure surfactants where mixed micelles form in only a few cases. At higher surfactant concentrations with boloform mole fractions greater than 0.5, mixed micelle formation is promoted and negative deviations from ideality are observed.

Mixtures of the di-cationic surfactant chlorhexidine digluconate with ordinary cationic surfactants have also been studied (77), as well as mixtures of different cationic bolaform and anionic surfactants including bolaforms with trimethylammonium bromide (78) and pyridinium head groups (79,80). Mixed micellar systems with bolaform surfactants are reviewed in this volume by Zana (81).

In contrast with the two head group bolaform surfactants, di-chain surfactants such as the dialkyl cationics have two hydrophobic chains or tails. CMCs of mixtures of monoalkyl and dialkyl cationic surfactants exhibit ideal mixing behavior, although there is evidence of synergism in surface tension reduction (82).

Finally, some very recent work has looked at surfactant mixtures of hydrocarbon surfactants with siloxane surfactants, for which the hydrophobic group consists of methylated Si-O structures. This work is reported on elsewhere in this volume (83).

Micelle Demixing. Evidence of demixing or the formation of more that one stable population of micelles in mixed systems has been observed or investigated by many different workers. Since mixing in micelles should be favored by entropy, the possibility of demixing seems to require a significant physical basis such as immisibility of surfactant hydrophobic groups in the core of the micelle, or steric effects which restrict mixing based on molecular geometries of the surfactants.

The strongest evidence for the phenomena of demixing in micelles has been in mixed fluorocarbon-hydrocarbon surfactant systems of like charge where the poor miscibility of fluorocarbons and hydrocarbons in the micellar core can provide a basis for phase separation. This is supported by observations of positive deviations from ideal mixing in the behavior of mixed CMCs (62-65,67-69), ultra-filtration experiments (47,48), and the results of NMR studies of self-diffusion (50), which are consistent with two populations of micelles. Additional results and discussion indicating that two type of micelles can co-exist in anionic-anionic fluorocarbon-hydrocarbon surfactant systems based on NMR (74) and pulse radiolysis (52) measurements appear in this volume.

The evidence for demixing in fluorocarbon-hydrocarbon surfactant systems is not unanimous however, with small angle neutron scattering results showing that at least some anionic-anionic mixtures form mixed micelles (59). As pointed out by Ottewill and co-workers (59), small angle neutron scattering measures spatial correlations and thus may provide the most direct means to observe micelle compositions. They have recently supplemented their neutron scattering results on this system with NMR studies which are reported in this volume (75).

In the case of mixtures of hydrocarbon surfactants, there is also evidence that mixed micelles fail to form in some systems. Perhaps the strongest case involves mixtures of bolaform surfactants with conventional surfactants where significant steric constraints on mixing in micelles exist. Here, the two cationic surfactant hydrophilic groups are connected by a hydrocarbon chain which limits separation between head groups in the micelle and thereby constrains packing of the alkyl chains. Results based on conductivity measurements show the formation of mixed micelles with conventional surfactants to be strongly correlated with the relative lengths of alkyl chains in the two surfactant types (76,81).

Evidence for the formation of two populations of micelles in more conventional mixed hydrocarbon surfactant systems has also been reported. This includes multiple breaks in surface tension versus concentration plots of zwitterionic-nonionic surfactant mixtures which are correlated with the relative chain lengths of the two species (84). This is discussed elsewhere in this volume by Ogino and Abe (85). QELS studies in the $C_{12}AO$-$C_{12}E_8$ system, also reported in this volume (43), show bimodal distributions correlated with the pH of the system and this has been interpreted as evidence that at low pH mixed micelles do not form.

Overall, there is substantial evidence that is consistent with demixing or the formation of more than one stable population of micelles in various surfactant systems. This indicates, that at least in some systems, sufficient immisibility of surfactant hydrophobic groups or steric effects based on molecular geometries of the surfactants can lead to the coexistence of two types of micelles in solution. However a degree of caution in interpreting the available evidence is warranted since some observations consistent with micellar demixing might arise from other effects.

Chemical Reactions in Mixed Micellar Systems. Chemical reaction rates can be significantly influenced by the presence of micelles and other surfactant aggregates. Reaction rates for various substrates in single surfactant systems can either be enhanced (micellar catalysis) or inhibited depending on the type of the reaction and the composition of the micelles. While no previous work on chemical reactions in mixed surfactant micelles appears in the literature, recent results for enzyme catalysed hydrolysis in anionic-nonionic mixed micelles (86), and of demethylation reactions in cationic-nonionic mixed micelles (87) are reported in this volume. These experimental results show that the presence and composition of mixed micelles can have a substantial effect on chemical reaction rates in both types of systems. Theoretical analysis of the observed reaction rates in mixed systems is also provided. Rubingh and Bauer (86) present a detailed theoretical model for the lipase catalysed rate of hydrolysis of surfactant substrates co-micellized with non-hydrolysable surfactants. Under limiting conditions, this takes the form of the Michaelis-Menten kinetic expression. Wright, Bunton and Holland (87) show that a simple pseudophase model can account for the decrease in the demethylation rate of methyl napthalene-2-sulfonate by bromide ion when nonionic surfactant is added to cationic micelles.

Modeling Mixed Micellar Systems

Typical models for treating mixed micellization are based on an equilibrium thermodynamic approach. They differ in how the micellar aggregate is modeled, whether counterions are taken into account (and how), and in the way nonideality is treated. Generally, the approach that is selected depends on the complexity of the system to be modeled, the properties to be described and the desired level of detail.

Pseudophase Separation Approach. The widely used pseudophase separation approach assumes that the mixed micelle or other surfactant aggregate can be treated much like a separate phase (88). While it is clear that micelles do not constitute a thermodynamic phase, they do mimic phase-like behavior in the sense that they can act as both a source and a sink for surfactant species in solution. This approach represents a limiting case where the aggregation number of the micelle is formally assumed to approach infinity. Mass-action models which explicitly take the number of molecules into account show this is a good approximation for micelles of about 50 or more molecules (89,90). Using the pseudophase separation approach greatly simplifies the modeling of complex mixed surfactant systems.

Ideal Mixing Models. The earliest approach for modeling mixed micellization was developed in the 1950s for binary surfactant systems by Lange (41) and Shinoda (42). This was based on the pseudophase separation approach with the assumption of ideal mixing in the micelles. Generally, ideal mixing models have been successful in describing the CMC behavior of binary ionic (28,39,41,42) and binary nonionic (22,91,92) surfactant mixtures, especially for surfactants with the same hydrophilic group. Clint has extended the ideal binary mixing model to explicitly treat surfactant monomer concentrations and micelle compositions (91).

A version of the binary ideal model has been developed by Moroi, et al. (37,38) to treat mixed ionic systems with different counterions including Na^+, K^+, Mg^{+2}, Mn^{+2}, Co^{+2}, Ni^{+2} and Cu^{+2}. Results show good agreement with experiment, and highlight the importance of charge density at the micellar surface due the presence of counterions.

Ideal mixing models have also been generalized for treating multicomponent surfactant mixtures with an arbitrary number of components (20,93,94). Treatment of multicomponent ideal mixtures has also been extended to two phase systems which contain an oil phase in addition to the surfactant solution (93,95).

Nonideal Mixing Models.   The regular solution approximation for treating nonideal mixing has been one of the most widely used developments in modeling mixed surfactant systems.   This approach was first applied by Rubingh to a broad range of different binary mixed surfactant systems (21).   While the form of this model was formally designed for treating nonionic surfactants, the approach was also found to be successful at describing the behavior of mixtures with ionic surfactants.   Over the years, this binary nonideal mixed micelle model has been used to model a wide variety of different micellar solutions (9-12,19-20,25,96-99), including situations as unusual as mixtures of cationic surfactants in concentrated sulfuric acid (99).   The model has also been used as the basis for developing treatments of synergism in binary mixed systems (7-14).   A basic introduction to the form of this model is presented elsewhere in this volume (27).

The regular solution approach for nonideal mixed micelles has been extended to multicomponent mixtures by developing a generalized pair-wise expression for activity coefficients based on the regular solution approximation and interaction parameters determined from binary mixtures (2,19,31,100,101). This allows nonideal behavior in multicomponent nonideal systems to be described without adjustable parameters. Multicomponent nonideal models are discussed elsewhere in this volume, including extended approaches for dealing explicitly with large numbers of polydisperse surfactant components (100) and the effects of nonideal interactions on the partitioning of polydisperse surfactants into a second (organic) phase (101).

A regular solution model which provides explicit treatment of counterion effects for binary ionic mixed micelles has been developed by Kamrath and Franses (90). This has also been extended to explicitly treat counterions in binary ionic-nonionic mixtures (102).

The regular solution approximation itself formally assumes that the entropy of mixing is zero and therfore that the excess heat of mixing dominates the mixing process.   Calorimetric measurements of micellar mixing have shown that this assumption does not hold in many systems (19,44,103,104) and the regular solution approximation has been criticized on fundamental grounds (105).   None the less, the functional form of the regular solution approach has been demonstrated to provide a robust and tractable basis for describing nonideal behavior in many mixed surfactant systems.   This has lead to a suggestion that the mixing parameter that arises in the regular solution approach be interpreted more generally as an excess free energy of mixing parameter which reduces to excess enthalpy only when the excess entropy is zero (2,19). It is interesting to note that studies of partial molar volumes show nonideal anionic-nonionic mixed micelles to exhibit zero volume of mixing behavior (106), justifying use of the regular solution approximation. Hall et al. have completed a detailed thermodynamic analysis of a nonideal cationic-sulfobetaine mixture which they find to conform well to regular solution theory (49). Further discussion and views on the regular solution approximation (27,107) are presented elsewhere in this volume.

Alternatives to the much-used regular solution approach for treating nonideality in mixed micelles have been developed. These include pseudophase separation models based on the Gibbs-Duhem equation which use slopes of the CMC as a function of monomer concentrations (23) or bulk composition (108). This approach offers the advantage that it does not assume a particular form for the free energy of mixing (as in the regular solution approach), although it also does not predict CMCs but uses them as input data to calculate other properties of the system.

Various multi-parameter approaches for treating nonideality have also been developed. These include models which explicitly account for the excess entropy of mixing (45,46) and counterion binding (101,109), and a group contribution model based on a summation of molecular interactions between surfactant functional groups (110).

Among the more interesting developments is a psuedophase model developed by Rathman and Scamehorn (29,111) which is based on electrostatic interactions only. This provides a reasonable description of nonideal binary mixed CMCs in ionic-nonionic systems without a nonideal interaction parameter using only single (unmixed) surfactant CMCs and fractional counterion binding on the pure ionic micelles. These results provide strong evidence of the importance of electrostatic interactions in nonideal mixing in ionic-nonionic systems.

Mass Action Models. In mass action models of mixed micellization the chemical potential depends on the aggregation number N of the micelle, and micellization is described as an equilibrium process where a range of different micelle sizes is present (102,112,113). Because with the mass action approach a finite number of micelles are formally present at all concentrations, it becomes necessary to define the CMC as a surfactant concentration at which only some (small) specified fraction of the total surfactant exists as micelles in the system. The models can be simplified by assuming the micelles to be monodisperse. Kamrath and Franses have developed mass action models for ideal mixing in binary nonionic-nonionic systems and ionic-ionic mixtures with the same polar group and ion (112). They have also developed a mass action model for nonideal mixing in ionic-nonionic mixtures (102). Wall and Elvingson have extended the mass action approach to treat the kinetics of micellization (113). Because of the number of parameters required and the additional complexity of mass action models compared to pseudophase separation models, these have only been applied in modeling a few systems.

Molecular Models. Molecular theories for *a priori* prediction of mixed micellization have been developed using molecular thermodynamic expressions for the free energy of micelle formation based on free energy contributions from a variety of different molecular interactions and properties. These can provide a way of relating the molecular geometry, size, and chemical nature of surfactant hydrophilic and hydrophobic groups to both macroscopic properties of the mixed system and the size of the resulting micellar aggregate. A variational principle approach to this problem was outlined by Stecker and Benedict (*114*). However this depended on phenomenological parameters for which no explicit molecular model was available. Statistical thermodynamic approaches for ideal mixed surfactant aggregates with different surfactant chain lengths have also been explored (*115*).

The first explicitly molecular approach for nonideal mixed micelles was taken by Nagarajan and has been demonstrated for both ideal and nonideal binary mixed systems (*116*). This approach is based on a thermodynamic expression for the free energy of mixed micelle formation which provides bulk and interfacial contributions in terms of molecular parameters. This permits calculation of CMCs, and both the distribution and average of micelle size and composition. Significant generalization and improvements to this molecular approach with extension to other aggregates including fluorocarbon-hydrocarbon mixed micelles is presented in this volume Nagarajan (*117*).

Important new developments in the molecular modeling of mixed micelles are also presented by Puvvada and Blankschtein in this volume (*118*). Their molecular-thermodynamic theory addresses a broad spectrum of behavior in mixed micellar systems including CMCs, distributions and average micellar size and composition, micellar shape, and phase transitions including the coexistence curve between micellar rich and micellar-poor phases in nonionic surfactant systems. Their approach also allows direct calculation of regular solution interaction parameters.

Surfactant Mixtures at Interfaces

The adsorption behavior of mixtures of soluble surfactants at interfaces and their consequent effect on interfacial properties varies significantly over a wide range of relative concentrations. At surfactant concentrations far below the mixed CMC, the interface is only sparsely covered by surfactant molecules and a significant amount of "bulk-like" water remains at the interface. As the concentration increases, surface tension decreases substantially and a "saturated" interfacial monolayer of adsorbed mixed surfactant molecules forms. Here, the average area per molecule at the interface becomes constant and the Gibbs equation can be used to determine molar areas from the constant slope of surface tension versus the logarithm of surfactant concentration. With further increases in the total surfactant concentration the surface tension levels off as the CMC, solubility limit, or other "phase boundary" is surpassed.

Experimental Measurements at Fluid Interfaces. Key properties for the description of surfactant mixtures adsorbed at fluid interfaces are interfacial tensions, molar areas for surfactant species at the interface, and composition of the adsorbed surfactant monolayer. Of these properties, the most experimentally accessible is the surface tension at the air-solution interface. Surface tension is typically measured using a tensiometer with either du Nouy ring or Wilhelmy plate, although the pendant or sessile-drop method is also widely used and has the advantage that the surface is renewed for each measurement. Surface tension measurements for many mixed systems have been reported in the literature. These have included detailed studies of adsorption behavior related to mixed micellization (*119-121*).

Fewer measurements have been reported for interfacial tensions at mixed surfactant solution-hydrocarbon interfaces (*35,122-124*). In this case the interfacial tensions between surfactant solution and hydrocarbon phase can be quite low, and the spinning-drop method is typically used.

Molar areas of surfactants at the interface can be determined from the slope of measured interfacial tension with the logarithm of surfactant concentration using the Gibbs equation. Since CMCs are often obtained using surface tension measurements, this information is frequently available. It can be shown that in the case of mixed surfactant systems, the area determined by the Gibbs equation is a weighted average area per mole at the interface based on interfacial composition.

The composition of surfactant at various interfaces in mixed systems can be determined using the Hutchinson method (*125*). Here, the concentration of each surfactant component in turn is varied while holding the concentration of the other component(s) constant. This allows molar areas for each component in the mixture to be determined from the slope of measured interfacial tensions with the logarithm of the individual surfactant component concentration. Taken together, these results allow the surfactant mole fractions at the interface to be calculated for a particular concentration and composition of the mixed surfactant system. Because of the number of measurements required to apply the Hutchinson method, this has been applied to relatively few mixed surfactant systems. Of these, the most useful and detailed studies have been reported by Gu and Rosen (*126*).

Radioisotopes have also been used in adsorption studies on mixed surfactant systems to provide information about surfactant composition at interfaces. Here, radiation detectors are used to quantify the amounts of surfactant adsorbed. Jayson and Thompson (*127*) have used this technique to study the adsorption of mixtures of tritiated SDS and C_8E_4 surfactants at the air-solution interface. Their results showed that surface compositions calculated using the Ingram model (see next section below) fail to predict measured surface excess mole fractions from surface tension and bulk composition. Application of this technique can thus provide a useful test of the validity of model predictions in mixed surfactant systems.

Surface pressures and molar areas of insoluble mixed surfactant monolayers at the air-solution interface can be directly studied using the Langmuir film balance. Surface pressures measured by this technique represent the pressure exerted by the monomolecular film, and are equivalent by definition with the difference between the pure solution surface tension and that in the presence of surfactant. The pressure-area isotherm for a typical insoluble monolayer as it is compressed in the film balance exhibits a low surface pressure region, followed by a sharp (often linear) increase in surface pressure as the monolayer is compressed, and finally a break in the slope at a "collapse pressure" where the monolayer is no longer stable.

There are three types of mixing problems associated with insoluble Langmuir monolayers. The first is the mixing of different insoluble surfactants in the monolayer. This is discussed for mixtures of different bile salts elsewhere in this volume (*128*). The second is mixing by penetration of the insoluble monolayer by soluble surfactant molecules from the solution phase.   Examples of this for methylamine oxides with anionic surfactants *(129)* and cholesterol with cationic surfactants *(130)* are given in a previous symposium volume, with others in an extensive recent review by Ter-Minassian-Saraga *(131)*. The third involves mixtures of different counterions associated with the monolayer. This problem is reviewed and a multicomponent mixed counterion model presented by Ahn and Franses *(132)* in this volume.

Experimental Measurements at Solid Interfaces. At solid-solution interfaces, the behavior of mixed surfactant systems can be characterized by adsorption isotherms. The usual approach for measuring adsorption onto a solid substrate is equilibration of the surfactant solution with the solid substrate at constant temperature, separation (usually by centrifugation), followed by chemical analysis of the surfactant solution to determined the amount of each species lost by adsorption. Examples of adsorption from mixed surfactant solutions onto solids include the adsorption of binary surfactant mixtures of anionic *(133)* and anionic-nonionic surfactants *(134)* onto alumina, anionic surfactants onto barite *(18)*, and anionic-nonionic surfactants onto polystyrene latex *(135)*.   Studies have also been made of adsorption from binary mixtures of fluorocarbon and hydrocarbon surfactants onto alumina *(136,137)* and ferric hydro sols *(138)*. Several studies are reported in this volume. These include calorimetric measurements and adsorption studies of anionic-nonionic surfactant mixture onto alumina by Fu, Somasundaran and Xu *(139)*, a study of the dynamics of adsorption of mixtures of anionic surfactants onto Al_2O_3 using an ATR-FTIR technique by Couzis and Gulari *(140)*, and a study of the adsorption of succinimide dispersants onto colloidal particles by Papke *(141)*.

Contact angles represent an experimentally measurable phenomenon which involves interfacial tensions at three different interfaces.   They are often closely related to practical applications since they provide a useful indication of the wettability of surfaces and emulsification of oils.   Contact angles are typically measured by observation of the angle formed by a droplet on a solid substrate using a instrument such as a goniometer. While few experimental studies of contact angles in mixed surfactant systems have been reported in the literature *(126,142)*, these provide a valuable test of theoretical models for surfactant mixtures at interfaces.

Approaches to Modeling Surfactant Mixtures at Interfaces. There are two fundamentally different approaches for treating surfactants at solution interfaces. These are the 2-D surface solution model *(143,144)* and the 2-D gas model *(145)*. Of these, the surface solution approach has most often been used for the modeling adsorption of soluble surfactants and the gas model for insoluble surfactant monolayers. While these are equivalent in the limit of low surfactant adsorption, the surface solution approach treats the interfacial region as a "surface phase" where a force field at the interface is used to distinguish the "surface phase" from bulk solution. This has been the primary approach used for treating mixed surfactant systems, and a number of variations on it have been developed.

Standard Surface Solution Approach. In the standard surface solution approach, the presence of solvent at the interface is explicitly taken into account *(143)*. This has been successfully applied to modeling binary mixed anionic-cationic mixtures below the CMC by Lucassen-Reynders (*143,146-148*) and Rodakiewicz-Nowak (*149-153*). For these systems the problem is simplified by the requirement of electroneutrality, the large difference in surface activity between the soluble surfactant and relatively insoluble ion pairs, and the formation of "saturated" monolayers at the interface. The extension of this approach to multicomponent (ternary) anionic-cationic surfactant systems has also been demonstrated by Rodakiewicz-Nowak (*152,153*).

For mixed surfactant solutions at or above the CMC, and for mixtures other than anionic-cationic surfactants, the standard surface solution approach becomes much less tractable because of the number of parameters involved, the requirement to formally account for water at the interface, and the effects of counterions. This has lead to the adoption of alternative approaches based on the pseudophase separation approach used successfully for treating nonideal mixed micellar systems.

Pseudophase Models. Ingram (*25,154*) was the first to consider treating nonideal surfactant mixtures adsorbed at interfaces using a regular solution approximation similar to that used for mixed micellization for nonideality. This approach employed submicellar surface tension functions and the nonideal mixed micelle model of Rubingh (*21*), and utilized a nonideal surface interaction parameter. Rosen and co-workers (*7-12,25,155*) have adopted this basic approach which uses a surface interaction parameter β^σ, surface tensions ratios in the surfactant mixture below the CMC, and does not explicitly include a force field term at the interface. In this approach changes in the molar areas of surfactant at the interface due to mixing are not taken into account. Also, because the calculations themselves depend on surface tensions or concentration ratios at given surface tensions below the CMC, there is no straight forward way to extend this approach for treating multiple components. They have subsequently modified and extended the method for treating synergism in surface tension behavior at other interfaces (*122,142,156*).

A related pseudophase model designed to require less experimental data to solve has been developed by Nguyen and Scamehorn (*157*). This is based on the Gibbs-Duhem equation and surface tension data, and does not require a nonideal interaction parameter as in the case of models using the form of the regular solution approximation.

An alternative approach which explicitly incorporates the force field term and extends the approach used for mixed micelle models to interfaces has been developed by Holland (*4,158,159*). Here, nonideal mixing is treated by using a form of the regular solution approximation with a surface interaction parameter β^s, but it should be noted that for a given system the magnitude of this parameter is *not* equivalent to that of β^σ used by Rosen and co-workers (*4*). As in the case for mixed micelles this is designed to be suitable for extension to multicomponent mixtures. Also, the derivation of this approach leads directly to considering the possibility of changes in the molar area of surfactants at the interface on mixing.

Changes in Molar Areas on Mixing. It is well known that the interfacial areas per surfactant molecule often change in going from pure to mixed systems. This is especially pronounced in the case of anionic-cationic surfactant mixtures. The importance of explicitly considering such changes in modeling mixed surfactant systems has been increasingly recognized (*4,124,126,158,159*). For example, the Gibbs-Duhem relation can be used to demonstrate the importance of taking changes in the areas per molecule at the interface on mixing into account if thermodynamic consistency is to be maintained (4).

These considerations have resulted in the explicit incorporation of changes in molar areas on mixing into interface models. In the approach developed earlier by Holland, the possibility of molar area changes arises directly in the derivation (158,159). When such changes are incorporated for anionic-cationic systems by using a simple overall ratio representing the average change in molar areas in going from the pure to mixed surfactant systems, good agreement with experimental results is seen together with a large reduction of the magnitude the surface interaction parameter β^s to near that of the micellar parameter β. This is discussed in detail elsewhere (4), and additional results and discussion are presented in this volume (160).

Rosen and co-workers have recently developed an alternative to their earlier model to account for molar area changes at the interface (126,142). This incorporates a force field term as in the 2-D surface solution approach, and assumes that molar areas after mixing depend on mole fractions in the monolayer and that the ratios of partial molar areas for surfactants in the mixed and pure system are constant. Their results with this method show good agreement with experimental surface tension results, and significantly improved agreement with surface mole fractions obtained using the Hutchinson method (126). The surface interaction parameters β^σ that result when molar areas are taken into account differ significantly from values from their earlier model. This modified approach and its application to understanding synergism in binary mixtures of surfactants at various interfaces is discussed by Rosen elsewhere in this volume (161).

Unsettled Issues. Unlike the case for mixed micelle models, the current state of the various modeling approaches for treating nonideality in mixed surfactant systems at interfaces seems to be unsettled. While a variety of different approaches (7-13,25,35,122,123,126,142,156-163) are all capable of providing useful descriptions of the behavior of binary mixtures at interfaces using a single nonideal interaction parameter, the magnitudes of the resulting parameters can vary significantly depending on the specific model and assumptions used. This makes it difficult to assign physical significance to the parameters and suggests that existing tabulations of them are of uncertain value. Of the key modeling issues, it now seems certain that allowing for changes in molar areas on mixing at interfaces will be necessary. The issue of how best to do this in a proper and unambiguous way is not yet resolved. Other modeling issues to be addressed include thermodynamic consistency, tractability, physical interpretation of nonideality parameters, and ways to generalise approaches for the treatment of multicomponent systems. In the case of adsorption of nonideal mixtures on to solid (hydrophilic) interfaces, the regular solution form for interaction parameters has been shown to fail in describing admicelles (133). This situation suggests that there are significant remaining issues to be resolved by future research on modeling nonideality at interfaces in mixed surfactant systems.

Phase Behavior of Mixed Systems

The phase behavior of surfactant systems is significantly more complex than that exhibited by non-amphiphilic compounds. This complexity is illustrated in Figure 2

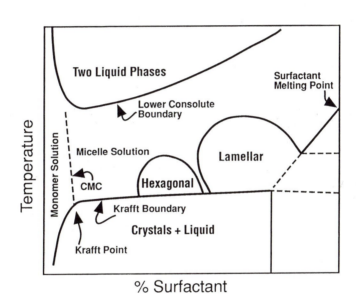

Figure 2. Typical surfactant-water phase diagram showing commonly encountered phase regions as a function of temperature and composition (see text).

with a hypothetical surfactant-water phase diagram, which shows commonly encountered phase regions. In this section, the phase behavior of mixed surfactant systems will be considered based on the properties of the individual surfactants composing the mixture.

The two phase region of surfactant crystals and water shown in Figure 2 indicates that below a characteristic temperature, known as the Krafft temperature, surfactant precipitates from solution. The boundary between this two phase region and the homogeneous solutions at higher temperature (or lower concentrations) is known as the Krafft boundary. The intersection of the Krafft boundary with the dashed line representing the CMCs at different temperatures is known as the Krafft point. If the phase rule is obeyed for micelles (i.e. if micellization can be treated as the formation of a separate phase), the Krafft boundary above the Krafft point should occur at constant temperature. The relative flatness of this line thus provides a good indication of the suitability of the phase separation approximation for a particular (single) surfactant. Since the Krafft boundary is relatively flat for most surfactants, the Krafft temperature of a surfactant is sometimes spoken of as if it is a unique characteristic of the surfactant (like a melting point) although in reality it is a (slowly changing) function of surfactant concentration.

Many surfactants exhibit two liquid phase regions. The alkyl ethoxylate nonionics and semi-polar surfactants, such as sulfoxides and phosphine oxides, show phase separation upon heating as is shown in Figure 2. The boundary between the two phase region and the micellar solution is known the lower consolute boundary in this case. In addition, certain zwitterionic surfactants show two liquid phases upon cooling and thus exhibit an upper consolute boundary. Since the surfactant solutions appear to become cloudy upon crossing the phase boundary, the temperature of phase separation is sometimes termed the cloud point temperature. In mixed surfactant systems it is important to distinguish between the cloud point curve and the coexistence curve. In single surfactant-water phase diagrams the compositions of the phases in equilibrium at a particular temperature are given by the point where the horizontal line at that temperature intersects the phase boundary. Generally, one of these phases is a very concentrated micellar solution while the other is nearly devoid of micelles.

Finally, we note the liquid crystalline phases commonly found in the middle composition range of the phase diagram. The two liquid crystal phases found in practically all surfactant-water systems are designated the Lamellar, which has a repeated bilayer morphology (or smectic type structure in thermotropic liquid crystal notation), and the Hexagonal. The Hexagonal phase is composed of unidirectionally oriented semi-infinite cylindrical micelles.

Solid-Liquid Phase Equilibria in Mixed Surfactant Systems. The precipitation of a solid phase from a micellar solution as the temperature is reduced below the Krafft temperature has been decribed in two ways. One paradigm has been that the Krafft temperature is the melting point of the hydrated surfactant phase (88), while the second identifies it as the temperature at which the solubility product is larger than the CMC so that micelles form rather than a solid precipitate (164). This latter definition has resulted in much progress recently in analysing the phase behavior in mixed surfactant systems by combining the calculation of monomer concentrations from mixed surfactant theory with the solubility product principle to define precipitation boundaries in surfactant mixtures. By analysing these systems where micelles and precipitating phase can coexist some general conclusions about the allowed micellar compositions can be drawn (*4*).

Isothermal Systems. Although a number of studies of the micellar properties of precipitating surfactant mixtures have been made (*165-167*), it was the work of Stellner et al. (*168*) which first clearly demonstrated that the solubility product principle was capable of accurately describing the solubility boundaries in the sodium dodecyl sulfate/dodecyl pyridinium chloride system. This work has been recently extended to include the effects of chain length, temperature and pH (*169*). Again, good agreement between experimental data and the theoretical prediction was found using this approach.

The precipitation of anionic surfactants with various metal ions has also been widely studied (*170-171*). The solubility product principle together with the appropriate ion concentration from mixed micelle theories can be employed to predict the precipitation boundaries in such systems as well. Stellner and Scamehorn (*172*) use such an approach to understand the increase in salinity tolerance of an anionic surfactant by the addition of nonionic. Because of its practical importance, the precipitation boundaries of Ca^{+2} ion with anionic surfactants have been frequently studied (*173-174*). The effect of alkyl ethoxylate nonionic surfactants in preventing the precipitation of calcium dodeyl sulfate has been modeled successfully using the regular solution approach to calculate monomer concentrations (*175*). This area continues to receive attention in the present volume where the effect of Ca^{+2} in anionic surfactant mixtures is modeled by Scamehorn (*176*).

Precipitation Temperatures in Surfactant Mixtures. When a surfactant with a high Krafft temperature is mixed with a surfactant with a lower Krafft temperature, the precipitation temperature of the mixture is generally lowered. Depending on the nature of the interaction with the solid phase different phase diagrams are possible. For example, eutectic behavior is exhibited by sodium dodecyl sulfate with other bivalent metal dodecyl sulfates (*177,178*). The phase implication of this eutectic behavior, namely that the precipitating phase is a single pure surfactant, has been verified in a number of these systems (*179*). Other binary surfactant mixtures form solid solutions as one proceeds from pure A to pure B. This type of behavior has been found for mixtures of octyl bezene sulfonate and dodecyl benzene sulfonate (*179*). Both eutectic behavior and solid solutions are observed in mixtures of alkyl sulfates depending on the chain length (*180*).

Alkyl ethoxylate nonionic surfactants often do not exhibit Krafft boundaries at normal atmospheric pressures. It has been shown by Nishikido (*181*) that the Krafft temperatures of these materials increase with increasing pressure. Thus the Krafft temperature of $C_{12}E_4$ which is below the freezing point of water is elevated to 31°C at pressures of 200 Mpa. Mixtures of $C_{12}E_4$ and $C_{12}E_6$ were found to form mixed solutions in the solid phase.

Finally, if very strong interactions occur between the two surfactants, such as those found in anionic-cationic and some zwitterionic-ionic mixtures, a double eutectic may be exhibited separated by a maximum in the Krafft temperature (*182,183*). Such behavior in non-amphiphilic systems is indicative of compound formation, suggesting that an analogous situation occurs in the above mixed surfactant systems.

A quantitative understanding of Krafft temperature lowering has been developed from both the perspective of the melting of a hydrated surfactant phase (88) or as the precipitation of a solid once the solubility product has been exceeded (4). With either approach the same equation for Krafft temperature lowering is found. The magnitude of this decrease is predicted to be proportional to the logarithm of the activity of the precipitating surfactant in the mixed micelle. Limited experimental studies appear to support this predicted dependence (4,177,184). It should be noted that the solubility product approach assumes the precipitation of a pure surfactant compound, and thus would not be expected to be applicable to the mixed solid solution case.

Liquid-Liquid Phase Separation in Mixed Surfactant Systems. Although the change in phase separation upon addition of a second surfactant to one exhibiting a cloud point has been studied for some time (4,185,186), it wasn't until recently that a quantitative theories for liquid-liquid separation in surfactant mixtures have been proposed (118,187). Here, modeling efforts for binary mixtures built upon earlier work to understand phase separtion in single component nonionic systems such as alkyl-ethoxylates, alkyl glucosides and an alkyl gylcerol ether (188-189). The resulting molecular-thermodynamic theory of Puvvada and Blankschtein (118) appears to provide an experimentally accessible approach to understanding both the general properties of surfactant mixtures and liquid-liquid phase separation.

   Nonionic-Nonionic Surfactant Mixtures. The liquid-liquid phase separation of the mixed systems $C_{12}E_6$-$C_{10}E_6$, $C_{12}E_6$-$C_{12}E_8$, $C_{12}E_6$-$C_{10}E_4$ have been studied in some detail (118,189). For these nonionic mixtures the cloud point is not changed dramatically upon addition of one nonionic to another. At temperatures near the critical temperatures, the two surfactant species partition equally between the two coexisting micellar phases implying that the cloud point curves and coexistence curves are identical in this range. Agreement between theoretically predicted cloud point temperatures and the experimentally measured values is good over the composition and concentration ranges studied. As the temperature above the critical temperature is increased significantly, the concentrated micellar phase becomes enriched in the low CMC component.

   Molecular thermodynamic theory also predicts significant micellar growth for both pure alkyl-ethoxylate nonionics and the mixtures at a characteristic temperature (T*) below the actual phase separation temperature (118,189). This transition is thought to occur at the onset of a sphere to cylinder shape in the micelle. The value of T* for mixed nonionics measured by QELS varies regularly with the mole fraction of the component surfactants and is successfully predicted by the model.

   Ionic-Nonionic Surfactant Mixtures. Ionic surfactants generally have a more pronounced effect on the cloud point temperature than do nonionic surfactants. Some of the ionic-nonionic surfactant systems which have been studied are; SDS-$C_{12}E_6$ (190), SDS-$C_{12}E_5$, DTAC-$C_{12}E_5$, SDS-$C_{12}E_5$, (191), SDS and SDeS with Triton X-100 (192), and various anionic-cationic mixtures with Triton X-100 (193). The addition of SDS to $C_{12}E_6$ resulted in a closed miscibility loop which shrinks to a smaller area upon increasing the mole ratio of SDS. At the relatively low mole ratio of 0.013, the miscibility loop disappears altogether (190). It has been suggested that this behavior is a result of increasing electrostatic interactions between mixed micelles due to the incorporation of charged surfactant. The importance of counterion condensation in modulating these electrostatic interactions in an anionic-nonionic system with added Cu^{+2} has also been investigated (*194*).

Modeling of the SDS-$C_{12}E_8$ system was able to show that the addition of a repulsive electrostatic term to a hard sphere potential can reproduce this shrinking miscibility loop behavior (*187*). It is interesting to note that molecular-thermodynamic theory also predicts a shrinking closed coexistence loop upon addition of an ionic surfactant to nonionic surfactants which show liquid-liquid separation behavior (*118*). It is suggested that in contrast to nonionic mixtures, ionic surfactants preferentially partition into the dilute phase to minimize this electrostatic repulsion.

Mixtures of cationic-nonionic surfactants have been studied by Holland and Rubingh (*4*). This showed that the relative effectiveness of different cationic surfactants in raising the cloud point temperature of $C_{12}E_5$ could be related to the mole fraction of cationic surfactant in the mixed micelle. This appears consistent with the idea that the extent of intermicellar electrostatic repulsive interaction controls the liquid-liquid separtion temperatures in these systems.

Anionic-Cationic Surfactant Mixtures.   The study of liquid-liquid phase separation in long chain surfactants of opposite charge is relatively recent. These systems appear to behave as a nonionic surfactact exhibiting a lower consulate boundary and hence have been called "pseudo nonionic" (*195*). Systems in which one or both of the ionic surfactants is also ethoxylated in the hydrophilic group frequently exhibit this type of behavior. The contribution of Mehreteab in this volume (*196*) discusses the effects of structure on the liquid-liquid solubility of these ethylated "ion-pair" type surfactants.

The presence of ethoxylation in the anionic-cationic pair is apparently not essential to realize liquid-liquid phase separation since mixtures of sodium N-lauroyl-N-methyl-β-alanine with C_{16}TAC and C_{18}TAC also exhibit such behavior (*182*). These systems also exhibit phase separation on heating. Although to date no quantitative comparison between theory and experiment has been made, some predictions of the expected behavior of such systems have been suggested.

Liquid-Liquid Crystal Phase Boundaries.   The most heavily studied mixed systems in the liquid crystal regions have been hydrocarbon-fluorocarbon surfactant mixtures (*197,198*). Some of the interest in these systems may stem from the fact that a number of such systems show evidence of two populations of micelles in dilute solution (see for example Fung et al., this volume). However, evidence for demixing has not been seen at the molecular level in x-ray studies in the lithium dodecyl sulfate-lithium perfluoro octane sulfonate in the hexagonal phase (*198*). Similar conclusions were drawn by Ravey et al. on mixed fluorocarbon-hydrocarbon nonionic surfactant systems even though the dilute solution behavior has suggested two population of micelles (*197*).

Mixtures of hydrocarbon surfactants in which liquid-liquid crystal phase transitions have been studied include CTAB with potassium laurate (KL) (*199*). In this case the effect of the CTAB-KL ratio on the thermal stability of the somewhat uncommon nematic phase was studied, and related to electrostatic interactions and surface area per head group.

Solubilization in Mixed Surfactant Systems. The incorporation of a water insoluble compound into a micelle is known as solubilization. From the standpoint of surfactant mixtures a question of interest is whether the surfactant mixture will solubilize more or less than the pure components based on a simple linear mole fraction weighting. Modeling by Treiner et al. (*200-202*) employed a regular solution approach to model the mixed micelle to answer this question. It was predicted that systems in which β was negative (many hydrocarbon-hydrocarbon surfactant mixtures) would show poorer solubilization efficiencies than the unmixed systems, while those exhibiting positive deviation (fluorocarbon-hydrocarbon) would show better solubilization (*200*). For the majority of surfactant mixtures and different solubilizates considered this behavior was observed (*202-204*), although a number of exceptions were also noted (*202*). Others have described the solubilization of nonpolar materials such as alkanes in terms of a Laplace pressure effect (*205,206*). A recent paper by Nishikido (*207*) presents thermodynamic equations on synergistic solubilization and provides a good starting point to understanding the literature on the theoretical approaches in this emerging area.

From the experimental standpoint the literature on solubilization is vast. Frequently large differences in solubilization behavior are observed depending on whether the solubilizate resides in the pallisade layer or the interior of the micelle. A important study of compounds falling in the latter category (micelle interior) is that of Smith et al. (*208*) which provided thermodynamic data employing a vapor pressure technique . A number of very interesting materials are expected to reside in the pallisade layer upon solubilization. These include phospholipids (*209*), di-chain amphiphiles (fabric softeners being important practical examples of this class), and alcohols (*210*). In the present volume, Dennis provides an overview of mixed micellization and solubilization between nonionic surfactants and phospholipids (*211*), while the paper by Marangoni et al. discusses the interaction of alcohols with anionic and cationic micelles (*212*).

Conclusion and Future perspectives

Future interest and progress in understanding of phenomena in mixed surfactant systems seems assured. This projection is based on the importance and increasingly sophisticated use of mixed surfactant systems in practical applications, and growing interest in the experimental and theoretical problems posed by these complex systems. We believe there will be several major themes in future developments toward understanding mixed surfactant systems.

One major area of activity will be the application of new experimental techniques to the study of mixed surfactant systems. In many cases this will involve adapting techniques currently applied to pure surfactant systems to mixed surfactant systems. Examples here are likely to include electron microscopy using new sample preparation techniques, kinetic studies of chemical reaction rates in mixed micelles and other aggregates, and the increased use of neutron scattering. The current gap of information on insoluble surfactant mixtures may also be filled by new temperature controlled optical microscopy techniques (*213*).

Another area will be expanding the generality and utility of thermodynamic models based on the psuedophase separation approach. Here, tractability for simultaneously treating larger numbers and more types of nonideal surfactant components in complex "real world" systems in the mixture will be important, while expanding the range of different phenomena treated. Some of the areas likely to be addressed include surfactant polydispersity, mixtures with unusual surfactant types, adsorption at various solution interfaces including both hydrophilic and hydrophobic solids and liquids, contact angles, and solubilization phenomena. Phase separation phenomena to form precipitates and liquid crystalline or other concentrated surfactant phases will also be important. Extension to dynamic processes such as chemical reactions, dynamic interfacial tension and contact angle changes is also a likely area for future work.

Finally, an important new area which should lead to more detailed understanding of mixed surfactant systems will be the development of sophisticated molecular models. This is likely to include molecular models for unusual surfactant types, micelle geometry (e.g. rod-like micelles), vesicles, solubilization phenomena and phase behavior. Some of the impetus behind these developments is recognition of the rapid improvements in computer technology which are beginning to put mainframe computer power onto desktops. This will potentially allow complex and computationally intensive models to be run routinely by a variety of different users, especially once these models have been coupled with databases containing background information and parameters for a wide range of different surfactant types.

Literature Cited

1. Scamehorn, J. F. In *Phenomena in Mixed Surfactant Systems;* Scamehorn, J. F. Ed.; ACS Symposium Series 311, American Chemical Society: Washington, DC, 1986; pp 1-27.
2. Holland, P. M., *Adv. Colloid Interface Sci.* **1986,** *26,* 111.
3. *Phenomena in Mixed Surfactant Systems;* Scamehorn, J. F. Ed.; ACS Symposium Series 311, American Chemical Society: Washington, DC, 1986.
4. Holland, P. M.; Rubingh, D. N. In *Cationic Surfactants: Physical Chemistry;* Rubingh, D. N.; Holland, P. M., Eds.; Surfactant Science Series 37; Marcel Dekker, Inc.: New York, NY, 1990; pp 141-187.
5. Tanford, C. *The Hydrophobic Effect;* John Wiley and Sons: New York, 1973.
6. Lucassen-Reynders, E. H.; Lucassen, J.; Giles, D. *J. Colloid Interface Sci.* **1981,** *82,* 150.
7. Hua, X. Y.; Rosen, M. J. *J. Colloid Interface Sci.* **1982,** *90,* 212.
8. Rosen, M. J.; Hua, X. Y. *J. Am. Oil Chem. Soc.* **1982,** *59,* 582.
9. Rosen, M. J.; Zhu, B. Y. *J. Colloid Interface Sci.* **1984,** *99,* 427.
10. Zhu, B. Y.; Rosen, M. J. *J. Colloid Interface Sci.* **1984,** *99,* 435.
11. Rosen, M. J. *Surfactants and Interfacial Phenomena, 2nd Ed.;* John Wiley & Sons: New York, 1989; Chapter 11, pp. 393-419.
12. Rosen, M. J. In *Phenomena in Mixed Surfactant Systems;* Scamehorn, J. F, Ed.; ACS Symposium Series 311, American Chemical Society: Washington, DC, 1986; pp 144-162.
13. Rosen, M. J. *Langmuir* **1991,** *7,* 885.
14. Jost, F.; Leiter, H.; Schwuger, M. J. *Colloid Polym. Sci.* **1988,** *266,* 554.
15. Rubingh, D. N.; Jones, T. *Ind. Eng. Chem. Prod. Res. Dev.* **1982,** *21,* 176.

16. Kurzendorfer, C. P.; Schwuger, M. J.; Lange, H. *Ber. Bunsenges. Phys. Chem.* **1978,** *82,* 962.
17. *Improved Oil Recovery by Surfactant and Polymer Flooding;* Shah, D. O.; Schechter, R. S., Eds.; Academic Press: New York, NY, 1977.
18. Dobiás, B. In *Phenomena in Mixed Surfactant Systems;* Scamehorn, J. F, Ed.; ACS Symposium Series 311, American Chemical Society: Washington, DC, 1986; pp 216-224.
19. Holland, P. M.; Rubingh, D. N. *J. Phys. Chem.* **1983,** *87,* 1984.
20. Holland, P. M. In *Structure/Performance Relationships in Surfactants;* Rosen, M. J., Ed.; ACS Symposium Series 253, American Chemical Society: Washington, DC, 1984; pp 141-151.
21. Rubingh, D. N. In *Solution Chemistry of Surfactants;* Mittal, K. L. ed.; Plenum Press, New York, NY, 1979, Vol. 3; pp 337-354.
22. Lange, H.; Beck, K. H. *Kolloid-Z.u.Z. Polymere* **1973,** *251,* 424.
23. Nguyen, C. M.; Rathman, J. F.; Scamehorn, J. F. *J. Colloid Interface Sci.* **1986,** *112,* 438.
24. Ingram, B. T.; Luckhurst, A. H. W. In *Surface Active Agents;* Soc. Chem. Ind.: London, 1979, pp 89-98.
25. Rosen, M. J.; Hua, X. Y. J. *J. Colloid Interface Sci.* **1982,** *86,* 164.
26. Corkhill, J. M.; Goodman, J. F.; Ogden, C. P.; Tate, J. R. *Proc. R. Soc. London Ser. A,* **1963,** *273,* 84.
27. Holland, P. M. In *Mixed Surfactant Systems;* Holland, P. M.; Rubingh, D. N., Eds.; ACS Symposium Series, American Chemical Society: Washington, DC, 1992; chapter 2.
28. Barry, B. W.; Morrison, J. C.; Russell, G. F. J. *J. Colloid Interface Sci.* **1970,** *33,* 554.
29. Rathman, J. F.; Scamehorn, J. F. *J. Phys. Chem.* **1984,** *88,* 5807.
30. Rathman, J. F.; Scamehorn, J. F. *Langmuir* **1986,** *2,* 354.
31. Graciaa, A.; Ben Ghoulam, M.; Marion, G.; Lachaise, J. *J. Phys. Chem.* **1989,** *93,* 4167.
32. Ishikawa, T.; Ogawa, M.; Esumi, K.; Meguro, K. *Langmuir* **1991,** *7,* 30.
33. Aratono, M.; Kanda, T.; Motomura, K. *Langmuir* **1990,** *6,* 843.
34. Puig, J. E.; Franses, E. I.; Miller, W. G. *J. Colloid Interface Sci.* **1982,** *89,* 441.
35. Rosen, M. J.; Murphy, D. S. *Langmuir* **1991,** *7,* 2630.
36. Mysels, K. J.; Otter, R. J. *J. Colloid Sci.* **1961,** *16,* 462.
37. Moroi, Y.; Motomura, K.; Matuura, R. *J. Colloid Interface Sci.* **1974,** *46,* 111.
38. Moroi, Y.; Nishikido, N.; Matuura, R. *J. Colloid Interface Sci.* **1975,** *50,* 344.
39. Shedlovsky, L.; Jacob, C. W.; Epstein, M. *J. Phys. Chem.* **1963,** *67,* 2075.
40. Treiner, C.; Fromon, M.; Mannebach, M. H. *Langmuir* **1989,** *5,* 283.
41. Lange, H. *Kolloid Z.* **1953,** *131,* 96.
42. Shinoda, K. *J. Phys. Chem.* **1954,** *58,* 541.
43. Xia, J; Dubin, P. L.; Zhang, H. In *Mixed Surfactant Systems;* Holland, P. M.; Rubingh, D. N., Eds.; ACS Symposium Series, American Chemical Society: Washington, DC, 1992; chapter 14.
44. Förster, T; von Rybinski, W.; Schwuger. M. J. *Tenside Surf. Det.* **1990,** *27,* 254.
45. Osborne-Lee, I. W.; Schechter, R. S.; Wade, W. H.; Barakat, Y. J. *J. Colloid Interface Sci.* **1985,** *108,* 60.

46. Osborne-Lee, I. W.; Schechter, R. S. In *Phenomena in Mixed Surfactant Systems;* Scamehorn, J. F, Ed.; ACS Symposium Series 311, American Chemical Society: Washington, DC, 1986; pp 30-43.
47. Asakawa, T.; Johten, K.; Miyagishi, S.; Nishida, M. *Langmuir* **1988,** *4,* 136.
48. Asakawa, T.; Miyagishi, S.; Nishida, M. *Langmuir* **1987,** *3,* 821.
49. Hall, D. G.; Meares, P.; Davidson, C.; Wyn-Jones, E.; Taylor, J. In *Mixed Surfactant Systems;* Holland, P. M.; Rubingh, D. N., Eds.; ACS Symposium Series, American Chemical Society: Washington, DC, 1992; Chapter 7.
50. Carlfors, J.; Stibs, P. *J. Phys. Chem.* **1984,** *88,* 4410.
51. Rathman, J. F.; Christian, S. D. *Langmuir* **1990,** *6,* 391.
52. Aoudia, M.; Hubig, S. M.; Wade, W. H.; Schechter, R. S. In *Mixed Surfactant Systems;* Holland, P. M.; Rubingh, D. N., Eds.; ACS Symposium Series, American Chemical Society: Washington, DC, 1992; Chapter 16.
53. Baglioni, P.; Dei, L.; Kevan, L.; Rivara-Minten, E. In *Mixed Surfactant Systems;* Holland, P. M.; Rubingh, D. N., Eds.; ACS Symposium Series, American Chemical Society: Washington, DC, 1992; Chapter 10.
54. Nishikido, N. *J. Colloid Interface Sci.* **1987,** *120,* 495.
55. Dubin, P. L.; Principi, J. M.; Smith, B. A.; Fallon, M. A. *J. Colloid Interface Sci.* **1989,** *127,* 558.
56. Tamori, K.; Esumi, K.; Meguro, K.; Hoffman, H. *J. Colloid Interface Sci.* **1991,** *147,* 33.
57. Bucci, S.; Fagotti, C.; Degiorgio, V; Piazza, R. *Langmuir* **1991,** *7,* 824.
58. Cummins, P. G.; Penfold, J.; Staples, E. *Langmuir* **1992,** *8,* 31.
59. Burkitt, S. J.; Ottewill, R. H.; Hagter, J. B.; Ingram, B. T. *Colloid Polym. Sci.* **1987,** *265,* 628.
60. Muto, Y; Esumi, K.; Meguro, K.; Zana, R. *J. Colloid Interface Sci.* **1987,** *120,* 162.
61. Meguro, K.; Muto, Y.; Sakurai, F.; Esumi, K. In *Phenomena in Mixed Surfactant Systems;* Scamehorn, J. F, Ed.; ACS Symposium Series 311, American Chemical Society: Washington, DC, 1986; pp 61-67.
62. Muckerjee, P.; Yang, A. Y. S. *J. Phys. Chem.* **1976,** *80,* 1388.
63. Smith, I. H.; Ottewill, R. H. In *Surface Active Agents;* Soc. Chem. Ind.: London, 1979, pp 77-87.
64. Funasaki, N.; Hada, S. *Chem. Lett.* **1979,** 717.
65. Shinoda, K.; Nomura, T. *J. Phys. Chem.* **1980,** *84,* 365.
66. Guo-Xi, Z.; Bu-Yao, Z. *Colloid Polym. Sci.* **1983,** *261,* 89.
67. Funasaki, N.; Hada, S. *J. Phys. Chem.* **1983,** *87,* 342.
68. Yoda, K.; Tamori, K.; Esumi, K.; Meguro, K. *J. Colloid Interface Sci.* **1985,** *104,* 279.
69. Zhao, G-X.; Zhu, B-Y. In *Phenomena in Mixed Surfactant Systems;* Scamehorn, J. F, Ed.; ACS Symposium Series 311, American Chemical Society: Washington, DC, 1986; pp 184-198.
70. Asakawa, T.; Miyagishi, S.; Nishida, M. *J. Colloid Interface Sci.* **1985,** *104,* 279.
71. Yoda, K.; Tamori, K.; Esumi, K.; Meguro, K. *J. Colloid Interface Sci.* **1989,** *131,* 282.
72. Yoda, K.; Tamori, K.; Esumi, K.; Meguro, K. *Colloids Surf.* **1991,** *58,* 87.

73. Mukerjee, P.; Handa, T. *J. Phys. Chem.* **1981,** *85,* 2298.
74. Fung, B. M.; Guo, W.; Christian, S. D.; Guzman, E. K. In *Mixed Surfactant Systems;* Holland, P. M.; Rubingh, D. N., Eds.; ACS Symposium Series, American Chemical Society: Washington, DC, 1992; Chapter 15.
75. Clapperton, R. M.; Ingram, B. T.; Ottewill, R. H.; Rennie; A. R. In *Mixed Surfactant Systems;* Holland, P. M.; Rubingh, D. N., Eds.; ACS Symposium Series, American Chemical Society: Washington, DC, 1992; Chapter 17.
76. Zana, R.; Muto, Y.; Esumi, K.; Meguro, K. *J. Colloid Interface Sci.* **1988,** *123,* 502.
77. Attwood, D.; Patel, H. K.; *J. Colloid Interface Sci.* **1989,** *129,* 222.
78. Ishikawa, M.; Matsumura, K.; Esumi, K.; Meguro, K. *J. Colloid Interface Sci.* **1991,** *141,* 10.
79. Moroi, Y.; Matuura, R.; Kuwamura, T.; Inokuma, S. *J. Colloid Interface Sci.* **1986,** *113,* 225.
80. Moroi, Y.; Matuura, R.; Tanaka, M.; Murata, Y.; Aikawa, Y.; Furutani, E.; Kuwamura, T.; Takahashi, H.; Inokuma, S. *J. Phys. Chem.* **1990,** *94,* 842.
81. Zana, R. In *Mixed Surfactant Systems;* Holland, P. M.; Rubingh, D. N., Eds.; ACS Symposium Series, American Chemical Society: Washington, DC, 1992; Chapter 19.
82. Weers, J. G.; Scheuing, D. R. *J. Colloid Interface Sci.* **1991,** *145,* 563.
83. Hill, R. M. In *Mixed Surfactant Systems;* Holland, P. M.; Rubingh, D. N., Eds.; ACS Symposium Series, American Chemical Society: Washington, DC, 1992; Chapter 18.
84. Abe, M.; Tsubaki, N.; Ogino, K. *J. Colloid Interface Sci.* **1985,** *107,* 503.
85. Ogino, K.; Abe M. In *Mixed Surfactant Systems;* Holland, P. M.; Rubingh, D. N., Eds.; ACS Symposium Series, American Chemical Society: Washington, DC, 1992; Chapter 8.
86. Rubingh, D. N.; Bauer, M. In *Mixed Surfactant Systems;* Holland, P. M.; Rubingh, D. N., Eds.; ACS Symposium Series, American Chemical Society: Washington, DC, 1992; Chapter 12.
87. Wright, S.; Bunton, C. A.; Holland, P. M. In *Mixed Surfactant Systems;* Holland, P. M.; Rubingh, D. N., Eds.; ACS Symposium Series, American Chemical Society: Washington, DC, 1992; Chapter 13.
88. Shinoda, K.; Hutchinson, E. *J. Phys. Chem.* **1962,** *66,* 577.
89. Benjamin, L. *J. Phys. Chem.* **1964,** *68,* 3575.
90. Kamrath, R. F.; Franses, E. I. *Ind. Eng. Chem. Fundam.* **1983,** *22,* 230.
91. Clint, J. *J. Chem. Soc.* **1975,** *71,* 1327.
92. Nishikido, N.; Morio, Y.; Matuura, R. *Bull. Chem. Soc. Japan* **1975,** *48,* 1387.
93. Harusawa, F.; Tanaka, M. *J. Phys. Chem.* **1981,** *85,* 882.
94. Warr, G. G.; Griese, F.; Healy, T. W. *J. Phys. Chem.* **1983,** *87,* 1220.
95. Graciaa, A.; Lachaise, J.; Bourrel, M.; Osborne-Lee, I. ; Schechter, R. S.; Wade, W. H. *Soc. Pet. Eng. J.;* **1987,** *2,* 305.
96. Scamehorn, J. F.; Schechter, R. S.; Wade, W. H. *J. Dispersion Sci. Technol.* **1982,** *3,* 261.
97. Zhu, D.-M.; Zhao, G.-X. *Colloids Surf.* **1990,** *49,* 269.
98. Jiding, X.; Yan, S.; Heyun, Z. In *Phenomena in Mixed Surfactant Systems;* Scamehorn, J. F, Ed.; ACS Symposium Series 311, American Chemical Society: Washington, DC, 1986; pp 287-311.

99. Müller, A. *Colloids Surf.* **1991**, *57*, 239.
100. Holland, P. M. In *Mixed Surfactant Systems;* Holland, P. M.; Rubingh, D. N., Eds.; ACS Symposium Series, American Chemical Society: Washington, DC, 1992; Chapter 6.
101. Graciaa, A.; Ben Ghoulam, M.; Schechter, R. S. In *Mixed Surfactant Systems;* Holland, P. M.; Rubingh, D. N., Eds.; ACS Symposium Series, American Chemical Society: Washington, DC, 1992; Chapter 9.
102. Kamrath, R. F.; Franses, E. I. In *Phenomena in Mixed Surfactant Systems;* Scamehorn, J. F, Ed.; ACS Symposium Series 311, American Chemical Society: Washington, DC, 1986; pp 44-60.
103. Hey, M. J.; MacTaggart, J. W. *J. Chem. Soc., Faraday Trans. 1* **1985**, *81*, 207.
104. Rathman, J. F.; Scamehorn, J. F. *Langmuir* **1988**, *4*, 474.
105. Hall, D. G.; Huddleston, R. W. *Colloids Surf.* **1985**, *13*, 209.
106. Nishikido, N.; Imura, Y.; Kobayashi, H.; Tanaka, M. *J. Colloid Interface Sci.* **1983**, *91*, 125.
107. Christian, S. D.; Tucker, E. E.; Scamehorn, J. F. In *Mixed Surfactant Systems;* Holland, P. M.; Rubingh, D. N., Eds.; ACS Symposium Series, American Chemical Society: Washington, DC, 1992; Chapter 3.
108. Motomura, K.; Yamanaka, M.; Aratono, M. *Colloid Polym. Sci.* **1984**, *262*, 948.
109. Moroi, Y.; Nishikido, N.; Saito, M.; Matuura, R. *J. Colloid Interface Sci.* **1975**, *52*, 356.
110. Asakawa, T.; Johten, K.; Miyagishi, S.; Nishida, M. *Langmuir* **1985**, *1*, 347.
111. Rathman, J. F.; Scamehorn, J. F. *Langmuir* **1987**, *3*, 372.
112. Kamrath, R. F.; Franses, E. I. *J. Phys. Chem.* **1984**, *88*, 1642.
113. Wall, S.; Elvingson, C. *J. Phys. Chem.* **1985**, *89*, 2695.
114. Stecker, M. M.; Benedek, G. B. *J. Phys. Chem.* **1984**, *88*, 6519.
115. Szleifer, I.; Ben-Shaul, A.; Gelbart, W. M. *J. Chem. Phys.* **1987**, *86*, 7094.
116. Nagarajan, R. *Langmuir* **1985**, *1*, 331.
117. Nagarajan, R. In *Mixed Surfactant Systems;* Holland, P. M.; Rubingh, D. N., Eds.; ACS Symposium Series, American Chemical Society: Washington, DC, 1992; Chapter 4.
118. Puvvada, S.; Blankschtein, D. In *Mixed Surfactant Systems;* Holland, P. M.; Rubingh, D. N., Eds.; ACS Symposium Series, American Chemical Society: Washington, DC, 1992; Chapter 5.
119. Motomura, K.; Ando, N.; Matsuki, H.; Aratono, M. *J. Colloid Interface Sci.* **1990**, *139*, 188.
120. Motomura, K.; Matsukiyo, H.; Aratono, M. In *Phenomena in Mixed Surfactant Systems;* Scamehorn, J. F, Ed.; ACS Symposium Series 311, American Chemical Society: Washington, DC, 1986; pp 163-171.
121. Aratono, M.; Kanda, T.; Motomura, K. *Langmuir* **1990**, *6*, 843.
122. Rosen, M. J.; Murphy, D. S. *J. Colloid Interface Sci.* **1986**, *110*, 224.
123. Rosen, M. J.; Murphy, D. S. *J. Colloid Interface Sci.* **1989**, *129*, 208.
124. Aveyard, R.; Binks, B. P.; Mead, J.; Clint, J. H. *J. Chem. Soc., Faraday Trans. 1* **1988**, *84*, 675.

125. Hutchinson, E. *J. Colloid Interface Sci.* **1948,** *3,* 413.
126. Gu, B.; Rosen, M. J. *J. Colloid Interface Sci.* **1989,** *129,* 537.
127. Jayson, G. G.; Thompson, G. *J. Colloid Interface Sci.* **1986,** *111,* 64.
128. Nagadome, S.; Shibata, O.; Miyoshi, H.; Kagimoto, H.; Ikawa, Y.; Igimi, H.; Suguhara, G. In *Mixed Surfactant Systems;* Holland, P. M.; Rubingh, D. N., Eds.; ACS Symposium Series, American Chemical Society: Washington, DC, 1992; Chapter 20.
129. Chang, D. L.; Rosano, H. L. In *Phenomena in Mixed Surfactant Systems;* Scamehorn, J. F, Ed.; ACS Symposium Series 311, American Chemical Society: Washington, DC, 1986; pp 116-132.
130. Alexander, D. M.; Barnes, G. T.; McGregor, M. A.; Walker, K. In *Phenomena in Mixed Surfactant Systems;* Scamehorn, J. F, Ed.; ACS Symposium Series 311, American Chemical Society: Washington, DC, 1986; pp 133-142.
131. Ter-Minassian-Saraga, E. In *Cationic Surfactants: Physical Chemistry;* Rubingh, D. N.; Holland, P. M., Eds.; Surfactant Science Series 37; Marcel Dekker, Inc.: New York, NY, 1990; pp 249-321.
132. Ahn, D.-J.; Franses, E. I. In *Mixed Surfactant Systems;* Holland, P. M.; Rubingh, D. N., Eds.; ACS Symposium Series, American Chemical Society: Washington, DC, 1992; Chapter 23.
133. Roberts, B. L.; Scamehorn, J. F.; Harwell, J. H. In *Phenomena in Mixed Surfactant Systems;* Scamehorn, J. F, Ed.; ACS Symposium Series 311, American Chemical Society: Washington, DC, 1986; pp 200-215.
134. Harwell; J. H.; Roberts, B. L.;Scamehorn, J. F. *Colloids Surf.* **1988,** *32,* 1.
135. Kronberg, B.; Lindstrom, M.; Stenius, P. In *Phenomena in Mixed Surfactant Systems;* Scamehorn, J. F, Ed.; ACS Symposium Series 311, American Chemical Society: Washington, DC, 1986; pp 225-240.
136. Esumi, K.; Otsuka, H.; Meguro, K. *J. Colloid Interface Sci.* **1991,** *142,* 582.
137. Esumi, K.; Otsuka, H.; Meguro, K. *Langmuir* **1991,** *7,* 2313.
138. Esumi, K.; Sakimoto, Y.; Yoshikawa, K.; Meguro, K. *Colloids Surf.* **1989,** *36,* 1.
139. Fu, E.; Somasundaran, P.; Xu, Q. In *Mixed Surfactant Systems;* Holland, P. M.; Rubingh, D. N., Eds.; ACS Symposium Series, American Chemical Society: Washington, DC, 1992; Chapter 25.
140. Couzis, A.; Gulari, E. In *Mixed Surfactant Systems;* Holland, P. M.; Rubingh, D. N., Eds.; ACS Symposium Series, American Chemical Society: Washington, DC, 1992; Chapter 24.
141. Papke, B. L. In *Mixed Surfactant Systems;* Holland, P. M.; Rubingh, D. N., Eds.; ACS Symposium Series, American Chemical Society: Washington, DC, 1992; Chapter 26.
142. Rosen, M. J.; Gu, B. *Colloids Surf.* **1987,** *23,* 119.
143. Lucassen-Reynders, E. H. In *Anionic Surfactants: Physical Chemistry of Surfactant Action,* Lucassen-Reynders, E. H., Ed.; Surfactant Science Series 11; Marcel Dekker, Inc.: New York, NY, 1981; pp 1-53.
144. Lucassen-Reynders, E. H. *Progress in Surface and Membrane Sci.* **1976,** *10,* pp 253-360.
145. Davies, J. T.; Rideal, E. K. *Interfacial Phenomena;* Academic Press: New York, NY, 1961.

146. Lucassen-Reynders, E. H. *Kolloid-Z.u.Z. Polymere.* **1972**, *250*, 356.
147. Lucassen-Reynders, E. H. *J. Colloid Interface Sci.* **1973**, *42*, 554.
148. Rodakiewicz-Nowak, J. *J. Colloid Interface Sci.* **1981**, *84*, 532.
149. Lucassen-Reynders, E. H. *J. Colloid Interface Sci.* **1982**, *85*, 178.
150. Rodakiewicz-Nowak, J. *J. Colloid Interface Sci.* **1982**, *85*, 586.
151. Rodakiewicz-Nowak, J. *J. Colloid Interface Sci.* **1983**, *91*, 368.
152. Rodakiewicz-Nowak, J. *Colloids Surf.* **1983**, *6*, 143.
153. Rodakiewicz-Nowak, J. *Z. Phys. Chemie Leipzig* **1985**, *266*, 997.
154. Ingram, B. T. *Colloid Polym. Sci.* **1980**, *25*, 191.
155. Rosen, M. J.; Zhu, Z. H. *J. Colloid Interface Sci.* **1989**, *133*, 473.
156. Rosen, M. J.; Gu, B.; Murphy, D. S.; Zhu, Z. H. *J. Colloid Interface Sci.* **1989**, *129*, 468.
157. Nguyen, C. M.; Scamehorn, J. F. *J. Colloid Interface Sci.* **1988**, *123*, 238.
158. Holland, P. M. In *Phenomena in Mixed Surfactant Systems;* Scamehorn, J. F, Ed.; ACS Symposium Series 311, American Chemical Society: Washington, DC, 1986; pp 102-115.
159. Holland, P. M. *Colloids Surf.* **1986**, *19*, 171.
160. Holland, P. M. In *Mixed Surfactant Systems;* Holland, P. M.; Rubingh, D. N., Eds.; ACS Symposium Series, American Chemical Society: Washington, DC, 1992; Chapter 22.
161. Rosen, M. J. In *Mixed Surfactant Systems;* Holland, P. M.; Rubingh, D. N., Eds.; ACS Symposium Series, American Chemical Society: Washington, DC, 1992; Chapter 21.
162. Huber, K. *J. Colloid Interface Sci.* **1991**, *147*, 321.
163. Góralczyk, D. *Colloids Surf.* **1991**, *59*, 361.
164. Moroi, Y.; Sugii, R.; Matuura, R. *J. Colloid Interface Sci.* **1984**, *98*, 184.
165. Chorro, M.; Kameka, N. *J. Chim. Phys.-Chim. Biol.* **1991**, *88*, 515.
166. Scheuing, D. R.; Weers, J. G. *Langmuir* **1990**, *6*, 665.
167. Kato, T.; Iwai, M.; Seimiya, T. *J. Colloid Interface Sci.* **1989**, *130*, 439.
168. Stellner, K. L.; Amante, J. C.; Scamehorn, J. F.; Harwell, J. H. *J. Colloid Interface Sci.* **1988**, *123*, 186.
169. Amante, J. C.; Scamehorn, J. F.; Harwell; J. H. *J. Colloid Interface Sci.* **1991**, *144*, 243.
170. Peacock, J. M.; Matejevic, E. *J. Colloid Interface Sci.* **1980**, *77*, 548.
171. Baviere, M.; Bazin, B.; Aude, R. *J. Colloid Interface Sci.* **1983**, *92*, 580.
172. Stellner, K. L.; Scamehorn, J. F. *J. Am. Oil Chem. Soc.* **1986**, *63*, 566.
173. Chou, S. J.; Bae, J. H. *J. Colloid Interface Sci.* **1983**, *96*, 192.
174. Gerbacia, W. E. F. *J. Colloid Interface Sci.* **1983**, *93*, 556.
175. Fan, X. J.; Stenius, P.; Kallay, N.; Matejevic, E. *J. Colloid Interface Sci.* **1988**, *121*, 571.
176. Scamehorn, J. F. In *Mixed Surfactant Systems;* Holland, P. M.; Rubingh, D. N., Eds.; ACS Symposium Series, American Chemical Society: Washington, DC, 1992; Chapter 27.
177. Hato, M.; Shinoda, K. *J. Phys. Chem.* **1973**, *77*, 378.
178. Moroi, Y.; Oyama, T.; Matuura, R. *J. Colloid Interface Sci.* **1977**, *60*, 103.
179. Tsujii. K.; Saito, N.; Takeuchi, T. *J. Phys. Chem.* **1980**, *84*, 2267.

180. Raison, M. *Proc. 2nd Int. Cong. Surface Activity,* 1957, p. 374.
181. Nishikido, N. *J. Colloid Interface Sci.* **1990,** *136,* 401.
182. Tsujii, K.; Okahashi, K.; Takeuchi, T. *J. Phys. Chem.* **1982,** *86,* 1437.
183. Nakama, Y.; Harusawa, F.; Murotani, I. *J. Am. Oil Chem. Soc.* **1990,** *67,* 717.
184. Rubingh, D. N., presented at Joint Central-Great Lakes ACS Regional Meeting, May, 1991.
185. Marsall, L. *Int. J. Pharm.* **1987,** *39,* 263.
186. Marsall, J. *Langmuir* **1988,** *4,* 90.
187. Meroni, A.; Pimpinelli, A.; Reatto, L. *Chem. Phys. Lett.* **1987,** *135,* 137.
188. Reatto, L.; Tau, M. *Chem. Phys. Lett.* **1984,** *108,* 292.
189. Puvvada, S.; Blankschtein, D. *J. Chem. Phys.* **1990,** *92,* 3710.
190. De Salvo Souza, L.; Corti M.; Cantu, L.; Degiorgio, V. *Chem. Phys. Lett.* **1986,** *131,* 160.
191. Nilsson, P. G.; Lindman, B. *J. Phys. Chem.* **1984,** *88,* 5391.
192. Valaulikar, B. S.; Manohar, C. *J. Colloid Interface Sci.* **1985,** *108,* 403.
193. Sadaghiana, A. S.; Khan, A. *J. Colloid Interface Sci.* **1991,** *144,* 191.
194. Trainer, C.; Fromon, M.; Mannebach, M. H. *Langmuir* **1989,** *5,* 283.
195. Mehreteab, A.; Loprest, F. J. *J. Colloid Interface Sci.* **1988,** *125,* 602.
196. Mehreteab, A. In *Mixed Surfactant Systems;* Holland, P. M.; Rubingh, D. N., Eds.; ACS Symposium Series, American Chemical Society: Washington, DC, 1992; Chapter 28.
197. Ravey, J. C.; Gherbi, A.; Stebe, M. J. *Progr. Coll. Polym. Sci.* **1989,** *79,* 272.
198. Tamori, K.; Esumi, K.; Meguro, K. *J. Colloid Interface Sci.* **1991,** *141,* 236.
199. Boonbrahm, R.; Saupe, A. *Mol. Cryst. Liq. Cryst.* **1984,** *109,* 225.
200. Treiner, C.; Bocquet, J. F.; Pommier, C. *J. Phys. Chem.* **1986,** *90,* 3052.
201. Treiner, C.; Khodja, A. A.; Fromon, M. *Langmuir* **1987,** *3,* 729.
202. Treiner, C.; Nortz, M.; Vaution, C. *Langmuir* **1990,** *6,* 1211.
203. Treiner, C.; Nortz, M.; Vaution, C. *J. Colloid Interface Sci.* **1988,** *125,* 251.
204. Yoda, K.; Tamori, K.; Esumi, K.; Meguro, K. *Colloids Surf.* **1991,** *58,* 87.
205. Weers, J. G. *J. Am. Oil Chem. Soc.* **1990,** *67,* 340.
206. Ward, A. J. I.; Quigley, K. *J. Dispersion Sci. Techn.* **1990,** *11,* 143.
207. Nishikido, N. *Langmuir* **1991,** *7,* 2076.
208. Smith, G. A.; Christian, S. D.; Tucker, E. E.; Scamehorn, J. F. *J. Colloid Interface Sci.* **1989,** *130,* 254.
209. Kamenka, N.; El Amrani, M.; Appell, J.; Lindheimer, M. *J. Colloid Interface Sci.* **1991,** *143,* 463.
210. Marangoni, D. G.; Kwak, J. C. T. *Langmuir* **1991,** *7,* 2083.
211. Dennis, E. A. In *Mixed Surfactant Systems;* Holland, P. M.; Rubingh, D. N., Eds.; ACS Symposium Series, American Chemical Society: Washington, DC, 1992; Chapter 29.
212. Marangoni, D. G.; Rodenhiser, A. P.; Thomas, J. M.; Kwak, J. C. T. In *Mixed Surfactant Systems;* Holland, P. M.; Rubingh, D. N., Eds.; ACS Symposium Series, American Chemical Society: Washington, DC, 1992; Chapter 11.
213. Laughlin, R. G.; Munyon, R. L. *J. Phys. Chem.* **1987,** *91,* 3299.

RECEIVED April 22, 1992

Chapter 2

Modeling Mixed Surfactant Systems
Basic Introduction

Paul M. Holland

General Research Corporation, Santa Barbara, CA 93111

A basic introduction to theory for modeling mixed surfactant systems is presented. This includes a demonstration of how equilibrium thermodynamic models can be developed to describe the behavior of mixed micellar solutions. The reader is provided with some simple tools for analysis and prediction in both multicomponent ideal and binary nonideal mixed micellar systems. Nonideal mixing is discussed in the context of activity coefficients and the regular solution approximation. Modeling of nonideal behavior in binary mixed micellar systems is compared with experimental results, and a tabulation of nonideal interaction parameters from a wide range of different binary mixed surfactant systems is provided.

Approaches to modeling mixed surfactant systems can vary from simple thermodynamic treatments which address only a few basic properties of the system to more complex molecular models which deal with properties such as micelle size and the composition distribution of mixed micellar aggregates. Generally the approach selected depends on the properties required and the extent to which one wants to emphasize either tractability for predicting the properties of complex systems such as multicomponent mixtures, or more detailed understanding of underlying phenomena in simpler mixed systems. In this introduction the emphasis will be on a simplified approach for modeling binary nonideal mixed micellar solutions.

Basic Modeling of Mixed Surfactant Systems

The models most widely applied in treating mixed surfactant systems use a simplified equilibrium thermodynamic approach which assumes that the the mixed micelle or other mixed surfactant aggregate can be treated much like a separate phase (1-4). This pseudophase separation approach represents a limiting case where the number of molecules in the aggregate becomes large (formally approaching infinity). Results from mass-action models which explicitly take the number of molecules in the

0097–6156/92/0501–0031$06.00/0

aggregate into account show this to be a good approximation for micelle sizes of about 50 or more molecules (5,6). Use of the pseudophase separation approach greatly simplifies the modeling of properties in complex mixed systems.

A typical starting point in developing such a model is to consider the chemical potentials of various surfactant species in solution. The chemical potential is defined as the gradient of the Gibbs free energy with concentration for a component at constant pressure and temperature while holding the concentration of all other components in the system constant. At equilibrium the chemical potential of a particular surfactant species is the same everywhere in the system. This provides a useful link between surfactant monomers in solution and those in mixed micelles or other surfactant pseudophases when developing a model.

In a formally nonionic surfactant system, the chemical potential of the ith surfactant component in solution can be expressed by

$$\mu_i = \mu_i^o + RT \ln C_i^m \tag{1}$$

in terms of a standard state chemical potential, μ^o, and the monomer concentration, C_i^m (see Legend of Symbols). At or above the CMC in a pure single-specie system a similar expression results

$$\mu_i^{Mo} = \mu_i^o + RT \ln C_i^* \tag{2}$$

where C_i^* is the CMC of the pure component. The chemical potential in the mixed micellar system can be expressed as

$$\mu_i^M = \mu_i^{Mo} + RT \ln f_i x_i \tag{3}$$

where f_i and and x_i are the activity coefficient and the mole fraction of the ith component. Since the chemical potential of the monomeric species must be equal to that in the mixed micelle at equilibrium, equations 1-3 can be combined to obtain (7)

$$C_i^m = x_i f_i C_i^* \tag{4}$$

which gives the monomer concentration in terms of the mixed micellar composition, an activity coefficient, and the CMC of the pure component.

It should be pointed out that while this approach does not explicitly include the effects of counterions it can be successfully applied to systems containing ionic surfactants. If one wishes to explicitly include counterions in a more sophisticated treatment (8), alternative relationships can be developed beginning with appropriate expressions for the chemical potentials.

Mixed CMCs. At the mixed CMC (C_{Mix}^*), the monomer concentration of a surfactant component is equal to its mole fraction as surfactant in the system (α_i) times the CMC. This can be combined with the constraint that the sum of mole fractions in the micelle must equal unity and equation 4, leading directly to a general result for the CMC of a mixed system (7)

$$\frac{1}{C_{Mix}^*} = \sum_{i=1}^{n} \frac{\alpha_i}{f_i C_i^*} \qquad (5)$$

This gives the mixed CMC in terms of the surfactant mole fractions of each component, the CMCs of the pure components, and activity coefficients in the mixed micelle. This expression has proven itself to be very useful in describing the behavior of a wide variety of mixed systems including those with ionic surfactants, even though it does not explicitly take counterions into account.

In the case of ideal systems, where the activity coefficients f_i are unity, calculation of the CMC in micellar systems with any number of components involves a simple summation of terms. For nonideal systems, one must first determine the activity coefficients.

Micellar Mole Fractions and Monomer Concentrations. Generalized expressions for both monomer concentrations and surfactant mole fractions in multicomponent nonideal mixed micellar systems can be readily developed. Mass balance considerations combined with equation 4 lead to (7)

$$\sum_{i=1}^{n} \frac{\alpha_i C}{C + f_i C_i^* - M} = 1 \qquad (6)$$

where M is the sum of the monomer concentrations

$$M = \sum_{i=1}^{n} C_i^m \qquad (7)$$

Solving Equation 6 for the total monomer concentration is straight forward for any number of components once the activity coefficients have been calculated, and the following expressions can be used to obtain individual monomer concentrations and mole fractions in the micelle, respectively.

$$C_i^m = \frac{\alpha_i f_i C_i^* C}{C + f_i C_i^* - M} \qquad (8)$$

$$x_i = \frac{\alpha_i C}{C + f_i C_i^* - M} \qquad (9)$$

For ideal mixing, the activity coefficients equal one by definition and the pseudophase model is easily solved for any number of surfactant components in the mixed micellar system. For nonideal systems, activity coefficients must first be determined.

Nonideal Mixing

Interactions between surfactant molecules in aggregates such as micelles arise from two sources. The first of these is based on the tendency for spontaneous self-association or the "hydrophobic effect" (9) which favors the formation of relatively large surfactant aggregates with the hydrophobic chains in the interior of the aggregate and the hydrophilic head groups at the surface in contact with aqueous solution. Since this does not specifically depend on the type of surfactant head group, this process tends to favor randomly mixed aggregates and represents the basis for ideal mixing of surfactants in micelles. The second of these involves interactions between unlike head groups in surfactant aggregates. Here electrostatic or other interactions can lead to significant nonideal effects on the properties of the mixed surfactant system, such as substantially lower CMCs and interfacial tensions compared those of the unmixed surfactants.

Activity Coefficients. From a thermodynamic point of view, nonideality in mixtures is best described using activity coefficients which represent the ratio of an effective over actual mole fraction of a given component. By definition, such activity coefficients become equal to one for either ideal mixtures and for pure components, and in the present case are defined with reference to a surfactant psuedophase such a mixed micelle. In nonideal mixtures of typical surfactants, the activity coefficients are usually less than one.

A generalized form for the activity coefficients in the micellar pseudophase for binary systems can be developed by considering the thermodynamics of mixing (1,10). As for mixtures of liquids, the excess free energy of mixing, G^E, can be expressed as

$$G^E = RT \left(x_1 \ln f_1 + (1 - x_1) \ln f_2 \right) \tag{10}$$

If one takes the partial derivative of this with respect to x_i, and eliminates some of the terms with the Gibbs-Duhem relation

$$\left[\frac{\partial G^E}{\partial x_1} \right]_{T,P} = RT \left(\ln f_1 - \ln f_2 \right) \tag{11}$$

results. Combining and rearranging these expressions provides the binary activity coefficients as a function of the excess free energy of mixing

$$\ln f_1 = \frac{1}{RT} \left[G^E + (1 - x_1) \frac{\partial G^E}{\partial x_1} \right] \tag{12}$$

$$\ln f_2 = \frac{1}{RT} \left[G^E + x_1 \frac{\partial G^E}{\partial x_1} \right] \tag{13}$$

The excess free energy of mixing itself is defined in terms of an excess heat (or enthalpy) of mixing H^E, and an excess entropy of mixing, S^E.

$$G^E = H^E - T S^E \tag{14}$$

Regular Solution Approximation. The regular solution approximation is introduced by assuming that the excess entropy of mixing, S^E, is zero. This allows the substitution of H^E in place of G^E. For binary mixtures this excess heat of mixing can be represented by

$$H^E = \beta \, x_1(1 - x_1) \, RT \tag{15}$$

where β is a dimensionless parameter which times RT represents a net difference in interaction energy between the mixed and unmixed systems. The form of this expression corresponds to the leading term in a lattice model description of liquid mixtures *(11)*. Since H^E should be constant, the parameter β will be temperature dependent. Substitution of the functional form of Equation 15 for G^E into the general expression for the activity coefficients directly leads to the activity coefficients of binary mixtures

$$f_1 = \exp \beta \, (1 - x_1)^2 \tag{16}$$

$$f_2 = \exp \beta \, x_1^2 \tag{17}$$

These are seen to be symmetric with respect to the surfactant composition of the aggregate. An alternative form for the interaction parameter, W, with the dimensions of energy, has been used by various other workers *(8,12)*. This is related to β by

$$\beta = \frac{W}{RT} \tag{18}$$

In the regular solution approximation where the excess entropy of mixing is defined to be zero, the parameter β can be formally interpreted as a parameter representing an excess heat of mixing. However, calorimetric measurements of excess heats of micellar mixing often show poor agreement with β values determined independently from CMC measurements, indicating that the assumptions of the the regular solution approximation do not hold for many binary surfactant mixtures *(10,13-15)*. It is interesting to note that similar observations are made in nonideal mixtures of liquids where the regular solution approach provides good descriptions of excess free energies of mixing, but results for heats of mixing are often poor. Hildebrand, Prausnitz and Scott consider this situation to result from a cancellation of errors, where the error in the heat of mixing is approximately canceled by the error in assuming the excess entropy of mixing to be zero *(16)*.

In the case of mixed surfactant systems, this situation has lead to the suggestion that the parameter β be interpreted more generally as an excess free energy of mixing parameter which only meets the formal criteria of the regular solution approximation when the excess entropy of mixing is zero (1,10). Here, the functional form of Equation 15 is empirically assumed to provide a reasonable description of excess free energy energies of mixing, yielding the same form for activity coefficients as before. It should be pointed out that the form of the regular solution approximation has proven itself to be remarkably successful in modeling the nonideal behavior of mixed surfactant systems. This includes extension (without adjustable parameters) to multicomponent nonideal mixtures using pairwise interaction parameters determined independently from binary systems (1,7,17).

To solve for the activity coefficients in binary mixed micellar systems using the above form of regular solution approximation one must first solve for surfactant micellar mole fractions. For a given value of β, these are readily obtained by iterative solution of the expression

$$\frac{\alpha_1 \, C_2^* \, f_2 \, (1-x_1)}{\alpha_2 \, C_1^* \, f_1 \, x_1} \;=\; 1 \qquad\qquad (19)$$

for x_1, followed by substitution into Equations 16 and 17.

The net interaction parameter β in a binary mixed surfactant system can be determined from a single measured value of the mixed CMC, and CMCs of the pure surfactant components. This is directly obtained following interative solution for x_1 in Equation 20, below.

$$\beta \;=\; \frac{\ln \, [\alpha_1 C_{Mix}^*/(x_1 C_1^*)]}{(1 - x_1)^2} \;=\; \frac{\ln \, [\alpha_2 C_{Mix}^*/((1 - x_1)C_2^*)]}{x_1^2} \qquad (20)$$

Example computer programs in BASIC for calculating of the mixed CMC and β parameter in binary nonideal mixed micellar systems are provided in Appendix I.

Results for Binary Mixtures

A substantial body of previous work (1-4,7,8,12,17-25) has shown that the above form of the regular solution approximation can be successfully used to model nonideality in a wide variety of different surfactant mixtures. This approach was first applied by Rubingh (18) and has been shown to provide useful descriptions of the behavior of mixed surfactant systems including nonionic/nonionic, ionic/ionic and nonionic/ionic surfactant mixtures.

A typical result for an anionic/nonionic binary surfactant mixture is shown in Figure 1, where the mixed CMC in the $C_{12}SO_4Na/C_8E_4$ system is plotted against the overall composition of the surfactant mixture. Here, experimental results for the CMC (points) are compared with the nonideal regular solution model (solid line) and the ideal model (dashed line). This mixture exhibits a significant deviation from ideality with mixed CMCs about a factor of two lower than those expected for ideal mixing in the micelles. It is also readily seen that a single parameter in the nonideal model, $\beta = -3.1$, provides a good description across a wide range of composition.

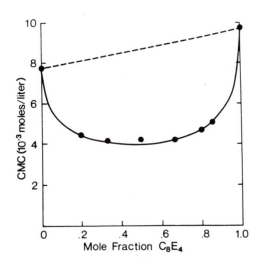

Figure 1.   CMCs in for binary anionic/nonionic mixtures of $C_{12}SO_4Na/C_8E_4$ at 25°C. The plotted points are experimental results, the solid line the prediction of the nonideal mixed micelle model with $\beta = -3.1$, and the dashed line is the prediction for ideal mixing.

Much larger deviations from ideality are seen in anionic/cationic mixed systems such as the $C_{10}SO_4Na/C_{10}TABr$ mixture in Figure 2, where mixed CMCs are an order of magnitude or more below those expected for ideal mixing. Here, the nonideal model (solid line) with $\beta = -13.2$ provides an excellent description of the mixed CMC behavior across a broad range of composition.

It is interesting to note that this simple pseudophase separation approach coupled with the regular solution approximation provides very useful descriptions of nonideality in mixed surfactant systems which include ionic surfactants even though the model is formally designed for nonionic surfactants and counterions are not explicitly taken into account. This has suggested that counterion effects are either relatively small or that they can be empirically accounted for by the interaction parameter of the model (7).

Measurements on a large number of different binary surfactant mixtures have been reported in the literature together with values of regular solution model interaction parameters. A representative selection of these are given in Tables I and II. Dimensionless micellar interaction parameters β for mixtures that contain only anionic and nonionic surfactants are given in Table I. Values for mixtures which include cationic or zwitterionic surfactants are given in Table II.

An overview of Tables I and II shows that for simple surfactant types nonideal interactions become progressively stronger (i.e. the dimensionless interaction parameter β becomes more negative) in going from mixtures of the same surfactant type to those of opposite charge. That is, nonionic/nonionic or cationic/cationic interactions are nearly ideal ($\beta \approx 0$), cationic/nonionic interactions are significantly stronger ($-5 \leq \beta \leq -1$) but somewhat weaker on average than anionic/nonionic interactions, and anionic/cationic interactions are much stronger ($\beta \leq -10$). In the case of zwitterionic surfactants, which have a more complex hydrophilic group structure, a greater range of nonideal behavior is observed, with zwitterionic/anionic interactions being comparable in strength to anionic/cationic interactions in some cases. More detailed examination of Tables I and II reveals a number of interesting effects on the magnitude of the interaction parameter β.

The effect of increasing ionic strength by adding salt can be seen in anionic/nonionic, cationic/nonionic, and anionic/cationic mixed systems. For example, in the $C_{12}SO_4Na/C_{12}E_8$ and $C_{12}PyrCl/C_{12}E_8$ systems, increasing ionic strength in going from water to 0.5 M NaCl changes β from -3.9 to -2.6, and -2.7 to -1.0, respectively. In the $C_{10}SO_4Na/C_{10}TABr$ system going from water to 0.05 M NaBr changes β from -18.5 to -13.2. Much of this observed decrease in the magnitude of the interaction parameter with increasing ionic strength can be attributed to added salt reducing the CMCs of the pure ionic surfactants while having little or no effect on the mixed CMC (this is apparent from examination of Equation 20).

In the case of surfactants which can accept a proton, pH can have a significant effect on the strength of the interaction parameter. For example, in unbuffered water the $C_{12}SO_4Na/C_{12}AO$ system behaves like a anionic/cationic system with $\beta = -16.5$, whereas in 0.5 mM Na_2CO_3 (pH $\approx$ 10) it behaves as a anionic/nonionic system with $\beta = -4.4$. In unbuffered water the zwitterionic/anionic system $C_{12}Np/C_{12}SO_4Na$ also behaves much like a cationic/anionic system with $\beta = -14.1$.

Structural effects on the magnitude of the interaction parameters can also be

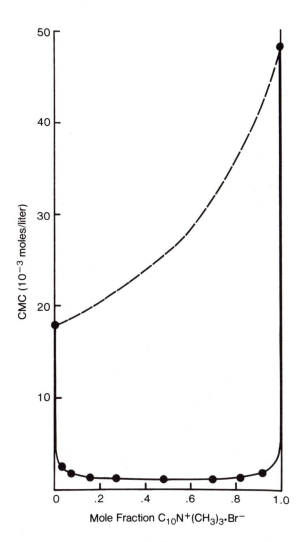

Figure 2. CMCs in for binary anionic/cationic mixtures of $C_{10}SO_4Na/C_{10}TABr$ in 0.05 M NaBr at 23°C. The plotted points are experimental results, the solid line the prediction of the nonideal mixed micelle model with $\beta = -13.2$, and the dashed line is the prediction for ideal mixing (From *Ref. 7*).

Table I. Interaction Parameters for Some Binary Mixed Micellar Systems

β	Type	Binary Mixture	Medium	T(°C)	Ref
-3.9	AN	$C_{12}SO_4Na/C_{12}E_8$	H_2O	25	19
-2.6	AN	$C_{12}SO_4Na/C_{12}E_8$	0.5 M NaCl	25	19
-3.6	AN	$C_{12}SO_4Na/C_{10}E_4$	0.5 mM Na_2CO_3	23	7
-4.1	AN	$C_{12}SO_4Na/C_8E_{12}$	H_2O	25	19
-3.4	AN	$C_{12}SO_4Na/C_8E_6$	H_2O	25	19
-3.1	AN	$C_{12}SO_4Na/C_8E_4$	H_2O	25	10
-1.6	AN	$C_{12}E_2SO_4Na/C_8E_4$	H_2O	25	10
-4.3	AN	$C_{15}SO_4Na/C_{10}E_6$	H_2O	25	19
-3.4	AN	$C_{12}SO_3Na/C_{12}E_8$	H_2O	25	3
-4.4	AN	$C_{12}SO_4Na/C_{12}AO$	0.5 mM Na_2CO_3	23	7
-3.7	AN	$C_{12}SO_4Na/C_{10}PO$	1.0 mM Na_2CO_3	24	7
-2.4	AN	$C_{12}SO_4Na/C_{10}MSO$	1.0 mM Na_2CO_3	24	7
-1.6	AN	$C_{12}SO_4Na/NPE_{10}$[†]	0.4 M NaCl	30	20
-3.2	AN	$C_{12}\Phi SO_3Na/NPE_{10}$[†]	0.15 M NaCl	30	20
-2.8	AN	$C_{12}\Phi SO_3Na/NPE_{10}$[†]	0.15 M NaCl	38	20
-2.1	AN	$C_{12}\Phi SO_3Na/NPE_{10}$[†]	0.15 M NaCl	46	20
-1.8	AN	$C_{12}\Phi SO_3Na/NPE_{10}$[†]	0.15 M NaCl	54	20
0.0	NN	$C_{10}PO/C_{10}MSO$	1.0 mM Na_2CO_3	24	7
-0.1	NN	$C_{10}E_3/C_{10}MSO$	H_2O	25	21
-0.4	NN	$C_{12}E_3/C_{12}E_8$	H_2O	25	22
-0.8	NN	$C_{12}AO/C_{10}E_4$	0.5 mM Na_2CO_3	23	7

Surfactant Types:

A= anionic N = nonionic C = cationic Z = zwitterionic

[†]Polydisperse surfactant mixture

Table II. Interaction Parameters for Some Binary Mixed Micellar Systems

β	Type	Binary Mixture	Medium	T(°C)	Ref
-16.5	AC	$C_{12}SO_4K/C_{12}AO$	H_2O	25	3
-10.5	AC	C_8SO_4Na/C_8TABr	H_2O	25	3
-18.5	AC	$C_{10}SO_4Na/C_{10}TABr$	H_2O	25	23
-13.2	AC	$C_{10}SO_4Na/C_{10}TABr$	0.05 M NaBr	23	7
-25.5	AC	$C_{12}SO_4Na/C_{12}TABr$	H_2O	25	24
-1.8	CN	$C_{10}TABr/C_8E_4$	0.05 M NaBr	23	7
-1.5	CN	$C_{14}TABr/C_{10}E_5$	H_2O	23	18
-2.4	CN	$C_{16}TACl/C_{12}E_5$	H_2O	23	18
-3.1	CN	$C_{16}TACl/C_{12}E_8$	0.1 M NaCl	25	19
-2.6	CN	$C_{18}TACl/C_{12}E_5$	2.4 mM NaCl	25	19
-4.6	CN	$C_{20}TACl/C_{12}E_8$	H_2O	25	19
-2.7	CN	$C_{12}PyrCl/C_{12}E_8$	H_2O	25	3
-1.4	CN	$C_{12}PyrCl^a/C_{12}E_8$	0.1 M NaCl	25	3
-1.0	CN	$C_{12}PyrCl/C_{12}E_8$	0.5 M NaCl	25	3
-1.3	CN	$C_{12}PyrCl/NPE_{10}{}^\dagger$	0.03 M NaCl	30	20
-1.7	CN	$C_{16}TACl/NPE_{10}{}^\dagger$	0.03 M NaCl	30	20
-0.2	CC	$C_{16}TACl/C_{12}PyrCl$	0.15 M NaCl	30	20
-10.6	ZA	$C_{12}Np^b/C_{10}SO_4Na$	H_2O	30	24
-14.1	ZA	$C_{12}Np/C_{12}SO_4Na$	H_2O	30	24
-15.5	ZA	$C_{12}Np/C_{14}SO_4Na$	H_2O	30	24
-5.0	ZA	$C_{12}BMG^c/C_{12}SO_4Na$	H_2O	25	25
-1.2	ZC	$C_{12}BMG/C_{12}TABr$	H_2O	25	25
-0.9	ZN	$C_{12}BMG/C_{12}E_8$	H_2O	25	25

$^a C_{12}Pyr = C_{12}(CH_3)-\overset{+}{N}C_5H_5$

$^b C_{12}Np = C_{12}-\overset{+}{NH_2}-CH2-CH2-COO^-$

$^c C_{12}BMG = C_{12}-\overset{+}{N}-CH_3-CH2-COO^-$
$\qquad\qquad\quad |$
$\qquad\qquad\ CH_2-C_6H_5$

observed. For example, a trend is seen with the number of units of exthoxylation for the system $C_{12}SO_4Na/C_8E_x$ in Table I, where as x increases from 4 to 12, β undergoes a relatively modest change from -3.1 to -4.1. A somewhat larger effect is seen when two units of exthoxylation are incorporated into $C_{12}SO_4Na$, causing the magnitude of β for the interaction with C_8E_4 to decrease from -3.1 to -1.6.

Structural effects are also apparent for variations in the length of the surfactant hydrocarbon chain. For example, a large effect is apparent for the C_xSO_4Na/C_xTABr system in Table II, where as x increases from 8 to 12, β changes dramatically from -10.2 to -25.5. The size of this effect is presumably exaggerated by the absence of added salt in these measurements coupled with the large variation in the CMCs of the pure components with x. It can be seen that with added salt the magnitude of β for the $C_{10}SO_4Na/C_{10}TABr$ system is -13.2, compared with -18.5 without. A similar structural effect is seen in the zwitterionic/anionic system $C_{12}Np/C_xSO_4Na$, where as x increases from 10 to 14, β changes from -10.6 to -15.5.

Finally, the effect of temperature on the strength of the micellar interaction parameter β is seen for the system $C_{12}\Phi SO_3Na/NPE_{10}$ in Table I. This shows that as the temperature increases from 30 to 54 °C, the magnitude of β decreases from -3.2 to -1.8. While the observed temperature dependence of β exhibits the expected trend it is much stronger than would be projected by Equation 15, thereby providing further evidence that the assumptions of the regular solution approximations do not hold for this system. It should also be pointed out that although there is some degree of polydispersity in the NPE_{10} used in these measurements, this is not expected to effect on the value of interaction parameters determined from CMC measurements (see chapter 6, this volume).

Summary

Useful thermodynamic models can be readily developed to treat nonideal mixed surfactant systems based on consideration of the chemical potentials of individual surfactant components in the mixture at equilibrium. Coupled with the pseudophase separation approach and a regular solution approximation for treating nonideal mixing in micelles, this provides a tractable means of treating nonideality and predicting behavior in a wide variety of different surfactant mixtures. Results for binary mixed micellar systems show that nonideal mixing is well described by this approach using a single parameter. Tabulated values of this nonideal interaction parameter for a wide range of different binary mixed systems are provided.

Legend of Symbols

C	total surfactant concentration
C_i^*	CMC of pure surfactant i
C_{Mix}^*	CMC of mixed surfactant system
C_i^m	monomer concentration of surfactant i
f_i	activity coefficient of surfactant i in mixed micelles
f_i^s	activity coefficient of surfactant i in mixed monolayer
R	gas constant
T	absolute temperature

x_i mole fraction of surfactant i in mixed micelles
α_i mole fraction of surfactant i in total surfactant
β dimensionless interaction parameter in mixed micelle

Appendix

Appendix I. BASIC computer programs for calculating (A) binary mixed CMC and (B) nonideal interaction parameter β in binary mixed micellar systems.   Input variables are β = BETA, α_1 = A1, C_1^* = CMC1, C_2^* = CMC2, and C_{Mix}^* = CMCM.

(A) Mixed CMC

```
 10   REM - CALCULATE BINARY MIXED CMC FROM MICELLAR BETA PARAMETER
 20   INPUT "INPUT BETA, A1, CMC1, CMC2 ", BETA, A1, CMC1, CMC2
 30   A2 = 1 - A1: X1 = 0: X2 = 1: G = A1*CMC2 / (A2*CMC1)
 40   FOR I = 1 TO 25
 50   XM1 = 0.5 * (X1 + X2): XM2 = 1 - XM1
 60   F1 = EXP(BETA*XM2*XM2): F2 = EXP(BETA*XM1*XM1)
 70   F = G * F2*XM2 / (F1*XM1)
 80   IF F > 1 THEN X1 = XM1 ELSE IF F < 1 THEN X2 = XM1 ELSE GOTO 100
 90   NEXT I
100   CMCM = 1/(A1/(F1*CMC1)+A2/(F2*CMC2)) : PRINT "MIXED CMC = " CMCM
```

(B) Micellar β Parameter

```
 10   REM - CALCULATE MICELLAR BETA FROM BINARY MIXED CMC
 20   INPUT "INPUT A1, CMC1, CMC2, CMCM ", A1, CMC1, CMC2, CMCM
 30   A2 = 1 - A1: G1 = CMCM * A1/CMC1: G2 = CMCM * A2/CMC2
 40   X1 = 0: X2 = 1
 50   FOR I = 1 TO 25
 60   XM1 = 0.5 * (X1 + X2): XM2 = 1 - XM1
 70   F1 = XM1*XM1 * LOG(G1/XM1): F2 = XM2*XM2 * LOG(G2/XM2)
 80   F = F1 - F2
 90   IF F > 0 THEN X1 = XM1 ELSE IF F < 0 THEN X2 = XM1 ELSE GOTO 110
100   NEXT I
110   BETA = LOG (G1/XM1) / (XM2 * XM2): PRINT "MICELLAR BETA = " BETA
```

Literature Cited

1. Holland, P. M. *Adv. Colloid Interface Sci.* **1986**, *26*, 111.
2. Scamehorn, J. F. In *Phenomena in Mixed Surfactant Systems;* Scamehorn, J. F. Ed.; ACS Symposium Series 311, American Chemical Society: Washington, DC, 1986; pp 1-27.
3. Rosen, M. J. In *Phenomena in Mixed Surfactant Systems;* Scamehorn, J. F, Ed.; ACS Symposium Series 311, American Chemical Society: Washington, DC, 1986; pp 144-162.
4. Holland, P. M.; Rubingh, D. N. In *Cationic Surfactants: Physical Chemistry;* Rubingh, D. N.; Holland, P. M., Eds.; Surfactant Science Series 37; Marcel Dekker, Inc.: New York, NY, 1990; pp 141-187.
5. Benjamin, L. *J. Phys. Chem.* **1964**, *68*, 3575.
6. Kamrath, R. F.; Franses, E. I. *J. Phys. Chem.* **1984**, *88*, 1642.

7. Holland, P. M.; Rubingh, D. N. *J. Phys. Chem.* **1983,** *87,* 1984.
8. Kamrath, R. F.; Franses, E. I. *Ind. Eng. Chem. Fundam.* **1983,** *22,* 230.
9. Tanford, C. *The Hydrophobic Effect;* John Wiley and Sons: New York, 1973.
10. Holland, P. M. In *Structure/Performance Relationships in Surfactants;* Rosen, M. J. Ed; ACS Symposium Series 253, American Chemical Society: Washington, DC, 1984; pp 141-151.
11. Münster, A. *Statistical Thermodynamics;* Springer-Verlag: Berlin, 1974; Vol. 2, p. 650.
12. Scamehorn, J. F.; Schechter, R. S.; Wade, W. H. *J. Dispersion Sci. Technol.* **1982,** *3,* 261.
13. Hey, M. J.; MacTaggart, J. W. *J. Chem. Soc., Faraday Trans. I* **1985,** *81,* 207.
14. Rathman, J. F.; Scamehorn, J. F. *Langmuir* **1988,** *4,* 474.
15. Förster, T.; von Rybinski, W.; Schwuger, M. J. *Tenside Surf. Det.* **1990,** *27,* 254.
16. Hildebrand, J. H.; Prausnitz, J. M.; Scott, R. L. *Regular and Related Solutions;* Van Nostrand: New York, NY, 1970; pp 96-7.
17. Graciaa, A.; Ben Ghoulam, M.; Marion, G.; Lachaise, J. *J. Phys. Chem.* **1989,** *93,* 4167.
18. Rubingh, D. N. In *Solution Chemistry of Surfactants;* Mittal, K. L. ed.; Plenum Press, New York, NY, 1979, Vol. 3; pp 337-354.
19. Lange, H.; Beck, K. H. *Kolloid-Z.u.Z. Polymere* **1973,** *251,* 424.
20. Nguyen, C. M.; Rathman, J. F.; Scamehorn, J. F. *J. Colloid Interface Sci.* **1986,** *112,* 438.
21. Ingram, B. T.; Luckhurst, A. H. W. In *Surface Active Agents;* Soc. Chem. Ind.: London, 1979, pp 89-98.
22. Rosen, M. J.; Hua, X. Y. J. *J. Colloid Interface Sci.* **1982,** *86,* 164.
23. Corkhill, J. M.; Goodman, J. F.; Ogden, C. P.; Tate, J. R. *Proc. R. Soc. London Ser. A,* **1963,** *273,* 84.
24. Zhu, B. Y.; Rosen, M. J. *J. Colloid Interface Sci.* **1984,** *99,* 435.
25. Rosen, M. J.; Zhu, B. Y. *J. Colloid Interface Sci.* **1984,** *99,* 427.

RECEIVED February 18, 1992

Chapter 3

Terminology for Describing Thermodynamic Properties of Nonideal Mixed Micellar Systems

Sherril D. Christian, E. E. Tucker, and John F. Scamehorn

Institute for Applied Surfactant Research, University of Oklahoma, Norman, OK 73019

The term 'regular solution theory' is frequently used to denote models for describing nonideality in aqueous mixed surfactant systems, particularly when activity coefficients of the components follow the Rubingh (1) one-parameter symmetrical activity coefficient relationships. However, this usage is inconsistent with that commonly employed in thermodynamic studies of nonelectrolyte solutions. Although the Gibbs-Duhem equation can with some justification be used to interrelate partial molar quantities for the components of the intramicellar 'pseudophase', the mixture should not be termed 'regular' simply because certain equations relating solute activities to intramicellar mole fractions are observed to apply. The regular solution label should be reserved for systems in which the entropy of formation of mixed micelles at constant volume does approximate that of an ideal solution. Mixed surfactant systems and micellar mixtures containing a single surfactant with an added solute seldom meet the ideal entropy requirement of regular solution theory, so describing these solutions as regular is usually inappropriate.

Several monographs provide information about the regular solution concept and its application to nonelectrolyte solutions. In 1929, Hildebrand proposed that a solution be termed *regular* if the entropy change is zero for transferring a small amount of one of its components from the actual solution to an ideal solution having the same composition (all at constant volume) (2). In other words, a solution is defined as regular if the partial molar entropy of a component in the mixture (at any given mole fraction) is the same as that of the component in an ideal solution *at the same mole fraction*.

0097–6156/92/0501–0045$06.00/0

The requirement that the volume of the mixture equal the sum of the volumes of the pure components was introduced by Hildebrand to compensate for non-specific enthalpic or energetic effects on the total entropy of solution (2-3). For example, hexane and perfluorohexane mix endothermically, with a positive volume change on mixing, and with a nonideal (positive) excess entropy of mixing. But if a solution of hexane and perfluorohexane, formed at a constant pressure of one atmosphere, is compressed isothermally back to its original volume, the total entropy of mixing quite closely approximates that of an ideal mixture (4). Such a mixture can therefore be described as (approximately) regular and deviations from ideality for such a system can be related to non-specific enthalpic (energy) contributions to the chemical potential. A major achievement of Hildebrand and his collaborators has been their use of the regular solution concept, together with solubility parameters to account for nonideal heats of mixing, to make quantitative or near-quantitative predictions of the Gibbs free energy and other properties of nonelectrolyte solutions (2-5). In systems where specific chemical complexes are absent, it is the rule rather than the exception that only the enthalpic contribution to the Gibbs free energy need be measured or predicted in order to predict the activities (or escaping tendencies) of all the components in a complicated nonelectrolyte mixture.

The prospect of using the regular solution concept to describe surfactant solutions has appealed to many investigators interested in mixed micelle properties(1,6-8). One problem encountered in testing the applicability of regular solution theory - - which in principle involves determining or calculating the entropy of mixing at constant volume - - is avoided in studies of many types of mixed micellar systems. It has been observed that even systems in which the mixed micelles form with large exothermic or endothermic heat effects exhibit quite small volume changes on mixing (at constant pressure) (9-10). In fact, even cationic-anionic surfactant mixtures, in homogeneous aqueous solution, form mixed micelles having total volumes almost exactly equal to the sum of the volumes of the separate one-component micelles (10). This is not to argue that the existence of larger volume effects would have precluded the use of regular solution theory, but the fact that volumes of mixing are so nearly ideal makes it unnecessary to consider nonideal entropy effects related to changes in volume.

The purpose of this short paper is to examine thermodynamic results pertaining to the formation of several types of mixed micelles to determine whether the designation 'regular solution' aptly applies to any of these systems. Except for the mixed micelles formed from surfactants which are so similar that the micelles form ideally, it will be concluded that virtually none of the various types of mixed micellar systems conform to the regular solution model. In other words, as in analogous studies of the thermodynamic properties of associating systems of many types, entropic effects pertaining to the formation of discrete complexes will be important, along with enthalpic and non-specific (ideal) entropy effects characteristic of other mixtures.

Empirical Representations of Activity Coefficients

In correlating the concentration-dependence of the activities of components of mixtures at a given temperature and pressure, it is useful to obtain equations relating the activity coefficients of species (based on pure-component standard states) to their mole fractions in solution. Thus it can be shown that in binary nonelectrolyte solutions at constant temperature and pressure, the natural logarithms of the activity coefficients of the components (designated by subscripts '1' and '2') can be expressed as:

$$\ln \gamma_1 = X_2^2[A + BX_1 + CX_1^2 + DX_1^3 + \ . \ . \ . \ ]$$
$$\ln \gamma_2 = X_1^2[(A - B/2) + (B - 2C/3)X_1 + (C - 3D/4)X_1^2 + \ . \ . \ . \ ] \quad (1)$$

where X_1 and X_2 are the mole fractions of species 1 and 2 in the mixture (adding up to 1) and where A, B, C, D, . . are empirical constants (*11-12*). The coefficients in the two series are related in such a way as to force conformity with the Gibbs-Duhem equation. The absence of terms in the expressions for $\ln \gamma_1$ and $\ln \gamma_2$ having a lower order than the square of the mole fraction of the opposite component is required by the choice of standard states (thus, the activity coefficient of a component must approach a value of 1 as the mole fraction of that component approaches 1) and the requirement that Raoult's law be obeyed as a limiting law (*3*).

Simple one-parameter forms of equations 1 given above, which are often called Margules equations (*13*), can be used to describe many nonelectrolyte systems (at constant temperature and pressure) in which enthalpy and volume changes are relatively small. The resulting symmetrical equations,

$$\ln \gamma_1 = AX_2^2 \quad \text{and} \quad \ln \gamma_2 = AX_1^2 \quad (2)$$

are convenient for fitting activity data for a variety of types of solutions, including many surfactant solutions in which mixed micelles form, treated by a 'pseudophase' equilibrium model (see below). Rubingh (*1*) was apparently the first investigator to use equations 2 to fit mixture cmc data for binary surfactant systems, so it is appropriate to call them *Rubingh* equations, or simply one-parameter Margules equations. Only rarely have forms of equation 1 been used with more than one parameter (*14*).

It can be shown that if equations 2 apply in the form $\ln \gamma_1 = wX_2^2/RT$ and $\ln \gamma_2 = wx_1^2/RT$, where the parameter w (which is equal to RTA) does not vary with either temperature or composition, then the entropy of mixing will be zero, and the solution may correctly be termed regular (*4,15*). Solutions may of course be regular without conforming to these simple equations, and this fact has led Hildebrand and his colleagues to argue strenuously against arbitrarily restricting the adjective 'regular' to apply to solutions having properties consistent with equations unrelated to the regular solution concept. On the other hand, one should also object to applying so useful a term as 'regular' to solutions that do not satisfy the ideal entropy requirement of regular solution theory. Studies of the temperature-dependence of activity coefficients for mixed micelles formed from dissimilar components have indicated that w can be a relatively strong function of temperature (*16*); therefore the adjective 'regular' should not be used to refer to systems simply because they have been shown to obey equations 2 at a single temperature. Given the specific nature of interactions in surfactant micelles (including association and repulsion terms involving the head groups in systems of mixed charge type), it is not surprising that specific entropy effects may come into play. Consequently, before a surfactant mixture can be termed regular, evidence should be presented that the system does in fact exhibit the ideal entropy of mixing behavior at constant volume required in the definition.

The Pseudo-Phase Equilibrium Model for Mixed Micelles

Numerous investigators have drawn analogies between micelle-monomer equilibrium and liquid-vapor equilibrium. In liquid-vapor equilibrium, a condensed phase (pure component or mixture) co-exists with a vapor phase in which the molecules may ordinarily be assumed to obey the ideal gas equation. At a given temperature, and at a nearly constant total pressure, the composition of the vapor is uniquely determined by the composition of the liquid phase. The activity of an individual component in the liquid mixture is taken to be equal to the ratio of the partial pressure of each component in the vapor to the vapor pressure of that pure component at the same temperature (and pressure). Similarly, micelles or mixed micelles can co-exist with the monomer(s) of the surfactant(s), and it is often a good approximation to assume that the monomer concentrations (and hence the thermodynamic activities of the components of the micelle) at a given temperature are uniquely determined by the composition of the micelle, provided the total concentration of the surfactant(s) in solution exceeds the critical micelle concentration (cmc). In fact, monomer concentrations of surfactants and of organic solutes, if present, in the extra-micellar solution (the 'bulk' aqueous phase) are ordinarily assumed to be proportional to the activities of these components; in other words, the monomers are assumed to be present in dilute enough concentrations to obey Henry's law.

For a pure single-component surfactant, in a given aqueous solution, the monomer concentration is equal to the critical micelle concentration, and both the activity and the activity coefficient of the surfactant are taken to be unity, as in the pure-component standard state for a liquid-vapor equilibrium system. (In most careful thermodynamic investigations of solutions involving ionic surfactants, a swamping electrolyte is added to maintain a nearly constant ionic strength and to avoid the effects of changes in the 'bulk' solution phase on solute activities.) Several studies of mixed micelle and organic solute-micelle systems have shown that the surfactant (and other solute) activities are the same for a given intramicellar composition (with constant added electrolyte), throughout moderate ranges of total surfactant composition (*17-19*).

If one accepts the analogy between liquid-vapor and micelle-monomer equilibrium, it may seem plausible to assume that the Gibbs-Duhem equation can be applied to the pseudo-phase, either in cases where two or more surfactants are present in the micelle or in micelle-organic solubilizate systems. This assumption may seem questionable, because the mixed micelle is not homogeneous in the usual chemical sense, because the size and shape of micelles can vary with composition, and because the presence of water or ions near the micellar surface is not explicitly taken into account. Nonetheless, in several studies it has been verified that activities of components in mixed micellar systems vary approximately as is required by the Gibbs-Duhem equation, and that various forms of equations (1) can be used to relate the activity coefficients of species in the micelle to the intramicellar mole fractions (denoted by X_1 and X_2, in the case of a binary pseudophase) (20). The presence of excess electrolyte, in systems involving ionic surfactants, undoubtedly helps to diminish the influence that variations in micellar charge, water activity, and ion activities might otherwise have on the validity of the assumed laws of bulk phase solutions. Treatment of mixed admicelles (adsorbed surfactant aggregates) as a pseudophase has been shown to result in consistency between thermodynamic data on the free energy of mixing and the Gibbs-Duhem equation (21). On the other hand, attempts to use the pseudophase equilibrium approximation to model activity coefficients of coacervate (22) and microemulsion (23) phases have not led to expressions consistent with the Gibbs-Duhem equation.

Nonideal Entropy Effects in Mixed Micelles

Neglecting volume changes on mixing, and given the assumptions of the pseudophase equilibrium model, one can decide on the basis of mixture cmc results (11) (which lead to Gibbs free energies of mixing) and heat of solution data whether or not nonideal entropy of mixing effects are important (24-25). Thus, the excess entropy of mixing at constant pressure and at a given temperature, T, is given by

$$S^E = (\Delta H^{mix} - G^E)/T \tag{3}$$

where ΔH^{mix} is the observed heat of mixing and G^E is the excess Gibbs free energy of mixing. Although studies providing both ΔH^{mix} and G^E results are rare, the general observation seems to be that in the case of exothermic mixing, ΔH^{mix} is larger numerically than G^E, indicating that S^E is negative (25). Such an effect conforms to commonly observed behavior in complexing systems, in that a negative entropy of association causes the entropy of mixing to be less positive than for an ideal solution (26).

A recent report gives calorimetric and cmc data for 10 binary surfactant mixtures of several different types (cationic-cationic, nonionic-nonionic, nonionic-anionic, and nonionic-cationic) (25). In almost all of these systems, ΔH^{mix} is found to be exothermic, although in the mixture of two cationic surfactants (N-hexadecyl-pyridinium chloride and hexadecyltrimethylammonium chloride) the effect is quite small. Mixed micelles of the cationic surfactant N-hexadecylpyridinium chloride and the zwitterionic dodecyldimethylphosphine oxide, surprisingly, form endothermically, but with positive values of S^E and negative G^E. Except for the nearly ideal cationic-cationic micelle system, all of the results indicate that S^E is significantly different from zero, and that TS^E is more negative than G^E for the systems which mix exothermically. Therefore, it is clear that departures from the ideal entropy of mixing required for conformity with regular solution theory are not only possible but typical of surfactant mixtures.

A similar situation applies to systems in which neutral organic molecules are solubilized by ionic surfactant micelles. Although direct calorimetric studies of such systems have been rare, precise solubilization 'isotherms' have been obtained for a number of volatile solutes in anionic, cationic, and neutral micelles at several temperatures (27-29). Thus, the entropies of solubilization of benzene and cyclohexane by N-hexadecylpyridinium chloride, sodium dodecylsulfate, and nonylphenylpoly(ethylene oxide) micelles are significantly nonideal, as is the entropy of solubilization of *tert*-butanol by N-hexadecylpyridinium chloride and sodium dodecylsulfate (28). The possible effects of volume changes on the thermodynamics of mixing in these systems have not been considered, but these effects are not apt to be great enough to cause S^E to approach zero if they are taken into account. In several cases, positive heats of formation of the micelle-solubilizate aggregates are accompanied by negative values of S^E.

Conclusions

Regular solutions are mixtures that form with ideal entropy of mixing at constant volume. By examining mixture cmc and heat of mixing results, one can determine whether regular solution theory can be used to describe these systems. The pseudophase equilibrium model can be applied to the formation of mixed micelles, and if volume changes are ignored (which seems justifiable for many types of surfactant mixtures), it is easy to estimate values of the excess entropy of mixing for forming mixed micelles. The fact that results for the mixed micelles apparently conform to the symmetrical activity coefficient equations, $\ln \gamma_1 = AX_2^2$ and $\ln \gamma_2 = AX_1^2$, does not help in determining if the formation of mixed micelles is regular. Only one test need be applied: whether S^E does or does not approximate zero. The available data make it clear that regular solution properties are rarely exhibited by mixed micelle systems, except for the trivial case of mixtures of similar surfactants of the same charge type, which are often nearly ideal. Micelles in which one or more surfactants exist with solubilized organic solutes also usually form with nonzero S^E values.

Acknowledgments

Financial support for this work was provided by the Department of Energy Office of Basic Energy Sciences Grant No. DE-FG05-84ER13678, National Science Foundation Grant No. CBT-8814147, U.S. Bureau of Mines Grant No. G1174140-4021, Oklahoma Center for the Advancement of Science and Technology, the Oklahoma Mining and Minerals Resources Research Institute, Aqualon Corp., Arco Oil and Gas Co., E.I. DuPont de Nemours & Co., Kerr-McGee Corp., Mobil Corp., Sandoz Chemical Co., Shell Development Co., Unilever Corp., and Union Carbide Corp.

Literature Cited

1. Rubingh, D. N. in *Solution Chemistry of Surfactants, Vol.1* , Mittal, K. L., Ed., Plenum Press: New York, 1979, p.337.
2. Hildebrand, J. H., *J. Am. Chem. Soc.*, **1929**, *51*, 66.

3. Hildebrand, J. H.; Scott, R. L. *Solubility of Nonelectrolytes*, 3rd. Edition, Reinhold: New York, 1950.

4. Hildebrand, J. H.; Scott, R. L. *Regular Solutions*, Prentice-Hall, Englewood Cliffs, N. J., 1962.

5. Hildebrand, J. H.; Prausnitz, J. M.; Scott, R. L. *Regular and Related Solutions*, Van Nostrand Reinhold, New York, 1970.

6. Holland, P. M., *Adv. Colloid and Interface Sci.*, **1986**, *26*, 111.

7. Scamehorn, J. F. in *Phenomena in Mixed Surfactant Systems*; Scamehorn, J. F., Ed.; ACS Symp. Ser., Vol. 311, American Chemical Society, Washington, D. C., 1986, p. 1.

8. Rosen, M. J. in *Phenomena in Mixed Surfactant Systems*; Scamehorn, J. F., Ed.; ACS Symp. Ser., Vol. 311, American Chemical Society, Washington, D. C., 1986, p. 144.

9. Nishikido, N.; Imura, Y.; Kobayashi, H.; Tanaka, M. *J. Colloid Interface Sci.*, **1983**, *91*, 125.

10. Lopata J. J.; Scamehorn, J. F.; in preparation.

11. Hansen, R. F.; Miller, F. A.; *J. Phys. Chem.*; **1954**, *58*, 193.

12. Christian, S. D.; *J. Chem. Educ.*; **1962**, *39*, 521.

13. Reid, R. C.; Prausnitz, J. M.; Poling, B. E.; *The Properties of Gases and Liquids*, Fourth Edition, McGraw-Hill: New York, 1987; p. 256.

14. Kamrath, R. F.; Franses, E. I.; *Ind. Eng. Chem. Fundam.*, **1983**, *22*, 230.

15. Guggenheim, E. A.; *Thermodynamics*; North-Holland: Amsterdam, 1949.

16. Nguyen, C. M.; Rathman, J. F.; Scamehorn, J. F.; *J. Colloid Interface Sci.* , **1986**, *112*, 438.

17. Rathman, J. F.; Christian, S. D.; *Langmuir*, **1990**, *6*, 391.

18. Christian, S. D.; Tucker, E. E.; Smith, G. A.; Bushong, D. S.; *J. Colloid Interface Sci.*, **1986**, *113*, 439.

19. Smith, G. A.; Ph. D. Dissertation, University of Oklahoma, 1986.

20. Osborne-Lee, I. W.; Schechter, R. S. in *Phenomena in Mixed Surfactant Systems*; Scamehorn, J. F., Ed.; ACS Symp. Ser., Vol. 311, American Chemical Society, Washington, D. C., 1986, p.30.

21. Lopata, J. J.; Harwell, J. H.; Scamehorn, J. F.; in preparation.

22. Yoesting, O. E.; Scamehorn, J. F., *Colloid Polym. Sci.*, **1986**, *264*, 148.

23. Haque, O.; Scamehorn, J. F., *J. Dispersion Sci. Technol.*, **1986**, 7, 129.

24. Holland, P. M. in *Structure/Performance Relationships in Surfactants*, Rosen, M., Ed., ACS Symp. Ser., Vol. 253, American Chemical Society, Washington, D. C., 1984, p. 141.

25. Rathman, J. F.; Scamehorn, J. F., *Langmuir*, **1988**, *4*, 474.

26. Foster, R.; *Organic Charge Transfer-Complexes*, Academic Press: London and New York, 1969, pp. 205-211.

27. Smith, G. A.; Christian, S. D.; Tucker, E. E.; Scamehorn, J. F. in *Use of Ordered Media in Chemical Separations*, Hinze, W. L.; Armstrong, D. W.; Eds., ACS Symp. Ser., Vol. 342, American Chemical Society, Washington, D.C., 1987, Ch. 10, p. 184.

28. Smith, G. A.; Christian, S. D.; Tucker, E. E.; Scamehorn, J. F., *J. Colloid Interface Sci.*, **1989**, *130*, 254.

29. Tucker, E. E.; Christian, S. D., *Faraday Symp. Chem. Soc.*, **1982**, *17*, 11.

RECEIVED January 21, 1992

Approaches to Modeling Mixed Surfactant Aggregates

Chapter 4

Micellization of Binary Surfactant Mixtures
Theory

R. Nagarajan

Department of Chemical Engineering, Pennsylvania State University, University Park, PA 16802

We develop a molecular theory of micelle formation in aqueous solutions of binary surfactant mixtures. The theory allows the prediction of the critical micelle concentration and the size and composition distribution of mixed micelles. Illustrative results are obtained for nonionic-nonionic, ionic-ionic, nonionic-ionic, anionic-cationic and hydrocarbon-fluorocarbon type of binary surfactant mixtures and compared against available experimental data. It is found that the theory predicts well the ideal and nonideal behavior exhibited by different mixtures. Of special interest are the predictions showing both the formation of miscible hydrocarbon-fluorocarbon mixed micelles and the coexistence of mutually immiscible hydrocarbon rich and fluorocarbon rich micelle populations. Finally, the theory allows one to identify the origin of mixture nonidealities in terms of various molecular interactions.

A molecular theory of mixed micelles was developed earlier by us [1], by extending the theory formulated by Nagarajan and Ruckenstein [2] for aggregation in single component surfactant systems. The theory predicted the size and the composition distribution of mixed micelles, thus allowing one to estimate the critical micelle concentration (cmc), the average micelle size and the average micelle composition, as a function of the composition of either the total surfactant mixture or the singly dispersed surfactants. The theory facilitated a molecular interpretation of the observed ideal and nonideal behavior of the mixed surfactant systems, in terms of the bulk and the surface interactions between surfactants constituting the micelles [1,3]. For hydrocarbon surfactant mixtures, the theory revealed that significant nonideality arises from the surface interactions.

0097–6156/92/0501–0054$11.50/0

The theory of mixed micelles was applied to surfactant-alcohol mixtures by Rao and Ruckenstein [4] who treated the short chain alcohols as nonionic surfactants. They showed that the decrease in cmc and the change in micellar aggregation number caused by the presence of alcohols, the extent of incorporation of alcohols inside the micelles and the alcohol partition equilibrium constant between the micellar and the aqueous phases are all predicted reasonably well for mixtures of some anionic, cationic, and nonionic surfactants with butanol, pentanol, hexanol and heptanol, respectively. Warr et al. [5] used fluorescence probe methods to measure the aggregation number of nonionic dodecyl maltoside (DM) and nonionic octaoxyethylene dodecyl ether ($C_{12}E_8$) mixed micelles and found that the composition dependence of the aggregation number of the micelle was not linear but possessed a distinctive s-shape, qualitatively similar to that predicted by our molecular theory. From small angle neutron scattering study of nonionic dodecyl maltoside (DM) and anionic sodium dodecyl sulfate (SDS) mixed micelles, Bucci et al. [6] observed that the micelle aggregation number exhibits a maximum when the micelle composition is altered in qualitative agreement with our theoretical predictions for nonionic-ionic mixed micelles [1]. More recently, our theory of mixed micelles was applied to binary mixtures of hydrocarbon and fluorocarbon surfactants by Asakawa et al. [7]. They showed that the theory correctly predicted the coexistence of two distinct, fluorocarbon rich and hydrocarbon rich populations of micelles as well as the growth in micellar size accompanying the composition changes. However, the increased region of immiscibility between the two coexisting micellar populations that was experimentally observed at high electrolyte concentrations was not quantitatively well predicted.

The principal goal of this paper is to improve upon various features of our earlier theory [1] with respect to the quantitative estimation of the free energy of micelle formation. In particular, the free energy model is made strictly predictive by eliminating some of the empirical expressions employed in our earlier treatment and replacing them with results from our recent theory for single component surfactants [8]. Firstly, we replace the empirical expression for calculating the free energy of conformational constraints on surfactant tails inside the micelles, by an analytical expression recently derived by Nagarajan and Ruckenstein [8]. The analytical free energy expression which was derived for single surfactant systems in ref. [8] is extended in this paper to treat surfactant mixtures having different hydrophobic tail lengths. Secondly, we replace the Debye-Huckel expression (with an empirical correction factor of 0.46) used in our earlier theory to compute ionic interactions in surfactant aggregates, with an approximate analytical solution to the Poisson-Boltzmann equation developed by Evans and Ninham [9]. Thirdly, we apply the solubility parameter theory of Hildebrand [10] to estimate the free energy contribution due to the mixing of surfactant tails inside the mixed aggregates. Fourthly, while treating spherocylindrical micelles, we allow the radii as well as the composition of the spherical ends of the micelles to be different from those of the cylindrical middle part as suggested by the work of Ben-Shaul et al. [*11*] on surfactant-alcohol mixtures.

The present work concentrates on binary surfactant mixtures but the approach developed here can readily be extended to solutions containing three or more surfactant components. The present theory does not take into account any interaggregate interactions and hence is strictly applicable only to dilute surfactant solutions. The interaggregate interactions, when present, have been shown to affect the aggregate size distribution by Ben-Shaul and Gelbart [*12*] in their study of single component surfactants. Blankschtein and coworkers [*13-15*], Goldstein [*16*] and Kjellander [*17*] have accounted for the interaggregate interactions in their treatments of phase separation in single component surfactant solutions. Indeed, both micelle formation and phase separation are simultaneously treated by Blankschtein and coworkers and by Goldstein. In their treatments, the form of interaggregate interaction energy assumed is such that the size distribution of aggregates is not affected by the interaggregate interactions. Recently, Puvvada and Blankschtein have extended the approach to binary mixtures of surfactants [*18*].

This paper is organized as follows. In the first section, we describe the general thermodynamic relations governing the molecular assembly of mixed surfactants. It is followed by a discussion of the geometrical characteristics of aggregates. Then we formulate the model for the free energy of aggregation. Finally, illustrative predictions for different types of surfactant mixtures are presented and wherever possible, compared with available experimental data. The details of the treatment relating to rodlike micelles are omitted from this text because of space restrictions, even though calculations exploring the formation of rods have been carried out for all the mixtures examined in this study. These details will be published elsewhere.

Size and Composition Distribution of Micelles

We consider a surfactant solution that consists of water molecules, singly dispersed surfactant molecules of type A and B and mixed micelles of various sizes and compositions. We denote by g the aggregation number of the mixed micelle containing g_A molecules of type A and g_B molecules of type B. The aggregate of a given size and composition is viewed as a distinct chemical component and the surfactant solution is treated as a multicomponent solution. At equilibrium, the total Gibbs energy of the surfactant solution is a minimum and this leads [*1*] to the size and composition distribution equation:

$$X_g = X_{1A}^{g_A} X_{1B}^{g_B} \exp - (\frac{g \, \Delta\mu_g^o}{kT}) \, ; \quad (\frac{g \, \Delta\mu_g^o}{kT}) = (\frac{\mu_g^o - g_A \, \mu_{1A}^o - g_B \, \mu_{1B}^o}{kT}) \quad (1)$$

In the above equation, μ_g^o is the standard chemical potential of a surfactant micelle containing g surfactant molecules while μ_{1A}^o and μ_{1B}^o are the standard chemical potentials of the singly dispersed A and B molecules, respectively. The standard chemical potentials are defined as those corresponding to infinitely dilute solution conditions. $\Delta\mu_g^o$ is the difference in the standard chemical potentials between g_A/g molecules of surfactant A and g_B/g molecules of surfactant B present in an aggregate of size g and the same molecules present in their singly dispersed states in water. X_{1A} and X_{1B} are the mole fractions of the singly dispersed surfactants A and B while X_g is the mole fraction of the aggregates of size g in the solution. The mole fraction X_g is dependent not only on the size g but also on the composition of the micelle. We define the mole fraction of components A and B in the singly dispersed surfactant mixture, in the mixed micelle and in the total surfactant solution, respectively, by the relations:

$$\alpha_{1A} = \frac{X_{1A}}{X_{1A} + X_{1B}} = \frac{X_{1A}}{X_1} \quad , \qquad \alpha_{1B} = \frac{X_{1B}}{X_{1A} + X_{1B}} = \frac{X_{1B}}{X_1} \tag{2}$$

$$\alpha_{gA} = \frac{g_A}{g_A + g_B} = \frac{g_A}{g} \quad , \qquad \alpha_{gB} = \frac{g_B}{g_A + g_B} = \frac{g_B}{g} \tag{3}$$

$$\alpha_{tA} = \frac{X_{1A} + \Sigma g_A X_g}{X_1 + \Sigma g X_g} \quad , \qquad \alpha_{tB} = \frac{X_{1B} + \Sigma g_B X_g}{X_1 + \Sigma g X_g} \tag{4}$$

In equation (4), the summation is over two independent variables, namely, the aggregation number $g = 2$ to ∞, and the micelle composition $\alpha_{gA} = 0$ to 1. For simplicity of notations, in this paper, a single summation is used in the various equations to account for all independent variables.

The size and composition distribution equation (1) can be rewritten in a form identical to that for a single surfactant system by using equations (2) and (3).

$$X_g = X_1^g \exp - \left(\frac{g \Delta\tilde{\mu}_g^o}{kT} \right) \tag{5}$$

where

$$\frac{\Delta\tilde{\mu}_g^o}{kT} = \frac{\Delta\mu_g^o}{kT} - [\alpha_{gA} \ln \alpha_{1A} + \alpha_{gB} \ln \alpha_{1B}] \tag{6}$$

From the size and composition distribution one can compute the average sizes of the aggregates via the relations

$$g_n = \frac{\Sigma\, g\, X_g}{\Sigma\, X_g}\,, \qquad g_w = \frac{\Sigma\, g^2\, X_g}{\Sigma\, g\, X_g} \tag{7}$$

where g_n and g_w denote the number and the weight average aggregation numbers. Similarly, the average composition of the mixed micelle can be calculated from

$$\overline{\alpha}_{gA} = \frac{\Sigma\, \dfrac{g_A}{g}\, X_g}{\Sigma\, X_g}\,, \qquad \overline{\alpha}_{gB} = \frac{\Sigma\, \dfrac{g_B}{g}\, X_g}{\Sigma\, X_g} \tag{8}$$

The critical micelle concentration X_{cmc} can be obtained by constructing a plot of one of the functions X_1, $\Sigma\, g\, X_g$ or $\Sigma\, g^2\, X_g$ (which are proportional to different experimentally measured properties of the surfactant solution) against the total concentration X of the surfactant in solution, $X = X_1 + \Sigma\, g\, X_g$. The cmc is estimated to be that value of the total surfactant concentration at which a sharp change in the plotted function (representing a physical property) occurs.

The critical micelle concentration, the average aggregation number and the average micelle composition can all be precisely estimated as specified above by computing the concentrations of aggregates of all possible sizes and compositions using the distribution equation (1). Such computations, though not complicated, are time consuming and are not always needed. A simpler approach yielding precise results is possible for systems displaying narrow size and composition dispersions. This is the case when micelles formed are spherical or globular with relatively small aggregation numbers (a condition met by most of the systems considered in this paper). In such systems, one can replace the entire size and composition distribution of aggregates with a single aggregate whose size and composition corresponds to the maximum in the distribution function, X_g. This approach is utilized in this paper for calculating the average composition and aggregation number for spherical or globular micelles. This is equivalent to approximating the size distribution or multiple equilibria model by the monomer-micelle equilibrium model. To estimate the cmc, we follow the suggestion of Hartley [19], $X_1 = g\, X_g(\text{maximum}) = X_{cmc}$, namely, the cmc is the concentration of the monomers when the amount of surfactant present as monomers is equal to that present as micelles. As mentioned earlier, the cmc can also be estimated by plotting one of the functions X_1, $\Sigma\, g\, X_g$ or $\Sigma\, g^2\, X_g$ against the total surfactant concentration. Another common method for estimating the cmc assumes it to be the total surfactant concentration at which about five percent of the surfactant is present as micelles, the remaining surfactant being singly dispersed. The different procedures give estimates for cmc varying from one another by at most five percent. Similar variations occur in the experimental determination of the cmc depending on which measurement technique is employed since different physical measurements sense the formation of micelles at somewhat different surfactant concentrations. Thus, the cmc estimated by any one of the procedures mentioned should be comparable to the experimentally determined cmc.

To proceed further and calculate the aggregation behavior of surfactants, a model for the standard free energy difference associated with micelle formation is necessary. This, in turn, requires the specification of the shapes of the aggregates and of their geometrical characteristics.

Geometrical Properties of Micelles

The hydrophobic interiors of the surfactant aggregates are constituted of the surfactant tails. Molecular packing requirements suggest [20,21] that no point within the aggregate can be farther than ℓ_{sA} or ℓ_{sB} (whichever is larger) from the aggregate core-water interface, where ℓ_{sA} and ℓ_{sB} refer to the extended lengths of the surfactant tails of molecules A and B. Keeping this constraint in view, one can assume that the small micelles are spherical in shape. The shapes of somewhat larger micelles have been examined by Israelachvili et al. [21] on the basis of local and overall molecular packing considerations and they have suggested shapes generated via ellipses of revolution. We note that for aggregation numbers that are up to 3 times as large as that of the largest spherical micelles, the average area per molecule for the ellipses of revolution suggested by Israelachvili et al. [21] is practically the same as for prolate ellipsoids. Therefore, the average geometrical properties of the non-spherical globular aggregates will be computed here as for prolate ellipsoids without implying however, that prolate ellipsoidal micelles form. When micelles are very large, we consider them to be rodlike with spherical endcaps and a cylindrical middle part. The two regions of the micelle are allowed to have differing radii. Figure 1 illustrates the different micellar shapes.

Spherical Micelles.   For spherical micelles containing g surfactant molecules, and having a radius R_s of the hydrophobic core, the total volume of the aggregate core, V_g, and the core surface area, A_g, are given by

$$V_g = \frac{4 \pi R_s^3}{3} = g \, v_s \; ; \; v_s = (\alpha_{gA} \, v_{sA} + \alpha_{gB} \, v_{sB}) \tag{9}$$

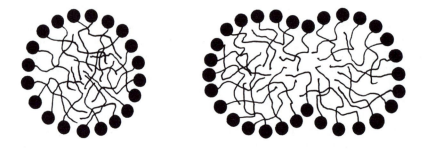

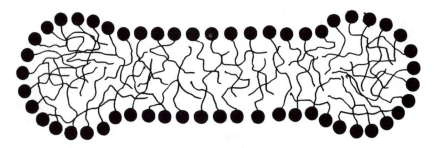

Figure 1. Schematic representation of spherical, globular and rodlike micelles. The geometrical characteristics of the globular micelles are calculated as those of prolate ellipsoids.

$$A_g = 4 \pi R_s^2 = g a \tag{10}$$

respectively. Here, v_{sA} and v_{sB} denote the volumes of the hydrophobic tails of A and B, while v_s is the micelle composition averaged volume of the hydrophobic tails. The variable, a, refers to the surface area of the hydrophobic core of the micelle per surfactant molecule. We define an area per molecule a_δ at a distance δ from the hydrophobic core of the micelle for latter use in the calculation of the ionic interaction energy at the micellar surface.

$$A_{g\delta} = 4 \pi (R_s + \delta)^2 = g a_\delta \tag{11}$$

We also define a geometrical ratio P that characterizes the average molecular packing in the aggregates for latter use in the calculation of surfactant tail deformation free energy.

$$P = \frac{V_g}{A_g R_s} = \frac{v_s}{a R_s} \tag{12}$$

Globular Micelles.   The micelles whose sizes are somewhat larger than those allowed by the spherical shape are called globular and their average geometrical characteristics are computed as for prolate ellipsoids. Consequently, one dimension of the prolate ellipsoid should be the largest radius allowed for the spherical micelles. This dimension is the semi-minor axis of the globular micelle and is denoted by R_s while the semi-major axis is denoted by b. The eccentricity E of the micelle is given by

$$E = [1 - (\frac{R_s}{b})^2]^{1/2} \tag{13}$$

The total volume of the hydrophobic core of the micelle is computed from

$$V_g = \frac{4 \pi R_s^2 b}{3} = g v_s \tag{14}$$

where v_s has already been defined as the composition averaged volume of the surfactant tails. The total surface area of the hydrophobic core is given by

$$A_g = 2 \pi R_s^2 [1 + \frac{\sin^{-1} E}{E (1 - E^2)^{1/2}}] = g a \tag{15}$$

The eccentricity of the micelle at a distance δ from the hydrophobic core is denoted by E_δ.

$$E_\delta = [1 - (\frac{R_s + \delta}{b + \delta})^2]^{1/2} \tag{16}$$

and the area of the micelle at that surface is calculated from

$$A_{g\delta} = 2 \pi (R_s + \delta)^2 [1 + \frac{\sin^{-1} E_\delta}{E_\delta (1 - E_\delta^2)^{1/2}}] = g a_\delta \tag{17}$$

Further, we define an equivalent radius R_{eq} of the globular micelle by considering the volume of the micelle to be the same as that of an equivalent sphere,

$$R_{eq} = (\frac{3 V_g}{4 \pi})^{1/3} \tag{18}$$

The packing factor P for the globular micelle defined as in equation (12) is given by

$$P = \frac{V_g}{A_g R_s} = \frac{v_s}{a R_s} \tag{19}$$

Free Energy of Formation of Mixed Micelles

We now formulate explicit expressions for the standard free energy difference between the surfactant molecules A and B present in a micelle and those present in the singly dispersed state in water. This free energy difference is composed of a number of contributions each of which has been described in our earlier work on single component surfactant systems [8]. Consequently, only the modifications warranted for the treatment of surfactant mixtures are discussed in detail here. For the derivation of the various equations one should consult reference [8].

Transfer Free Energy of the Surfactant Tail.   When micelles form, the hydrophobic tail of the surfactant is transferred from its contact with water to the hydrophobic core of the micelle. The contribution to the free energy from this transfer process is estimated by considering the micelle core to be like a liquid hydrocarbon (For fluorocarbon micelles, we simply change the description to that of a liquid fluorocarbon). The fact that the micelle core differs somewhat from a liquid hydrocarbon gives rise to an additional free energy contribution that is evaluated separately. The transfer free energy of the surfactant tail from water to a liquid hydrocarbon state is estimated from independent experimental data regarding the solubility of hydrocarbons in water [22,23]. On this basis, we have estimated the transfer free energy for a methylene group in an aliphatic tail as a function of temperature to be

$$\frac{(\Delta\mu^o_g)_{tr}}{kT} = 5.85 \ln T + \frac{896}{T} - 36.15 - 0.0056\,T \tag{20}$$

For a methyl group in the aliphatic chain, the temperature dependent transfer free energy is

$$\frac{(\Delta\mu^o_g)_{tr}}{kT} = 3.38 \ln T + \frac{4064}{T} - 44.13 + 0.02595\,T \tag{21}$$

In the above two relations, the temperature T is expressed in °K.

For surfactants containing two hydrocarbon chains, one can estimate the free energy contributions from the methyl and the methylene groups using the above equations. However, one expects considerable intramolecular interactions between the two chains in their singly dispersed state. Hence, the contribution to the transfer free energy of the two chains would be smaller than the estimate obtained by assuming there are two independent single chains. Tanford [20] estimated that the second chain of a dialkyl molecule contributes a transfer free energy that is sixty percent of an equivalent single chain molecule. In the absence of more detailed thermodynamic studies on such double chain molecules, we will assume Tanford's estimation to be adequate and use it in our free energy calculations.

For fluorocarbons, only limited thermodynamic transfer free energy data are available. Mukerjee and Handa [24] measured interfacial tension changes at the perfluorohexane-water interface caused by the adsorption of sodium perfluorobutyrate, sodium perfluorooctanoate and sodium perfluorodecanoate. From these interfacial tension changes, they estimated the free energy of transfer of the perfluoroalkyl chains from water to perfluorohexane. Using these data, we estimate the transfer free energy at 25°C to be

$$\frac{(\Delta\mu_g^o)_{tr}}{kT} = -6.20 \ (\text{for } CF_3), \ -2.25 \ (\text{for } CF_2) \tag{22}$$

Transfer free energies at other temperatures are presently unavailable.

For a mixed micelle having the composition $(\alpha_{gA}, \alpha_{gB})$, the transfer free energy per surfactant molecule is given by

$$\frac{(\Delta\mu_g^o)_{tr}}{kT} = \alpha_{gA} \frac{(\Delta\mu_g^o)_{tr,A}}{kT} + \alpha_{gB} \frac{(\Delta\mu_g^o)_{tr,B}}{kT} \tag{23}$$

The transfer free energy estimates from equation (20) to (22) do not include any contributions arising from the mixing of hydrocarbon-hydrocarbon, fluorocarbon-fluorocarbon, or hydrocarbon-fluorocarbon chains inside the micellar core. Such a contribution is taken into account separately, below.

Deformation Free Energy of the Surfactant Tail. The surfactant tails inside the hydrophobic core of the aggregate are not in a state identical to that in liquid hydrocarbons (or fluorocarbons). This is because, one end of the surfactant tail in the aggregate is constrained to remain at the micelle-water interface, while the entire tail has to assume a conformation consistent with the maintenance of an uniform density equal to that of liquid hydrocarbon in the micelle core. Consequently, the formation of micelles is associated with a positive free energy contribution stemming from the deformation of the anchored surfactant tail. We have used a lattice picture for the micelle core to develop simple analytical expressions for the chain deformation free energy as explicit functions of the micelle aggregation number and micellar shape [8]. The use of a lattice requires the specification of the size of the molecular segment which can be placed on the lattice without any orientational constraints. As suggested by Dill and Flory [25], for aliphatic hydrocarbon chains, a suitable segment is that which contains about 3.6 methylene groups. Correspondingly, the linear dimension of a lattice site, denoted by L, is taken equal to about 4.6Å. This linear dimension also represents the typical spacing between alkane molecules in the liquid state, namely, L^2 is equal to the cross-sectional area of a polymethylene chain. Consequently, a surfactant tail with an extended length of ℓ_s is considered as made up of N segments, where $N = \ell_s/L$. One end of the surfactant tail, namely that attached to the polar head group is constrained to be located at the micelle-water interface. The other end (the terminal methyl group) is free to occupy any position in the entire volume of the aggregate as long as an uniform segment density can be maintained within the micelle core. Obviously, the chains will be locally deformed in order to satisfy both the packing and the uniform density constraints. The conformational free energy per surfactant tail can be determined by calculating the integral of this local deformation energy over the entire volume of the micelle.

When the surfactants A and B have different hydrophobic tail lengths, segments of both molecules may not be present everywhere in the micellar core. Such a chain packing problem has been addressed by Szleifer et al. [26] and we will make use of some of their analysis. Let us assume that surfactant A has a longer tail than surfactant B, $\ell_{sA} > \ell_{sB}$. If the mixed micelles have the radius (semiminor axis in the case of globular micelles) R_s that is less than both ℓ_{sA} and ℓ_{sB}, then even the tail of the short chain surfactant B can reach everywhere within the core of the micelle. On the other hand, if $\ell_{sA} > R_s > \ell_{sB}$, then clearly the inner region of the micellar core, of dimension $(R_s - \ell_{sB})$ can only be reached by the A tails and not by the B tails, since the B tails cannot extend beyond their fully extended length, ℓ_{sB}. In generalizing our earlier result [8] for the free energy contribution due to the deformation of the tails, we take into account the different extents to which the A and the B tails are stretched for the two situations described above. Consequently, we can write

$$\frac{(\Delta\mu_g^o)_{def}}{kT} = B_g \left[\alpha_{gA} \frac{R_s^2}{N_A L^2} + \alpha_{gB} \frac{Q_g^2}{N_B L^2}\right], \quad B_g = \left(\frac{9\,P\,\pi^2}{80}\right) \tag{24}$$

where

$$Q_g = R_s \text{ if } R_s < \ell_{sA}, \ell_{sB}; \quad Q_g = \ell_{sB} = N_B L \text{ if } \ell_{sA} > R_s > \ell_{sB} \tag{25}$$

N_A and N_B refer to the number of segments in the tails of surfactants A and B, respectively. P is the packing factor defined before in equation (12) for spheres and equation (19) for globular aggregates.

For the condition, $\ell_{sA} > R_s > \ell_{sB}$, the innermost region of the micelle is not accessible to surfactant B. Therefore, the micelle must contain sufficient number of the A surfactants to completely fill up the inner region. This factor provides another condition for chain packing inside the micelles besides the more obvious criterion of $R_s \le \ell_{sA}$. To establish this additional packing condition, we define by f the fraction of A tail that can reach the inner region of the micellar core. To a first order approximation, $f = (\ell_{sA} - \ell_{sB}) / \ell_{sA}$. Therefore, for spherical micelles one has to satisy the condition

$$\frac{4\pi}{3} (R_s - \ell_{sB})^3 \le g\,\alpha_{gA}\,v_{sA}\,f \tag{26}$$

This in conjunction with equation (9) for the volume of the hydrophobic core yields the packing condition

$$\left(\frac{R_s - \ell_{sB}}{R_s}\right)^3 \le f\, \eta_A \,, \quad \eta_A = \frac{\alpha_{gA}\, V_{sA}}{\alpha_{gA}\, V_{sA} + \alpha_{gB}\, V_{sB}} = 1 - \eta_B \tag{27}$$

In this equation, η_A and η_B denote the volume fractions of A and B tails in the micelle core. Equation (27) defines the largest allowed value for R_s for a given micelle composition α_{gA}, or, the smallest allowed value for the micelle composition α_{gA} for a given micelle radius R_s. This equation was first derived by Szleifer et al. [26] in their study of chain packing in mixed surfactant aggregates. The parameter f is only defined to a first order approximation above, using its maximum possible value. Since there is no simple way to estimate f, we introduce a more restrictive packing criterion, that the radius be less than the composition averaged value of the tail lengths of the two surfactants, $R_s \le (\eta_A\, \ell_{sA} + \eta_B\, \ell_{sB})$. If this criterion is satisfied, then equation (27) is automatically satisfied.

For surfactants with two chains in their hydrophobic part, the deformation free energy can be calculated following the same approach adopted for single chain surfactants. One can show that the deformation free energy calculated using equation (24) for a single chain should be multiplied by a factor of two in order to obtain the deformation free energy per double tailed surfactant molecule. Therefore, when a double tailed surfactant is present in the mixture (as surfactant A and/or B), the term accounting for its deformation free energy in equation (24) is simply multiplied by two to account for both tails. The molecular packing constraint given by equation (27) remains unaffected whether the surfactant is with a single or a double tail.

Fluorocarbon chains have different geometrical characteristics compared to hydrocarbon chains. Consequently, the definition of a lattice site for micelles involving fluorocarbon surfactants has to be different. Lattice models have no simple means of accomodating such differences in the geometrical properties of molecules that are to be located on the same lattice. This is a common problem in lattice theories of solutions when the same solute is mixed with solvent molecules of differing sizes. At the present time, the ambiguity in the definition of the lattice site for mixed systems cannot be easily resolved. Since most examples studied in this work pertain to hydrocarbon surfactants, we retain the lattice definition given earlier as a general one and apply it also to systems containing fluorocarbon surfactants. Specifically, we will estimate the number of lattice segments N for fluorocarbon chains via the length ratio, $N = \ell_s/L$. A more fundamental approach to this problem is not possible without considerable mathematical complexity and is not undertaken here.

Aggregate Core-Water Interfacial Free Energy.   The formation of micelle generates an interface between the hydrophobic core region consisting of the surfactant tails and the surrounding water medium. The free energy associated with the formation of this interface has been taken in our previous calculations [1,2,8] as equal to the product of the area of the interface and the macroscopic interfacial tension characteristic of this interface. Such an expression is generalized for the case of binary mixtures.

$$\frac{(\Delta \mu_g^o)_{int}}{kT} = \frac{\sigma_{agg}}{kT} [a - \alpha_{gA} a_{oA} - \alpha_{gB} a_{oB}] \tag{28}$$

Here, σ_{agg} is the macroscopic micelle core-water interfacial tension, a is the surface area of the hydrophobic core per surfactant molecule defined earlier, and a_{oA} and a_{oB} are the areas per molecule of the core surface shielded from contact with water by the polar head groups of surfactants A and B.

The interfacial tension σ_{sw} between the surfactant tail (s may refer to A or B) and the surrounding water (w) is calculated in terms of the surface tensions σ_s of the surfactant tail and σ_w of water via [27] the relation

$$\sigma_{sw} = \sigma_s + \sigma_w - 2.0 \psi (\sigma_s \sigma_w)^{1/2} \tag{29}$$

where ψ is a constant with a value of about 0.55 [27,28] for aliphatic hydrocarbon tailed surfactants. For such tails, it has also been shown [8] that

$$\sigma_s = 35.0 - 325 M^{-2/3} - 0.098 (T - 298) \tag{30}$$

where M is the molecular weight of the surfactant tail, T is in °K, and σ_s is expressed in dyne/cm. The surface tension of water is correlated [8] by the expression

$$\sigma_w = 72.0 - 0.16 (T - 298) \tag{31}$$

where the surface tension is given in dynes/cm and the temperature in °K.

Since the interfacial tension against water of various hydrocarbon tails of surfactants are close to one another as shown by the above equations, we approximate the aggregate core-water interfacial tension σ_{agg} by the micelle composition averaged value:

$$\sigma_{agg} = \eta_A \sigma_{Aw} + \eta_B \sigma_{Bw} \tag{32}$$

where σ_{Aw} and σ_{Bw} are calculated using equation (29). The composition averaging in equation (32) is carried out using the volume fractions of surfactant tails in the micellar core, η_A and η_B, defined in equation (27).

For fluorocarbon systems, water-perfluorohexane interfacial tension has been experimentally determined [24] at 25°C to be 56.45 dyne/cm. Temperature dependence of this interfacial tension is presently not available. Since this interfacial tension value is comparable to water-hydrocarbon interfacial tensions discussed above, even for the hydrocarbon-fluorocarbon mixed surfactants, we will use equation (32) for estimating the magnitude of σ_{agg}.

The areas a_{oA} and a_{oB} that appears in equation (28), depend on the extent to which the polar head group shields the cross-sectional area of the surfactant tail. As discussed in reference [8], if the polar head group of surfactant A has a cross-sectional area a_{pA} larger than L^2, (for double tailed surfactants, $2L^2$) then for such a surfactant, a_{oA} is taken equal to L^2 (for double tailed surfactants, $2L^2$). If the polar head group area a_{pA} is less than L^2 (for double tailed surfactants, $2L^2$), then the head group shields only a part of the cross-sectional area of the tail from the contact with water and a_{oA} is taken equal to a_{pA}. Identical considerations apply to surfactant B in the determination of a_{oB}.

Head Group Steric Interactions.   The formation of micelle brings the polar head groups of the surfactant molecules to the surface of the aggregate where they are crowded when compared to their isolated states as singly dispersed surfactant molecules. This generates steric repulsions among the head groups. The associated repulsive free energy has been estimated in our earlier work [1,2,8] using an excluded area concept borrowed from the van der Waals equation of state. This expression is generalized for binary surfactant mixtures as

$$\frac{(\Delta\mu_g^0)_{steric}}{kT} = -\ln\left[1 - \frac{(\alpha_{gA}a_{pA} + \alpha_{gB}a_{pB})}{a}\right] \tag{33}$$

The above expression is inadequate if the polar head groups cannot be considered as compact such as in the case of nonionic surfactants having polyoxyethylene chains as head groups. An alternate treatment for head group interactions in such systems was developed in our recent work [8]. However, the quantitative predictions from this approach are as yet unsatisfactory. Hence, given its simplicity, we retain equation (33) as the basis for calculating steric interactions even for polyoxyethylene surfactants.

Head Group Dipole Interactions.   For zwitterionic surfactants, one has to consider the mutual interactions between the permanent dipoles of the polar head groups. In general, the dipole-dipole interactions are dependent on the orientation of the dipoles. Because of the chain packing in micelles, one expects to find the dipoles stacked such that the poles of the dipoles are located on parallel surfaces. The dipole-dipole interactions in such a case are repulsive and they can be estimated by visualizing the arrangement of the poles of the dipoles as constituting an electrical capacitor [21]. The distance between the planes of the capacitor is equated to the distance of charge separation on the zwitterionic head group. Consequently, the dipole-dipole interactions for dipoles having a charge separation d can be computed for spherical aggregates from [2]

$$\frac{(\Delta\mu^o_g)_{dipole}}{kT} = \frac{2\pi e^2 R_s}{\varepsilon\, a_{dipole}\, kT} \left[\frac{d}{d + R_s} \right] \alpha_{g\,dipole} \tag{34}$$

In the above relation, e denotes the electronic charge, ε is the dielectric constant of the solvent, R_s is the radius of the spherical micelle and $\alpha_{g\,dipole}$ is the fraction of surfactant molecules in the micelle having a dipolar head group. The same equation is employed for globular aggregates. If both A and B are zwitterionic surfactants with the same headgroup then

$$\alpha_{g\,dipole} = \alpha_{gA} + \alpha_{gB} = 1 \;, \; a_{dipole} = a. \tag{35}$$

If surfactant A is zwitterionic while surfactant B is nonionic or ionic, then

$$\alpha_{g\,dipole} = \alpha_{gA} \;, \; a_{dipole} = a/\alpha_{g\,dipole}. \tag{36}$$

The dipole-dipole interactions may be relevant even when surfactants do not possess zwitterionic head groups. Such a situation occurs when the surfactant mixture consists of an anionic and a cationic surfactant. The two oppositely charged surfactants may be visualized as forming ion pairs. Depending upon the location of the charges on the two surfactant head groups, these ion pairs may be assigned with a dipole moment. The variable d in equation (34) which stands for the distance of charge separation in the zwitterionic head groups, now refers to the distance between the locations of the anionic and the cationic charges, perpendicular to the micelle core surface. Thus, for such systems

$$\alpha_{g\,dipole} = \text{Smaller of} \, (\alpha_{gA}\,,\,\alpha_{gB}\,)/2 \;, \; a_{dipole} = a/\alpha_{g\,dipole} \;, \; d = |\delta_A - \delta_B| \tag{37}$$

The factor 2 in the above equation accounts for the fact that a dipole is associated with two surfactant molecules, treated as a pair. The distance δ refers to the location of the charge from the hydrophobic core surface as explained below. In equation (34), the dielectric constant ε is taken to be that of pure water. The dielectric constant as a function of temperature is correlated [8] by the relation

$$\varepsilon = 87.74 \, \exp\left[-0.0046\,(T - 273)\right] \tag{38}$$

where T is expressed in °K.

Head Group Ionic Interactions. Ionic interactions arise at the micellar surface if the surfactant has a charged head group. An approximate analytical expression to calculate this free energy contribution has been derived by Evans and Ninham [9] assuming that the surfactant molecules are completely dissociated.

$$\frac{(\Delta\mu_g^0)_{ionic}}{kT} = 2\left[\ln\left(\frac{S}{2} + \{1 + (\frac{S}{2})^2\}^{1/2}\right) - \frac{2}{S}\left(\{1 + (\frac{S}{2})^2\}^{1/2} - 1\right)\right.$$

$$\left. - \frac{2C_g}{\kappa S}\,\ln\left(\frac{1}{2} + \frac{1}{2}\{1 + (\frac{S}{2})^2\}^{1/2}\right)\right]\alpha_{g\,ion} \tag{39}$$

where

$$S = \frac{4\pi e^2}{\varepsilon\,\kappa\,a_{\delta\,ion}\,kT}\;,\qquad \kappa = \left(\frac{8\pi n_o e^2}{\varepsilon\,kT}\right)^{1/2},\qquad n_o = \frac{(C_{1\,ion} + C_{add})}{10^3}\,N_{Av} \tag{40}$$

The area per molecule a_δ which appears in the above equation is evaluated at a distance δ from the hydrophobic core surface. This distance δ is estimated [8] as the distance from the hydrophobic core surface to the surface where the center of the counterion is located. κ is the reciprocal Debye length, n_o is the number of counterions in solution per cm^3, $C_{1\,ion}$ is the molar concentration of the singly dispersed ionic surfactant molecules, C_{add} is the molar concentration of the salt added to the surfactant solution and N_{Av} is the Avogadro's number. The last term in the right hand side of equation (39) is the curvature correction term where C_g is given by

$$C_g = \frac{2}{R_s + \delta}\;;\quad \frac{2}{R_{eq} + \delta} \tag{41}$$

for spherical and globular micelles, respectively. If both A and B are ionic surfactants with the same kind of charged head groups, then

$$\alpha_{g\,ion} = \alpha_{gA} + \alpha_{gB} = 1\;,\quad a_{\delta\,ion} = a_\delta\;,\quad \delta = \alpha_{gA}\,\delta_A + \alpha_{gB}\,\delta_B\;. \tag{42}$$

If A is ionic while B is nonionic or zwitterionic, then

$$\alpha_{g\,ion} = \alpha_{gA} \;, \quad a_{\delta\,ion} = a_{\delta}/\alpha_{g\,ion} \;, \quad \delta = \delta_A \;. \tag{43}$$

If A and B are both ionic but oppositely charged, i.e. one is anionic while the other is cationic, then

$$\alpha_{g\,ion} = |\alpha_{gA} - \alpha_{gB}| \;, \quad a_{\delta\,ion} = a_{\delta}/\alpha_{g\,ion} \;, \quad \delta = \alpha_{gA}\,\delta_A + \alpha_{gB}\,\delta_B \;. \tag{44}$$

Free Energy of Mixing of Surfactant Tails.   This is the only contribution not present in the free energy model for single component surfactant solutions. This contribution accounts for the entropy and the enthalpy of mixing of the surfactant tails of molecules A and B in the hydrophobic core of the micelle, with respect to the reference states of pure A or pure B micelle cores. We calculate this free energy contribution using the Flory-Huggins expression for the entropy of mixing and the solubility parameter based expression for the enthalpy of mixing. It is known that the free energy expression based on the solubility parameter theory of Hildebrand may not be adequate for precise phase equilibria calculations, but does provide reasonable estimates of solution nonidealities while describing liquid mixtures [29]. Moreover, the Hildebrand equation is characterized by simplicity of functional form. Hence, we employ the Hildebrand theory here for calculating the free energy of mixing inside the micelle core.

$$\frac{(\Delta\mu_g^o)_{mix}}{kT} = [\,\alpha_{gA}\,\ell n\,\eta_A + \alpha_{gB}\,\ell n\,\eta_B\,] +$$

$$[\,\alpha_{gA}\,V_{sA}\,\{\,\delta^H_A - \delta^H_{mix}\,\}^2 + \alpha_{gB}\,V_{sB}\,\{\,\delta^H_B - \delta^H_{mix}\,\}^2\,]/kT \tag{45}$$

In the above equation, the first term is the Flory-Huggins expression for the entropy of mixing of the surfactant tails inside the micelle core. η_A and η_B are the volume fractions of A and B tails, defined in equation (27). The second term represents the enthalpy of mixing written in the framework of the Hildebrand theory [10]. δ^H_A and δ^H_B refer to the solubility parameters of the tails of surfactants A and B (not of the surfactant as a whole, but only of the tail) while δ^H_{mix} is the volume fraction averaged solubility parameter of all the components within the micelle core, $\delta^H_{mix} = \eta_A\,\delta^H_A + \eta_B\,\delta^H_B$.

Free Energy Model and Mixture Nonideality.   All the free energy contributions can be calculated as a function of the aggregation number and the composition of the mixed micelles using equations provided above. The free energy of formation of mixed micelles appearing in equation (1) is, thus

$$(\Delta\mu_g^o) = (\Delta\mu_g^o)_{tr} + (\Delta\mu_g^o)_{def} + (\Delta\mu_g^o)_{int} + (\Delta\mu_g^o)_{steric}$$

$$+ (\Delta\mu_g^o)_{dipole} + (\Delta\mu_g^o)_{ionic} + (\Delta\mu_g^o)_{mix} \qquad (46)$$

In the framework of this free energy model, a molecular interprtetation of the ideal and nonideal behavior exhibited by surfactant mixtures can be provided.

For ideal mixed micelles, the free energy of formation of mixed micelle is the composition averaged value of the free energies of formation of pure component micelles plus the ideal entropy of mixing of the two types of surfactants.

$$(\Delta\mu_g^o)_{A+B} = \alpha_{gA} (\Delta\mu_g^o)_A + \alpha_{gB} (\Delta\mu_g^o)_B + (\Delta\mu_g^o)_{mix}(\text{Ideal}) \qquad (47)$$

Therefore, if the free energy of formation of mixed micelles, excluding the ideal entropy of mixing term is a nonlinear function of composition, then the surfactant mixture will exhibit nonideal behavior. Further, if the free energy of mixing contribution deviates from the ideal entropy of mixing, then also the mixed surfactant system will be nonideal. If the magnitude of the free energy of formation of the mixed micelles is larger than the ideal value, then we characterize the system as displaying a negative deviation from ideality. Obviously, positive deviation from ideality occurs when the free energy of formation of mixed micelles has a magnitude smaller than the ideal value.

Considering the various contributions to the free energy of micellization in equation (46), we note that they can be decomposed into bulk and surface contributions besides a free energy of mixing. (Note that although the free energy of mixing is a part of the bulk component, we view it as a separate term).

$$(\Delta\mu_g^o)_{A+B} = (\Delta\mu_g^o)_{A+B} (\text{Bulk}) + (\Delta\mu_g^o)_{A+B} (\text{Surface}) + (\Delta\mu_g^o)_{mix} \qquad (48)$$

The transfer free energy of the surfactant tail and the tail deformation free energy provide the bulk contributions while the aggregate-water interfacial free energy and the steric, dipolar and ionic interaction energies between the head groups provide the surface contributions.

The transfer free energy of the surfactant tails is a composition averaged value of the two pure component micelles. The tail deformation free energy is also a composition averaged value if the two surfactants have equal tail lengths. For such conditions, the bulk component of the free energy of mixed micelles does not give rise to nonidealities of the surfactant mixture. However, if the surfactant tail lengths differ from one another, then the tail deformation free energy is a nonlinear function of the aggregate composition. In such a case, the nonideality of the surfactant mixture can originate from the bulk component.

If we consider the surface component of the free energy of micellization, all the contributions depend upon the area a per molecule of the aggregate. When the surfactant tails are identical in size, the area a is composition independent. However, all the surface contributions have a nonlinear composition dependence because of the functional forms of the various free energy contributions. When the surfactant tails differ in their sizes, the area a is a nonlinear function of the aggregate composition. This additional nonlinearity simply adds on to the already existing nonlinear dependence of the surface free energy contributions on the micelle composition. Thus, in all situations, the surface component of the free energy of mixed micelles contributes to nonideal behavior.

Finally, the free energy of mixing contribution is estimated taking into account the size differences between the hydrophobic tails. Also a enthalpy of interaction is introduced. Thus this contribution deviates from the ideal entropy of mixing except when the two surfactants have identical tails. Therefore, the free energy of mixing contribution can also give rise to nonideal behavior of the surfactant mixture.

In summary, the nonideal behavior of mixed micelle originates from the nonlinear composition dependence of the various free energy contributions as well as due to the deviation of the free energy of mixing from the ideal entropy of mixing. Depending upon the relative importance of the various free energy contributions, the nonidealities may be pronounced or only slight, as will be seen below. Further, both positive and negative deviations from ideal behavior will be observed for different surfactant mixtures. These general observations are examined through calculations for different kinds of binary surfactant mixtures below.

Predictions of Molecular Theory

For spherical or globular micelles, the size and composition dispersion was found to be narrow based on our earlier theory [1]. Consequently, one can consider the average aggregation number and the average micelle composition to be those values of g and α_{gA}, respectively, for which the micelle concentration X_g has a maximum. The computations in this paper are carried out as follows. For a given value of α_{1A}, we search for the values of the mixed micelle composition α_{gA} and the radius R_s (for spherical micelles) or the semimajor axis b (for the ellipsoidal micelles) for which X_g has a maximum, as a function of the concentration of singly dispersed surfactants, X_1. As noted earlier, the semiminor axis of the ellipsoidal micelles is the largest radius allowed for the spherical micelles for any given micelle composition. The critical micelle concentration reported in this paper is that value of X_1 for which the total amount of the surfactant in the micellized form is equal to that in the singly dispersed form, namely $X_{cmc} = X_1 = g\,X_g$(maximum) [19]. The average aggregation numbers and the average micelle compositions reported in this paper (unless otherwise specified) correspond to the condition that the total concentration of the surfactant present is equal to the cmc. The search for the parameter values that maximize the aggregate concentration X_g was carried out using a standard IMSL (International Mathematical and Statistical Library) subroutine ZXMWD. This subroutine is designed to carry out the search for the global extremum of a function of q independent variables subject to any specified constraints on the variables.

Molecular Parameters. The surfactants considered in this study are represented in Figure 2 (along with symbols used to refer to them in the text) and the various molecular constants used in the calculations are summarized in Table I. For hydrocarbon surfactants, the molecular volume of the surfactant tail containing n_c carbon atoms is calculated from the group contributions of $(n_c - 1)$ methylene groups and the terminal methyl group [8].

$$v\,(CH_3) = 54.6 + 0.124\,(\,T - 298\,)\;\text{Å}^3\;;$$

$$v\,(CH_2) = 26.9 + 0.0146\,(\,T - 298\,)\;\text{Å}^3 \tag{49}$$

where T is in °K. The extended length of the surfactant tail ℓ_s is taken to be temperature independent and is calculated using a group contribution of 1.265 Å for the methylene group and 2.765 Å for the methyl group.

For fluorocarbon surfactants, extensive volumetric data are not available in order to estimate the temperature dependent group molecular volumes for the CF_3 and CF_2 groups. Using data available [30-32] at 25°C, we estimate that $V(CF_3) = 1.67\,V(CH_3)$ and $V(CF_2) = 1.44\,V(CH_2)$. The ratios between the volumes of the fluorocarbon and the hydrocarbon groups are taken to be the same at all temperatures in the absence of extensive experimental data. The extended length of the fluorocarbon chain is estimated using the same group contributions as for hydrocarbon tails, namely, 2.765Å for the CF_3 group and 1.265Å for the CF_2 group [7].

The solubility parameters for the hydrocarbon and fluorocarbon tails of surfactants can be estimated using a group contribution approach on the basis of the tabulated data available [10,29]. The solubility parameter estimated for the fluorocarbon tail of the surfactant SPFO on this basis is $12.3\,\text{MPa}^{1/2}$. This approach utilizes only the pure component properties for the determination of solubility parameters. Since, the solubility parameters estimated on this basis have been found inadequate for quantitative description of hydrocarbon-fluorocarbon mixture properties, an alternate

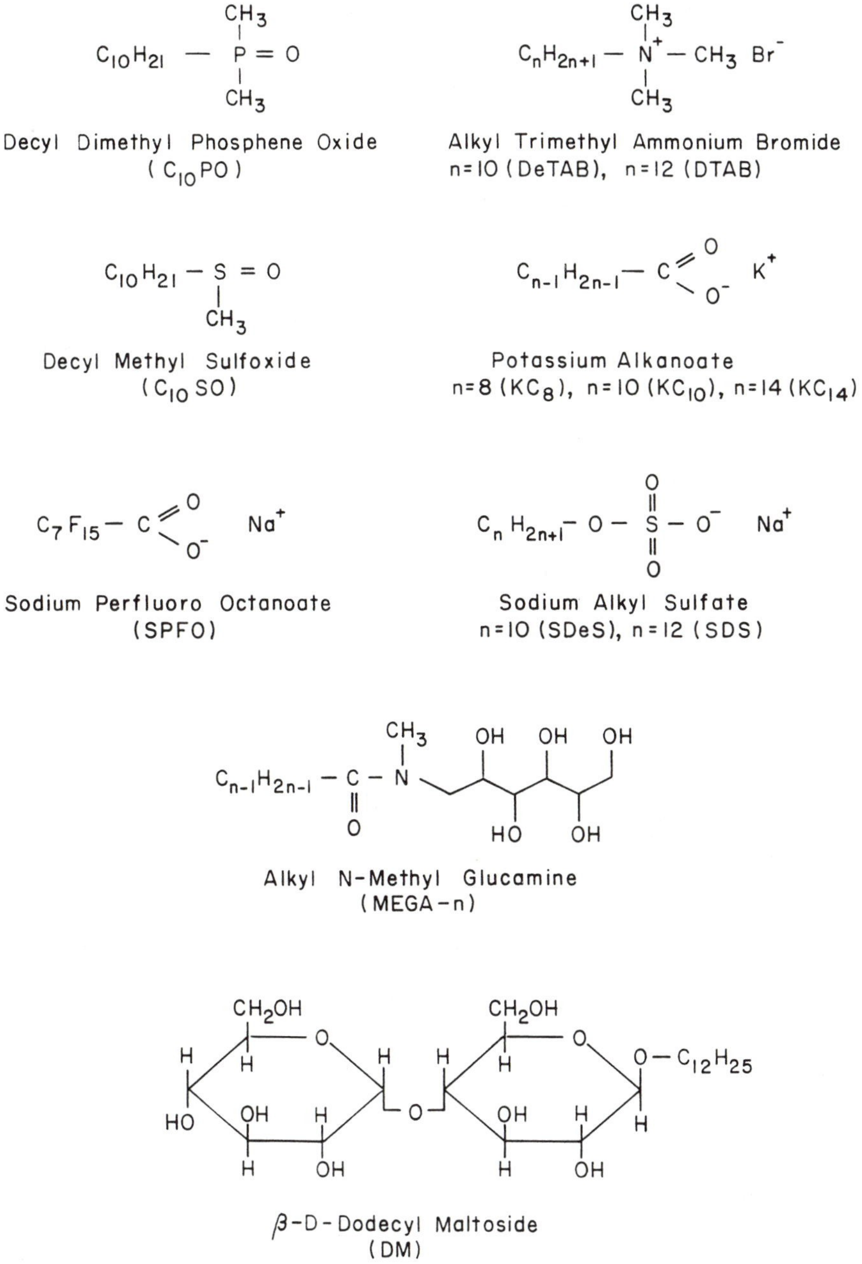

$$C_{10}H_{21} - \overset{\overset{\displaystyle CH_3}{|}}{\underset{\underset{\displaystyle CH_3}{|}}{P}} = O$$

Decyl Dimethyl Phosphene Oxide
($C_{10}PO$)

$$C_nH_{2n+1} - \overset{\overset{\displaystyle CH_3}{|}}{\underset{\underset{\displaystyle CH_3}{|}}{N}}{}^+ - CH_3 \quad Br^-$$

Alkyl Trimethyl Ammonium Bromide
n=10 (DeTAB), n=12 (DTAB)

$$C_{10}H_{21} - \overset{\overset{\displaystyle }{}}{\underset{\underset{\displaystyle CH_3}{|}}{S}} = O$$

Decyl Methyl Sulfoxide
($C_{10}SO$)

$$C_{n-1}H_{2n-1} - C \overset{\displaystyle \nearrow O}{\underset{\displaystyle \searrow O^-}{}} \quad K^+$$

Potassium Alkanoate
n=8 (KC$_8$), n=10 (KC$_{10}$), n=14 (KC$_{14}$)

$$C_7F_{15} - C \overset{\displaystyle \nearrow O}{\underset{\displaystyle \searrow O^-}{}} \quad Na^+$$

Sodium Perfluoro Octanoate
(SPFO)

$$C_nH_{2n+1} - O - \overset{\overset{\displaystyle O}{\|}}{\underset{\underset{\displaystyle O}{\|}}{S}} - O^- \quad Na^+$$

Sodium Alkyl Sulfate
n=10 (SDeS), n=12 (SDS)

Alkyl N-Methyl Glucamine
(MEGA - n)

β-D-Dodecyl Maltoside
(DM)

Figure 2. Chemical structure of the surfactant molecules considered in this study and the symbols used to refer to them in the text.

Table I. Molecular Constants for Surfactants

Surfactant	v_s	ℓ_s	a_o	a_p	δ	d	δ^H
	Å^3	Å	Å^2	Å^2	Å	Å	$\text{MPa}^{1/2}$

Nonionic Hydrocarbon Surfactants

$C_{10}SO$	296	14.1	21	39	-	-	16.63
$C_{10}PO$	296	14.1	21	48	-	-	16.63
DM	350	16.7	21	43	-	-	16.76
MEGA-8	217	10.4	21	34	-	-	16.31
MEGA-9	244	11.6	21	34	-	-	16.44

Anionic Hydrocarbon Surfactants

SDeS	296	14.1	17	17	5.45	-	16.63
SDS	350	16.7	17	17	5.45	-	16.76
KC_8	217	10.4	11	11	6.00	-	16.31
KC_{10}	271	12.9	11	11	6.00	-	16.54
KC_{14}	378	17.9	11	11	6.00	-	16.81

Cationic Hydrocarbon Surfactants

DeTAB	296	14.1	21	54	3.45	-	16.63
DTAB	350	16.7	21	54	3.45	-	16.76

Anionic Fluorocarbon Surfactants

SPFO	325	10.4	11	11	5.55	-	12.30
							9.5

approach is to back calculate the solubility parameter values by fitting the mixture properties. Mukerjee and Handa [24] have estimated group contributions to the solubility parameters by fitting critical solution temperature data on hydrocarbon-fluorocarbon mixtures. On the basis of these group contributions, we calculate the solubility parameter of the SPFO tail to be approximately $9.5\text{MPa}^{1/2}$. Computed results are presented later utilizing both these estimates. It may be mentioned that improved accounting of the interactions between hydrocarbon and fluorocarbon chains would be necessary to improve the quantitative accuracy of our predictions.

Of the molecular constants needed for the calculations, there can be some small uncertainty in estimating a_p for nonionic, a_p and δ for ionic and a_p and d for zwitterionic surfactants. Reasonable estimates for these constants can be obtained from knowledge of various bond lengths and atomic or ionic radii. To improve the estimates, information about conformation and orientation of the polar heads will be necessary. It is this factor which is responsible for the small uncertainty in the estimated molecular constants. Although the estimates employed in our calculations (Table I) could be revised, the molecular constants are not adjustable parameters. Indeed, the requirement that the same value of the molecular constant be used to predict the cmc, micelle size and the influence of salt and temperature for pure micelles, and also the micelle composition of mixed micelles, solubilization behavior, etc., provides an unusually stringent test.

Nonionic Hydrocarbon-Nonionic Hydrocarbon Mixtures. The aggregation behavior of binary mixtures of decyl methyl sulfoxide, $C_{10}SO$ and decyl dimethyl phosphene oxide, $C_{10}PO$ at 24°C has been examined and the results are presented in Figures 3 and 4. In Figure 3, the cmc values are plotted against the composition of the micelles and the composition of the singly dispersed surfactant. Note that when the total surfactant concentration is equal to the cmc, the amount of micelles present is small and hence, one can practically equate the composition of the monomers to the composition of the total surfactant. Also shown for comparison are the experimental data obtained by Holland and Rubingh [33] based on surface tension measurements. Figure 4 presents the average aggregation numbers predicted by the theory as a function of the composition of the mixed micelle and that of the monomers. No experimental data are available for comparison. The size of the mixed micelle is approximately linear with composition but one can discern a slight s-shaped curve as was observed in the experiments of Warr et al. [5] for another binary mixture of nonionic surfactants.

It has been shown that the cmc of this mixed surfactant system could be calculated from the cmc values of the individual surfactants by assuming the phenomenological ideal mixed micelle model [33]. In the framework of the free energy model presented here, nonidealities in this binary mixture can arise from the fact that the two surfactants have somewhat differing head group cross sectional areas while possessing identical tails. Since there are no volume differences between the hydrophobic tails of the two surfactants, for any aggregation number, the area per molecule of the mixed micelle is independent of the micelle composition. However, the steric interactions between head groups at any aggregation number is a nonlinear function of the micelle composition. This is the source of mixture nonideality. Nevertheless, the deviation from ideal mixing conditions are rather small, thus allowing the phenomenological ideal mixed micelle model

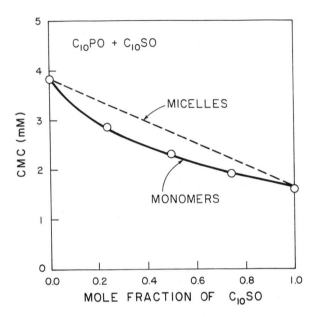

Figure 3. The cmc of $C_{10}PO + C_{10}SO$ mixtures as a function of the composition of micelles (dotted line) and that of singly dispersed surfactants (continuous line) at 24°C. The experimental data are from ref. [33].

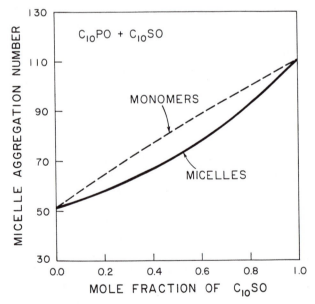

Figure 4. The average aggregation number of $C_{10}PO + C_{10}SO$ mixed micelles at the cmc as a function of the composition of micelles (continuous line) and that of the singly dispersed surfactants (dotted line) for the conditions specified in Figure 3.

to predict the mixture cmc satisfactorily. The small nonideality, however, is reflected in the micelle aggregation number in terms of the slight nonlinear dependence of the aggregation number on the micelle composition.

Ionic Hydrocarbon-Ionic Hydrocarbon Mixtures. The micellization behavior of mixtures of two anionic surfactants having the same head group has been examined. The surfactants are sodium dodecyl sulfate, SDS and sodium decyl sulfate, SDeS which differ from one another in their hydrocarbon tail lengths. The predicted cmc values are plotted against the composition of the singly dispersed surfactant in Figure 5. Also shown are the experimental data obtained by Mysels and Otter [34] using conductivity measurements and the data of Shedlowsky et al. [35] based on e.m.f. measurements. The predicted mixed micelle compositions as a function of the composition of the singly dispersed surfactants are compared in Figure 6 with the experimental data obtained by Mysels and Otter [34] using conductivity measurements.

In these binary mixtures, one source of nonideality arises from the volume differences between the hydrophobic tails of the two surfactants. Consequently, at any given aggregation number, the area per molecule of the mixed micelle is a nonlinear function of the micelle composition and hence, all the free energy contributions reflect this nonlinear dependence. The other source of nonideality is the change in ionic strength of the solution as the composition is modified. In the absence of any added salt, the ionic strength is determined by the concentration of the singly dispersed surfactants. This concentration changes with composition, thus modifying the ionic interactions at the micelle surface nonlinearly with respect to composition. Given the importance of the ionic interactions to the free energy of micellization, the nonideality is more perceptible in these binary mixtures. The predicted micelle aggregation numbers shown in Figure 7 reflect the nonideality arising from the change in ionic strength of the surfactant solution. Since the ionic strength has a significant influence on the ionic interactions between the head groups, the aggregation numbers are larger for the mixed micelles than for the pure component micelles over most of the composition.

Somewhat contrasting behavior is shown by the mixtures of cationic surfactants, dodecyl trimethyl ammonium bromide, DTAB and decyl trimethyl ammonium bromide, DeTAB. This mixture is similar to the SDS-SDeS mixture with respect to the tail lengths of the two surfactants. However, the trimethyl ammonium bromide head group has a very large area a_p compared to that of the anionic sulfate head group. The predicted cmc values as a function of the micelle and the monomer compositions are presented in Figure 8. For comparison, experimental cmc data obtained by Garcia-Mateos et al. [36] using electrical conductivity measurements are also presented. One can observe the large cmc changes in DTAB-DeTAB mixtures as the composition is altered which also affects the ionic strength of

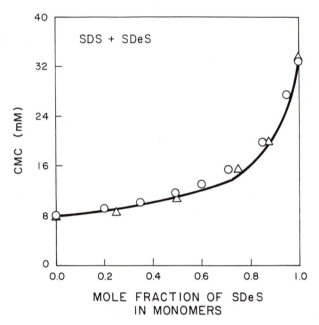

Figure 5. The cmc of SDS + SDeS mixtures as a function of the composition of singly dispersed surfactants. The experimental data shown by circles are from ref. [34] while the data shown by triangles are from ref. [35].

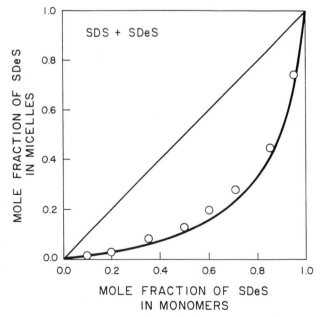

Figure 6. The composition of SDS + SDeS mixed micelles as a function of the composition of singly dispersed surfactants. The experimental data shown by circles are from ref. [34].

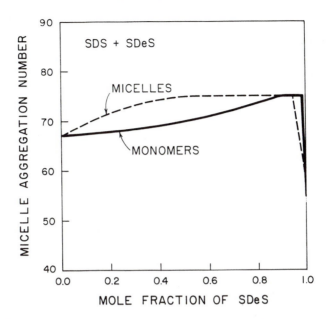

Figure 7. The average aggregation number of SDS + SDeS mixed micelles as a function of the composition of the micelles (dotted line) and that of singly dispersed surfactants (continuous line).

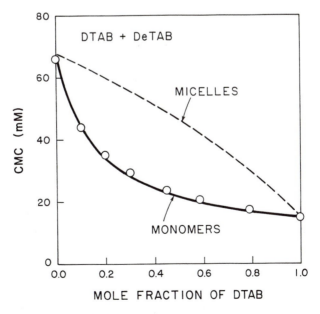

Figure 8. The cmc of DTAB + DeTAB mixtures as a function of the composition of the micelles (dotted line) and that of singly dispersed surfactants (continuous line). The experimental data are from ref. [36].

the solution. Nevertheless, the predicted micelle aggregation numbers plotted in Figure 9 do not show any growth above that of pure component micelles. This differing behavior is a direct consequence of the larger steric repulsions in the case of these cationic surfactants with their bulky head groups when compared to the sulfate head groups of SDS-SDeS.

To explore the consequences of significant variations in the hydrophobic tail lengths, we have computed the micellization behavior of mixtures of anionic potassium alkanoates, namely, potassium tetradecanoate, $KC_{14}-$ potassium octanoate, KC_8 and potassium decanoate, $KC_{10}-$ potassium octanoate, KC_8 at 25°C. The calculated cmc, micelle composition and micelle aggregation numbers are shown in Figures 10 to 12 as a function of the composition of the monomers. Also shown in Figure 6 are the experimental cmc data [37] obtained by Shinoda using dye solubilization measurements. One can observe from Figure 11 that the less hydrophobic KC_8 is almost completely excluded from the micelles in $KC_8 + KC_{14}$ mixtures because of the comparatively much stronger hydrophobicity of KC_{14}. This contrasts against the behavior displayed by $KC_8 + KC_{10}$ mixtures.

Ionic Hydrocarbon-Nonionic Hydrocarbon Mixtures.   In contrast to the two kinds of binary mixtures considered above where nonidealities were relatively small, mixtures of ionic and nonionic surfactants display significant nonideal behavior. We consider mixtures of anionic sodium dodecyl sulfate, SDS and nonionic decyl methyl sulfoxide, $C_{10}SO$ at 24°C and in the presence of 1 mM Na_2CO_3. The calculated cmc values are shown in Figure 13 while the aggregation numbers are presented in Figure 14, both as functions of the monomer composition. The experimental cmc data based on surface tension measurements [33] are also provided for comparison. The results show that considerable nonideality is exhibited by this binary mixture. The cmc values of the mixed system are substantially smaller than those expected for ideal mixed micelles. Thus, a negative deviation from ideal behavior is observed.

The two surfactants differ somewhat in their hydrophobic tail lengths and in the sizes of the polar head groups. Similar variations occurred in the case of nonionic-nonionic mixtures considered before but by themselves, they do not give rise to significant nonidealities. However, in the present case, one component is ionic while the other is nonionic. This results in a large variation in the ionic interaction energy at the micelle surface as the micelle composition is modified. This free energy contribution is thus responsible for most of the nonideal behavior exhibited by this system.

We also present the model predictions for binary mixtures of anionic sodium dodecyl sulfate, SDS and nonionic $\beta-$ dodecyl maltoside, DM that has been experimentally investigated by Bucci et al. [6]. In this system, the two surfactants have identical hydrophobic parts and hence, all the

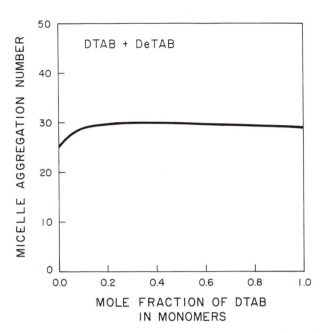

Figure 9. The average aggregation number of DTAB + DeTAB mixed micelles as a function of the composition of singly dispersed surfactants.

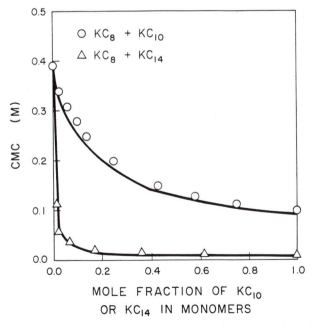

Figure 10. The cmc of $KC_8 + KC_{10}$ mixtures and $KC_8 + KC_{14}$ mixtures as a function of the composition of singly dispersed surfactants. The experimental data are from ref. [37].

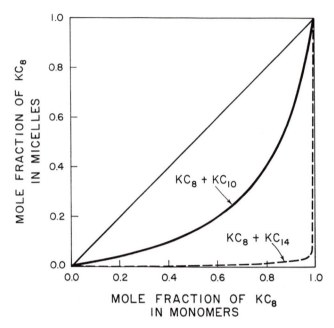

Figure 11. The composition of $KC_8 + KC_{10}$ mixed micelles and $KC_8 + KC_{14}$ mixed micelles as a function of the composition of singly dispersed surfactants.

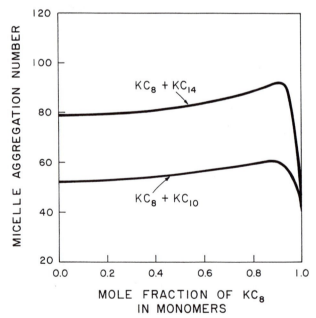

Figure 12. The average aggregation number of $KC_8 + KC_{10}$ mixed micelles and $KC_8 + KC_{14}$ mixed micelles as a function of the composition of singly dispersed surfactants.

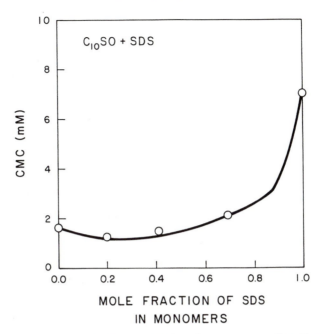

Figure 13. The cmc of SDS + C$_{10}$SO mixtures as a function of the composition of singly dispersed surfactants. The experimental data are from ref. [*33*].

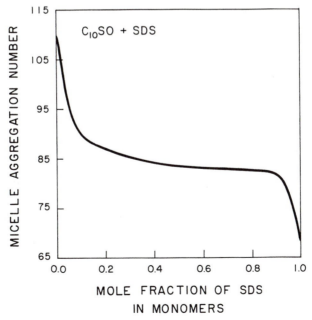

Figure 14. The average aggregation number of SDS + C$_{10}$SO mixed micelles as a function of the composition of monomers.

nonideality emerges from the differences in the size and charge of the two head groups. The calculated average aggregation numbers and those estimated from neutron scattering measurements [6] are shown in Figure 15 in solutions with and without added salt. The average aggregation numbers and the average micelle compositions are estimated at a total surfactant concentration of 50 mM and at 25°C. Figures 16 and 17 present respectively, the cmc and micelle composition, as functions of the composition of the singly dispersed surfactant at two concentrations of added electrolyte, NaCl. One can see the preferential incorporation of nonionic DM molecules in the micelles over most of the composition range. In the absence of any added electrolyte, this preference is stronger since the presence of anionic SDS in micelles will give rise to a positive free energy contribution. Obviously, in the presence NaCl, the electrostatic repulsions between ionic head groups are reduced and hence a larger number of SDS molecules are incorporated into the mixed micelles.

Anionic Hydrocarbon-Cationic Hydrocarbon Mixtures.   Anionic and cationic surfactants when present together are expected to form ion pairs with no net charge and thereby undergo a decrease in their aqueous solubility resulting in precipitation. This phenomenon is widely observed [38,39]. But depending upon their hydrophobic parts, these surfactant mixtures can also exhibit the formation of mixed micelles or mixed spherical bilayer vesicles in certain concentration and composition domains. As discussed earlier, depending upon the location of the charges on the anionic and the cationic surfactants, one may associate a permanent dipole moment with each ion pair. Consequently, these surfactant mixtures can behave as part ionic single chain molecules and a part zwitterionic paired chain molecules. We have calculated the aggregation properties of binary mixtures of decyl trimethyl ammonium bromide, DeTAB and sodium decyl sulfate, SDeS and the results are presented in Figures 18 and 19. The calculated and experimental [33] cmc values are shown in Figure 18 while information on the micelle composition is found from Figure 19. The formation of rodlike mixed micelles is indicated over the entire composition range. The micelle composition data show that the mixed micelles contain approximately equal numbers of the two types of surfactants over the entire composition space. Indeed, the small deviation from the micelle composition value of 0.5 arises because of the differing sizes of the polar head groups. The formation of spherical bilayer vesicles from binary mixtures of this type will be examined in a latter paper.

Anionic Fluorocarbon-Nonionic Hydrocarbon Mixtures.   We consider the aggregation behavior of nonionic alkyl-N-methyl glucamines (MEGA-n) and anionic sodium perfluorooctanoate (SPFO) mixtures that have been experimentally studied by Wada et al. [40]. In such mixtures, the nonideality associated with the mixing of hydrocarbon and fluorocarbon surfactant tails

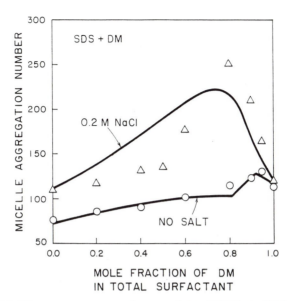

Figure 15. The averge aggregation number of SDS + DM mixed micelles as a function of the total surfactant composition at a total surfactant concentration of 50 mM. The experimental data are from ref. [6] where the circles refer to micelle sizes in the absence of any added salt while the triangles correspond to a 0.2 M concentration of added NaCl.

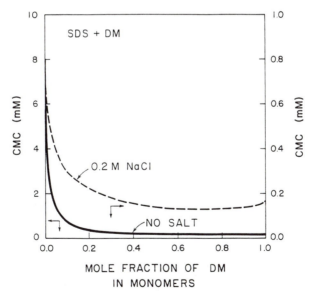

Figure 16. The cmc of SDS + DM mixtures as a function of the composition of monomers. The conditions correspond to those in Figure 15.

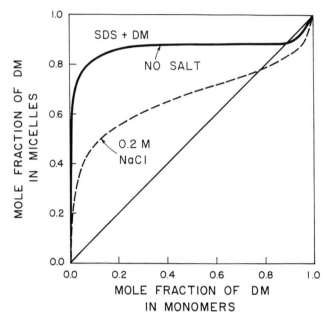

Figure 17. The composition of mixed micelles of SDS + DM as a function of the composition of monomers. The conditions correspond to those in Figure 15.

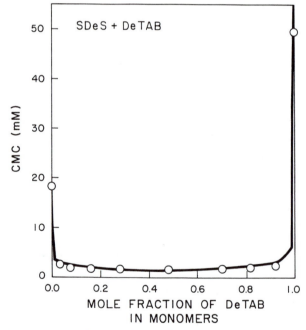

Figure 18. The cmc of SDeS + DeTAB mixtures as a function of the composition of singly dispersed surfactants. The experimental data are from ref. [33].

is superimposed on another type of nonideality associated with the mixing of anionic and nonionic headgroups. For such polar head group mixtures in hydrocarbon surfactants, our earlier calculations revealed considerable negative deviations from ideality. For mixtures of hydrocarbon and fluorocarbon tails we anticipate strong positive deviations from ideality. Thus, for the mixtures under study, both negative and positive deviations from ideality occur, partially compensating one another. As a result, these binary mixtures exhibit reduced nonideality. Figures 20 and 22 present the calculated cmc values as a function of the composition of the monomers. The calculations have been performed with two alternate values for the solubility parameter of SPFO, one determined from pure fluorocarbon properties ($12.3 MPa^{1/2}$), and the other determined from fluorocarbon-hydrocarbon mixture properties ($9.5 MPa^{1/2}$). The cmc values calculated using these two solubility parameter estimates are found to provide bounds for the measured cmc. We note that the cmc values of pure MEGA-n surfactants determined by Wada et al. [40] using surface tension measurements are 10 to 20 percent lower than those estimated from their light scattering measurements. All the mixture measurements have been made using the surface tension technique and it is therefore, conceivable that the actual cmc values may be somewhat larger than those indicated by the experimental points in Figures 20 and 22. Both MEGA-8-SPFO and MEGA-9-SPFO mixtures display the same qualitative behavior. Miscibility between the two surfactants is promoted because of the head group interactions and a single kind of mixed micelles exists in solution. This is seen in the micelle composition data shown in Figures 21 and 23 for MEGA-8-SPFO and MEGA-9-SPFO mixtures, respectively.

Anionic Hydrocarbon-Anionic Fluorocarbon Mixtures.   We have calculated the micellization behavior of sodium perfluorooctanoate, SPFO and sodium decyl sulfate, SDeS mixtures and the results are shown in Figures 24 and 25. The cmc values are compared with available experimental data [40] in Figure 24 where the composition of the mixed micelle is also shown. One can observe positive deviations in cmc in contrast to the negative deviations seen for all the binary mixtures discussed previously. This positive deviation is a direct consequence of the nonideality in the interactions between the hydrocarbon and fluorocarbon tails. This positive deviation which was also present in SPFO-MEGA-n mixtures was compensated partially by the negative deviations due to head group interactions in those systems. Such compensating negative head group interactions are absent in SPFO-SDeS mixtures. The calculated micelle composition data show the interesting feature that over a composition domain two types of micelles coexist. One can observe the hydrocarbon rich and the fluorocarbon rich micelles with the average compositions of $\alpha_{gA} = 0.32$ and 0.79, in coexistence. Phenomenological models of mixed micelles employing regular solution concept are not capable of predicting such asymmetry in the compositions of

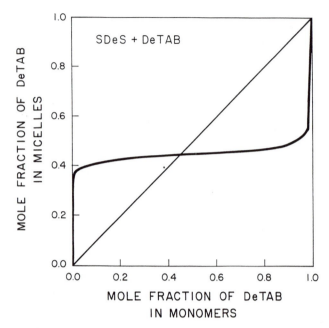

Figure 19. The average composition of SDeS + DeTAB mixed micelles as a function of the composition of monomers.

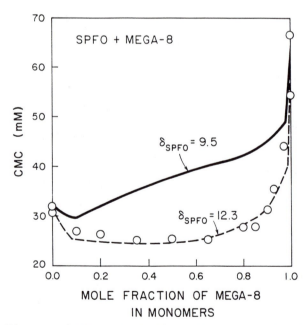

Figure 20. The cmc of MEGA-8 + SPFO mixtures as a function of the composition of singly dispersed surfactants. The calculated results are presented for two limiting values of the solubility parameter of the SPFO tail. The experimental data are from ref. [40].

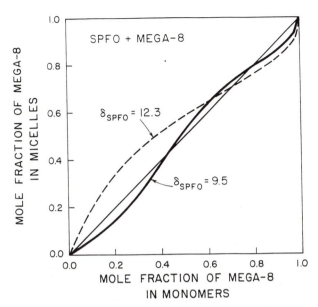

Figure 21. The average composition of MEGA-8 + SPFO mixed micelles as a function of the composition of monomers. The conditions correspond to those described in Figure 20.

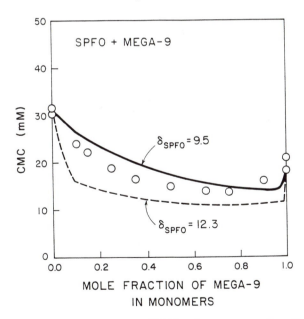

Figure 22. The cmc of MEGA-9 + SPFO mixtures as a function of the composition of singly dispersed surfactants. The calculated results are presented for two limiting values of the solubility parameter of the SPFO tail. The experimental data are from ref. [40].

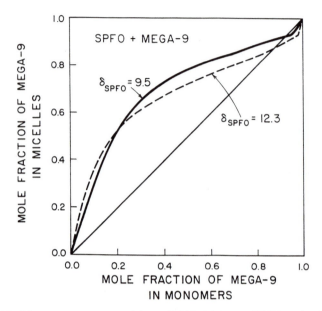

Figure 23. The average composition of MEGA-9 + SPFO mixed micelles as a function of the composition of monomers. The conditions correspond to those described in Figure 22.

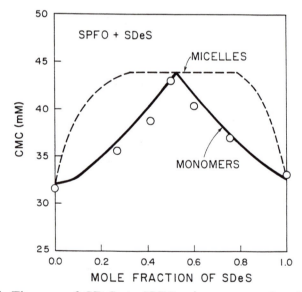

Figure 24. The cmc of SDeS + SPFO mixtures as a function of the composition of micelles (dotted line) and that of singly dispersed surfactants (continuous line). The experimental data are from ref. [40]. The experimental and calculated results indicate the coexistence of two micelle populations.

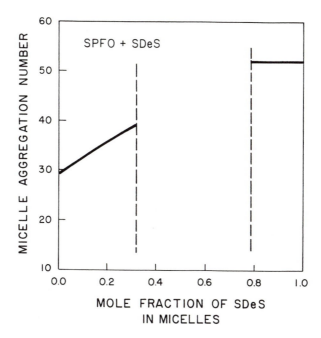

Figure 25. The average aggregation number of SDeS + SPFO mixed micelles as a function of the composition of micelles. Two distinct micelle sizes corresponding to the two coexisting micelle populations are indicated.

the coexisting micelle populations. The average aggregation numbers of the mixed micelles have been predicted and these results are shown in Figure 25. The micelles that are fluorocarbon and hydrocarbon rich are found to have distinctly different aggregation numbers.

Conclusions

A molecular thermodynamic treatment of binary surfactant mixtures in aqueous media is developed in this paper. The model allows the prediction of the size and the composition distribution of spherical, globular and rodlike mixed micelles. Central to the model are the expressions for various contributions to the free energy of formation of mixed micelles. These expressions are obtained by extending the theory for single surfactants developed recently by us.

The present model improves upon our earlier treatment of mixed micelles in a number of ways. Of special importance are the analytical expressions for the free energy of deformation of the surfactant tails associated with the molecular packing requirements within the micelle core. Also, ionic interactions are calculated using an approximate analytic solution to the Poisson-Boltzmann equation available in the literature. The free energy of mixing of surfactant tails within the micelle core is estimated on the basis of Hildebrand's solubility parameter approach. Further, the free energies of transfer of the surfactant tails from water into micelles and other variables are given as explicit functions of temperature. The model is now strictly predictive in the sense that the empirical factors present in our earlier treatment of mixed micelles are all entirely replaced.

The theory is used to predict the micellization behavior of various binary surfactant mixtures differing in both the hydrophobic tails and the polar head groups. Negative and positive deviations from ideal behavior are observed for the surfactant systems considered. For hydrocarbon-hydrocarbon surfactant mixtures, close to ideal behavior is observed for nonionic-nonionic mixtures. Negative deviations from ideality become significant in the order ionic-ionic, ionic-nonionic and anionic-cationic mixtures. For hydrocarbon-fluorocarbon surfactants, with ionic-ionic head groups, positive deviation from ideal behavior is observed. Of interest is the observed coexistence of two populations of mixed micelles one rich in fluorocarbon and the other in hydrocarbon. In contrast, when ionic-nonionic head groups are considered, the micelles become miscible with a single type of mixed micelle being present.

The thermodynamic model allows an interpretation of the observed ideal and nonideal behavior of mixed micelles in terms of bulk and surface components of the free energy of micelle formation as well as the free energy of mixing. In general, the surface component leads to negative deviations from the ideal behavior. The bulk component contributes weakly or not at all to nonideality. The free energy of mixing contributes to a positive deviation from ideality for hydrocarbon-fluorocarbon mixtures.

Literature Cited

1. Nagarajan,R. *Langmuir* 1985, *1*, 331.
2. Nagarajan,R.; Ruckenstein,E. *J. Colloid Interface Sci.* 1979, *71*, 580.
3. Nagarajan,R. *Adv. Colloid Interface Sci.* 1986, *26*, 205.
4. Rao,I.V.; Ruckenstein,E. *J. Colloid Interface Sci.* 1986, *113*, 375.
5. Warr,G.G.; Drummond,C.J.; Grieser,F.; Ninham,B.W.; Evans,D.F. *J. Phys. Chem.* 1986, *90*, 4581.
6. Bucci,S.; Fagotti,C.; Degiorgio,V.; Piazza,R. *Langmuir* 1991, *7*, 824.
7. Asakawa,T.; Fukita,T.; Miyagishi,S. *Langmuir* 1991, *7*, 2112.
8. Nagarajan,R.; Ruckenstein,E. *Langmuir* 1991, *7*, 2934.
9. Evans,D.F.; Ninham,B.W. *J. Phys. Chem.* 1983, *87*, 5005.
10. Hildebrand,J.H.; Prausnitz,J.M.; Scott,R.L. *Regular and Related Solutions* Van Nostrand Reinhold Company: New York, 1970.
11. Ben-Shaul,A.; Rorman,D.H.; Hartland,G.V.; Gelbart,W.M. *J. Phys. Chem.* 1986, *90*, 5277.
12. Gelbart,W.M.; Ben-Shaul,A.; McMullen,W.E.; Masters,A. *J. Phys. Chem.* 1984, *88*, 861.
13. Blankschtein,K.; Thurston,G.M.; Benedek,G.B. *Phys. Rev. Lett.* 1985, *54*, 955.

14. Blankschtein,D.; Thurston,G.M.; Benedek,G.B. *J. Chem. Phys.* 1986, *85*, 7268.
15. Puvvada,S.; Blankschtein,D. *J. Chem. Phys.* 1990, *92*, 3710.
16. Goldstein,R.E. *J. Chem. Phys.* 1986, *84*, 3367.
17. Kjellander,R. *J. Chem. Soc. Faraday Trans. 2* 1982, *78*, 2025.
18. Puvvada,S.; Blankschtein,D. *This Book* 1992.
19. Hartley,G.S. *Aqueous Solutions of Paraffin Chain Salts* Hermann; Paris, 1936.
20. Tanford,C. *The Hydrophobic Effect* Wiley: New York, 1973; Second Edition, 1980.
21. Israelachvili,J.N.; Mitchell,D.J.; Ninham,B.W. *J. Chem. Soc. Faraday Trans. 2* 1976, *72*, 1525.
22. Abraham,M.H. *J. Chem. Soc. Faraday Trans. 1* 1984, *80*, 153.
23. Abraham,M.H.; Matteoli,E. *J. Chem. Soc. Faraday Trans. 1* 1988, *84*, 1985.
24. Mukerjee,P.; Handa, T. *J. Phys. Chem.* 1981, *85*, 2298.
25. Dill,K.A.; Flory,P.J. *Proc. Natl. Acad. Sci. USA* 1980, *77*, 3115.
26. Szleifer,I.; Ben-Shaul,A.; Gelbart,W.M. *J. Chem. Phys.* 1987, *86*, 7094.
27. Girifalco,L.A.; Good,R.J. *J. Phys. Chem.* 1957, *61*, 904.
28. Good,R.J.; Elbing,E. *Ind. Eng. Chem.* 1970, *62*, 54.
29. Prausnitz,J.M. *Molecular Thermodynamics of Fluid-Phase Equilibria* Prentice-Hall: Englewoodcliffs, N.J., 1969.
30. Dunlop,R.D.; Scott,R.L. *J. Phys. Chem.* 1962, *66*, 631.
31. Hildebrand,J.H.; Dymond,J. *J. Chem. Phys.* 1967, *46*, 624.
32. Funasaki,N.; Hada,S. *J. Phys. Chem.* 1983, *87*, 342.
33. Holland,P.M.; Rubingh,D.N. *J. Phys. Chem.* 1983, *87*, 1984.
34. Mysels,K.J.; Otter,R.J. *J. Colloid Sci.* 1961, *16*, 462.
35. Shedlowsky,L.; Jakob,C.W.; Epstein,M.B. *J. Phys. Chem.* 1963, *67*, 2075.
36. Garcia-Mateos,I.; Velazquez,M.M.; Rodriguez,L.J. *Langmuir* 1990, *6*, 1078.
37. Shinoda,K. *J. Phys. Chem.* 1954, *58*, 541.
38. Malliaris,A.; Binana-Limbele,W.; Zana,R. *J. Colloid Interface Sci.* 1986, *110*, 114.
39. Stellner,K.L.; Amante,J.C.; Scamehorn,J.F.; Harwell,J.H. *J. Colloid Interface Sci.* 1988, *123*, 186.
40. Wada,Y.; Ikawa,Y.; Igimi,H.; Makihara,T.; Nagadome,S.; Sugihara,G. *Fukuoka University Science Reports* 1989, *19*, 173.

RECEIVED April 22, 1992

Chapter 5

Molecular–Thermodynamic Theory of Mixed Micellar Solutions

Sudhakar Puvvada[1] and Daniel Blankschtein

Department of Chemical Engineering and Center for Materials Science and Engineering, Massachusetts Institute of Technology, Cambridge, MA 02139

We review a recently formulated molecular-thermodynamic theory to describe and predict micellization, phase behavior and phase separation of mixed micellar solutions. The theoretical formulation consists of blending a thermodynamic theory of mixed micellar solutions, which captures the salient features of these complex fluids at the macroscopic level, with a molecular model of mixed micellization, which captures the essential driving forces for micellization at the molecular level. The molecular-thermodynamic theory is then utilized to predict a broad spectrum of mixed micellar solution properties as a function of surfactant type, surfactant composition, and solutions conditions such as temperature. The predicted properties include (i) the critical micellar concentration (CMC), (ii) the average mixed micellar composition at the CMC, (iii) the micellar size and composition distribution and its characteristics, (iv) the micellar shape, and (v) the coexistence curve bounding the two-phase region of the phase diagram. The broad spectrum of theoretical predictions compares very favorably with the experimentally measured properties.

Surface-active compounds used in commercial applications typically consist of a mixture of surfactants because they can be produced at a relatively lower cost than that of isomerically pure surfactants. In addition, in many surfactant applications, mixtures of dissimilar surfactants often exhibit properties superior to those of the constituent single surfactants due to synergistic (attractive)

---

[1]Current address: Center for Bio/Molecular Science and Engineering, Naval Research Laboratory, Washington, DC 20375–5000

0097–6156/92/0501–0096$06.00/0
© 1992 American Chemical Society

interactions between the different surfactant species (*1,2*). Indeed, in solutions containing mixtures of surfactants, the tendency to form aggregated structures (mixed micelles) can be substantially different than in solutions containing only the constituent single surfactants. For example, the critical micellar concentration (CMC) of a mixture of anionic and cationic surfactants in aqueous solution is considerably lower than the CMC's of each individual surfactant in aqueous solution (*3*). On the other hand, antagonistic (repulsive) interactions, in mixtures of hydrocarbon-based and fluorocarbon-based surfactants in aqueous solution, result in mixture CMC's that can be considerably higher than the CMC's of the constituent single surfactants in aqueous solution (*4,5*). In general, specific interactions (synergistic or antagonistic) between surfactants result in solutions of surfactant mixtures having micellar and phase behavior properties which can be significantly different from those of solutions containing the constituent single surfactants. Consequently, understanding specific interactions between the various surfactant species present in the solution is of central importance to the surfactant technologist. Indeed, in order to tailor surfactant mixtures to a particular application, the surfactant technologist has to be able to predict and manipulate (i) the tendency of surfactant mixtures to form monolayers, micelles, and other self-assembling aggregates in solution, (ii) the properties of the formed aggregates such as their shape and size, (iii) the distribution of the various surfactant species between monomers and mixed micelles, and (iv) the phase behavior and phase equilibria of solutions containing surfactant mixtures.

In spite of their considerable practical importance, as well as the challenging theoretical issues associated with the description of these complex fluids, solutions of surfactant mixtures have not received the full attention that they deserve. In particular, previous theoretical studies of mixed micellar solutions have evolved along two very different, seemingly unrelated, fronts. On the one hand, significant efforts have been devoted to understand the mixture CMC (*6-11*), as well as the micellar size and composition distribution (*12,13*). On the other hand, very little effort has been devoted to understand the solution behavior at higher surfactant concentrations where intermicellar interactions become increasingly important and control the phase behavior and phase separation phenomena (*14*). In view of this, it is quite clear that there is an immediate need to develop a theoretical description of mixed micellar solutions capable of *unifying* the previously disconnected treatments of micellization and phase behavior, including phase separation, into a *single coherent computational framework*.

To contribute to this much desired *theoretical unification*, we have recently formulated (*15,16*) a molecular-thermodynamic theory of mixed micellar solutions which consists of blending a *molecular model of mixed micellization*, which captures the essential physico-chemical factors which control mixed micelle formation and growth, with a *thermodynamic theory*, which captures the salient features of these complex fluids at the macroscopic level. The molecular model of mixed micellization utilizes information, which is readily available experimentally or can be estimated using *statistical mechanics* and/or *scaling-type arguments* (*16,17*), about surfactant molecules and water. This information includes the chemical nature and size of the surfactant hydrophilic moiety and hydrophobic moiety (typically an hydrocarbon chain), values of interfacial tensions between hydrocarbon and water, and values of the transfer free energies of hydrocarbons from water to bulk hydrocarbon. In addition, the value of the free-energy change associated with hydrocarbon-chain packing inside the micellar core is needed.

In this paper, we review the central elements of the molecular-thermodynamic theory, and present examples of its qualitative and quantitative predictive capabilities, including a comparison with experimentally measured mixed micellar solution properties. For complete details and explicit mathematical derivations, the reader is referred to Refs. 15, 16 and 17.

Thermodynamic Theory of Mixed Micellar Solutions

The thermodynamic theory of mixed micellar solutions consists of evaluating the total Gibbs free energy of the solution G, and subsequently analyzing G using the methods of equilibrium thermodynamics (*16,17*). For the sake of clarity, we present a theoretical description of aqueous solutions containing *binary* surfactant mixtures. Consider a solution containing N_A surfactant A molecules, N_B surfactant B molecules, and N_w water molecules in thermodynamic equilibrium at temperature T and pressure P. The total surfactant mole fraction X and the composition of the mixture α_{soln} are equal to $(N_A + N_B)/(N_w + N_A + N_B)$ and $N_A/(N_A + N_B)$, respectively. If X exceeds the critical micellar concentration (CMC), the various surfactant molecules will self-assemble to form a distribution of mixed micelles $\{N_{n\alpha}\}$, where $N_{n\alpha}$ is the number of mixed micelles having aggregation number n and composition α. Note that in such a mixed micelle, there are $n\alpha$ surfactant A molecules and $n(1-\alpha)$ surfactant B molecules, such that, $N_A = \sum_{n\alpha} n\alpha N_{n\alpha}$, and $N_B = \sum_{n\alpha} n(1-\alpha)N_{n\alpha}$. Note also that mixed micelles of different sizes and compositions are treated as distinct chemical species in chemical equilibrium with each other and with the free monomers in solution.

The Gibbs free energy of the mixed micellar solution G is modelled as the sum of three distinct contributions: the free energy of formation G_f, the free energy of mixing G_m, and the free energy of interaction G_i, that is, $G = G_f + G_m + G_i$. These three contributions are chosen to provide a heuristically appealing identification of the various factors responsible for mixed micelle formation and growth, on the one hand, and for phase behavior and phase separation, on the other.

The free energy of formation is expressed as (*16,17*)

$$G_f = N_w\mu_w^o + N_A\mu_A^o + N_B\mu_B^o + \sum_{n\alpha} nN_{n\alpha}g_{mic}(sh,n,\alpha,l_c) , \qquad (1)$$

where $\mu_w{}^\circ$, $\mu_A{}^\circ$, and $\mu_B{}^\circ$ are the standard-state chemical potentials, and $g_{mic}(sh, n, \alpha, l_c)$ is the free energy of mixed micellization which captures the free-energy change (per monomer) associated with transferring $n\alpha$ surfactant A monomers and $n(1-\alpha)$ surfactant B monomers from water into a mixed micelle of shape sh, aggregation number n, micellar composition α, and core minor radius l_c. The numerical magnitude of g_{mic}, which reflects the propensity of a mixed micelle to form and subsequently grow, summarizes the many complex physico-chemical factors responsible for micelle formation, such as, the hydrophobic effect, interfacial effects, conformational effects associated with restricting the hydrocarbon chains in the micellar core, steric and electrostatic interactions between the hydrophilic surfactant moieties, and an entropic contribution associated with mixing the two surfactant species in the mixed micelle (*15-17*).

The free energy of mixing the formed mixed micelles, free monomers, and water is modelled by an expression of the form (*15-17*)

$$G_m = kT \left[N_w \ln X_w + \sum_{n\alpha} N_{n\alpha} \ln X_{n\alpha} \right], \qquad (2)$$

where $X_w = N_w/(N_w + N_A + N_B)$ is the mole fraction of water, and $X_{n\alpha} = N_{n\alpha}/(N_w + N_A + N_B)$ is the mole fraction of micelles having aggregation number n and composition α. $-G_m/T$ is an entropic contribution which reflects the number of ways in which the distribution of mixed micelles, the free monomers, and the water molecules can be positioned in the solution as a function of the solution concentration and composition.

The free energy of interaction reflects interactions between mixed micelles, water molecules, and free monomers in the solution. At the level of a mean-field type quadratic expansion, this contribution takes the following form (*15*)

$$G_i = -\frac{1}{2} C_{eff}(\alpha_{soln})(N_A + N_B)(\phi_A + \phi_B), \qquad (3)$$

where ϕ_A and ϕ_B are the volume fractions of surfactants A and B, respectively, and $C_{eff}(\alpha_{soln})$ is an effective mean-field interaction parameter for the mixture which is related to the single-surfactant interaction parameters, C_{AW} and C_{BW}, and a specific interaction parameter C_{AB} through

$C_{eff}(\alpha_{soln}) = C_{AW}\alpha_{Soln} + C_{BW}(1-\alpha_{soln}) - C_{AB}\alpha_{soln}(1-\alpha_{soln})\sqrt{\gamma_A \gamma_B}/\gamma_{eff}$, where γ_A (γ_B) is the ratio of the molecular volumes of surfactant A (surfactant B) and water, and $\gamma_{eff} = \alpha_{soln}\gamma_A + (1-\alpha_{soln})\gamma_B$.

All thermodynamic properties of the mixed micellar solution are governed by the proposed Gibbs free energy model through the chemical potential of water μ_w, and the chemical potential of a mixed micelle, having aggregation number n and composition α, $\mu_{n\alpha}$, which are obtained from G by simple differentiation, that is, $\mu_w = (\partial G/\partial N_w)_{T,P,N_{n\alpha}}$ and $\mu_{n\alpha} = (\partial G/\partial N_{n\alpha})_{T,P,N_w,N_{n'\alpha'}}$. Note that the chemical potentials of a surfactant A monomer μ_A and of a surfactant B monomer μ_B are obtained by setting n=1, and $\alpha=1$ (for A) or 0 (for B), respectively, in the expression for $\mu_{n\alpha}$.

When the mixed micellar solution is in thermodynamic equilibrium, the chemical potential $\mu_{n\alpha}$ is related to the chemical potentials μ_A and μ_B by the condition of chemical equilibrium, that is, $\mu_{n\alpha} = n\alpha\mu_A + n(1-\alpha)\mu_B$. Using the expressions for $\mu_{n\alpha}$, μ_A, and μ_B obtained from G in this condition, the following expression for the micellar size and composition distribution $\{X_{n\alpha}\}$ is obtained

$$X_{n\alpha} = \frac{1}{e}X_1^n e^{-n\beta g_m(n,\alpha,\alpha_1)} , \qquad (4)$$

where $\beta = 1/kT$, $\beta g_m(\alpha,\alpha_1) = \beta g_{mic}(\alpha) - 1 - \alpha\ln\alpha_1 - (1-\alpha)\ln(1-\alpha_1)$ is a modified dimensionless free energy of mixed micellization, $X_1 = X_{1A} + X_{1B}$ is the total mole fraction of free monomers in the solution, and $\alpha_1 = X_{1A}/X$ is the composition of free monomers. Eq. (4) indicates that a delicate balance between two opposing factors determines the magnitude of $X_{n\alpha}$. The first is the Boltzmann factor $\exp(-n\beta g_m)$ reflecting the energetic advantage of forming a micelle (n-mer) from n free monomers. The second factor X_1^n reflects the large entropic disadvantage associated with localizing n surfactant molecules in a single micelle.

To determine $X_{n\alpha}$ using Eq. (4), one needs to know X_1, α_1, and g_m (or equivalently g_{mic}). The first two, X_1 and α_1, are obtained by solving the material balance equations for surfactants A and B, $X_A = \alpha_{soln}X = \alpha_1 X_1 + \sum_{n,\alpha} n\alpha X_{n\alpha}$, and $X_B = (1-\alpha_{soln})X = (1-\alpha_1)X_1 + \sum_{n,\alpha} n(1-\alpha)X_{n\alpha}$, respectively. The third one, g_{mic}, is calculated using the molecular model of mixed micellization to be presented in the next section. Subsequently, the values of X_1 and α_1, so deduced, are inserted back into Eq. (4) to compute the entire micellar size and composition distribution $\{X_{n\alpha}\}$.

Once $\{X_{n\alpha}(X,\alpha_{soln},T,P)\}$ is known, in the context of the Gibbs free-energy formulation, all equilibrium properties of the mixed micellar solution can be evaluated. These properties include the CMC, characteristic features of the distribution such as the weight-average mixed micelle aggregation number, and the phase behavior including phase separation (see the Section on Results and Discussion).

Molecular Model of Mixed Micellization

As discussed in the previous section, the free energy of mixed micellization g_{mic} constitutes a central element in the evaluation of micellar solution properties such as the CMC, the micellar size and composition distribution, and the phase behavior including phase separation of mixed micellar solutions. We have also shown (*15*) that g_{mic} captures, at the micellar level, specific interactions between the two surfactant species and thus determines the observed nonidealities in the mixture CMC as well as in other micellar properties. The free energy of micellization $g_{mic}(sh,n,\alpha,l_c)$ represents the free-energy change (per monomer) associated with creating a micelle having shape sh, aggregation number n, composition α, and core minor radius l_c, from nα A-type and n(1-α) B-type surfactant molecules. As described earlier, the magnitude of g_{mic} reflects many complex physico-chemical factors such as the hydrophobic effect, interfacial effects, conformational free-energy changes associated with restricting the hydrocarbon chains in the micellar core, steric and electrostatic interactions between the hydrophilic moieties at the micellar core-water interface, and intramicellar entropic contributions associated with mixing the two surfactant species in a mixed micelle.

   To evaluate the free-energy contributions associated with the various physico-chemical factors mentioned above, we have developed (*16,17*) a thought process to *visualize* the reversible formation of a mixed micelle having shape sh, aggregation number n, composition α, and core minor radius l_c (final state) from nα surfactant A monomers and n(1-α) surfactant B monomers (initial state) in water, as shown in Figure 1. Since the numerical magnitude of the free-energy change associated with a reversible process should be independent of the path connecting the initial and final states, we have chosen a series of convenient intermediate states to calculate g_{mic}. Each individual step, connecting these intermediate states, contributes to g_{mic} a free-energy change that can be evaluated using available experimental data or statistical-thermodynamics, and simple scaling-type arguments. Below, we describe (*16,17*) the various steps of the thought process used to estimate g_{mic} for a mixed micelle having shape sh, aggregation number n, composition α, and core minor radius l_c. As described in Ref. 17, the CH_2 group adjacent to the hydrophilic moiety lies within the hydration shell of this moiety, and therefore does not possess any hydrophobic properties. Accordingly, in the remainder of this paper, we denote by *head*, the hydrophilic moiety and the CH_2 group adjacent to it, and by *tail* the rest of the surfactant molecule.

   In the first step of the thought process, the heads, if charged, are discharged along with the counterions. Subsequently, in the second step, the bond between the head and the tail of each surfactant molecule is broken. In the third step, the nonpolar hydrocarbon tails of surfactants A and B are transferred from water to a mixture of hydrocarbons A and B whose composition is equal to the micellar composition α. In the fourth step, an hydrocarbon droplet having shape sh and core minor radius l_c is created from the hydrocarbon mixture having composition α. That is, in this step, an interface

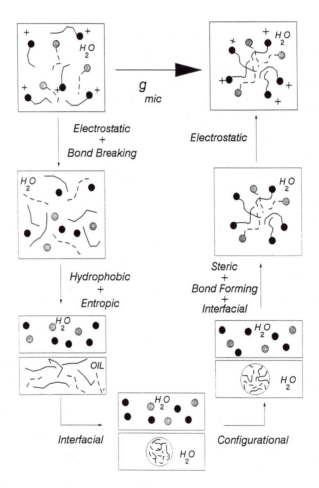

Figure 1. Schematic representation of thought process to *visualize* the various physico-chemical factors involved in the formation of a mixed micelle. Surfactant A is represented by a black head and a full tail, and surfactant B by a grey head and a dashed tail.

separating the hydrocarbon mixture from water is created. Note that, within this hydrocarbon droplet, the surfactant tails are unrestricted and can move freely. However, in a micelle, one end of the tail is attached to the head and therefore is restricted to lie close to the micellar core-water interface. Accordingly, in the fifth step, this restriction is imposed on the tails. Therefore, at the end of the fifth step, the creation of the micellar core has been completed. The creation of the micellar corona of heads follows next. Accordingly, in the sixth step, the discharged heads are reattached to the tails at the micellar core-water interface. This involves three operations: recreating the bond between the head and the tail, screening part of the micellar core-water interface from contact with water, and introducing steric repulsions between the heads. Note that the free-energy change associated with reforming the bond is assumed to be equal and opposite in sign to that associated with breaking the bond in the second step, and consequently the two contributions cancel each other. Finally, in the seventh step, the heads, if charged, are recharged along with the associated counterions which completes the creation of the micellar corona, and thus of the entire mixed micelle.

The free-energy contributions associated with each step of the thought process can be related (*16,17*) to the various physico-chemical factors associated with micellization. These contributions include (*16,17*):

(i) The hydrophobic free energy $g_{w/hc}$ associated with transferring the hydrocarbon tails from water to an hydrocarbon mixture in the third step. Note that $g_{w/hc}$ is obtained from available solubility data (*20*). In addition, group-contribution methods (*21*), and molecular simulations (*22*) can also be used to estimate $g_{w/hc}$.

(ii) The interfacial free energy g_σ associated with creating the micellar core-water interface in the fourth step, as well as with shielding part of that interface in the sixth step. Note that g_σ is obtained from available hydrocarbon-water interfacial tensions, and the interfacial area per monomer. In the calculation of g_σ, the effect of interfacial curvature on the interfacial tensions is captured using the Tolman equation (*23*), where the Tolman distance is computed using simple scaling-type arguments.

(iii) The configurational (packing) free energy $g_{hc/mic}$, arising from the loss in conformational degrees of freedom in the fifth step, is evaluated using a mean-field single-chain statistical-mechanical approach (*16,17,24*). Specifically, the calculations involve (a) generating the various conformations of a single central tail, subject to a mean-field generated by the other tails, using the rotational isomeric state approximation, (b) excluding all those conformations in which the tail is exposed to water outside the micellar core, and (c) evaluating the resulting partition function.

(iv) The steric free energy g_{st} associated with repulsive steric interactions between the heads in the sixth step. The heads at the interface are modelled as a localized monolayer, and the free energy associated with creating this monolayer is evaluated using techniques from statistical mechanics and Monte-Carlo simulations. Important inputs to the computation of g_{st} are the average head cross-sectional areas a_h's of the two surfactant species, which are estimated from the known molecular structures of the surfactant heads, supplemented by simple scaling-type arguments. In addition, for chain-like hydrophilic moieties, the a_h's can be estimated more rigorously by generating the conformations of the chain-like heads, in the context of the rotational isomeric state approximation (25), and subsequently determining the corresponding average cross-sectional areas.

(v) The electrostatic free energy g_{elec}, associated with the first and the seventh steps, is obtained by solving the Poisson-Boltzmann equation for the electric field strength and the distribution of ions around the mixed micelle.

The total free energy of mixed micellization g_{mic} is then computed by summing the five contributions described above, that is,

$$g_{mic}(sh,n,\alpha,l_c) = g_{w/hc} + g_\sigma + g_{hc/mic} + g_{st} + g_{elec} . \tag{5}$$

Note that the various contributions to g_{mic} in Eq. (5) are evaluated for a particular micellar shape sh, micellar composition α, and micellar-core minor radius l_c. The resulting *optimum* values sh^*, α^*, and l_c^* are then determined by a minimization procedure (16,17). Note also that the value of sh^*, so deduced, determines whether the mixed micelles exhibit *one-dimensional, two-dimensional, or no growth* (16,17).

Results and Discussions

The molecular-thermodynamic theory described in the previous two sections is utilized below to predict a broad spectrum of mixed micellar solution properties as a function of surfactant molecular structure, surfactant composition, and solution conditions such as temperature. The predicted properties include (i) the CMC, (ii) the average mixed micellar composition at the CMC, (iii) the weight-average mixed micelle aggregation number, and (iv) the liquid-liquid phase separation coexistence curves, including the compositions of the two coexisting micellar-rich and micellar-poor phases.

First, we present both *qualitative and quantitative predictions* based on a simpler phenomenological analytical model for g_{mic} (not presented in this paper, see Ref. 15) rather than on the more accurate and detailed molecular model of micellization presented in the previous section. Although most of these predictions are qualitative in nature, they enable us to obtain *analytical* expressions for many of the useful mixed micellar solution properties, as well as to shed light on the physical basis of some of the observed experimental trends. In the context of the simpler model for g_{mic}, predictions are made for aqueous solutions containing binary mixtures of nonionic-nonionic, and nonionic-ionic surfactants.

Subsequently, we utilize the detailed molecular model of mixed micellization presented in the previous section to *numerically* compute g_{mic} in order to provide more accurate *quantitative* predictions of mixed micellar properties of aqueous solutions containing binary mixtures of nonionic surfactants belonging to the alkyl polyethylene oxide (C_iE_j) family. The broad spectrum of theoretical predictions compares very favorably with the experimentally measured properties.

Predictions Based on Simple Analytical Model for g_{mic}. We have recently shown *(15)* that by summing the various contributions to g_{mic} one can, to leading order in α and l_c, express g_{mic} as $B + (A-B)\alpha + C\alpha^2 + kT[\alpha\ln\alpha + (1-\alpha)\ln(1-\alpha)]$, where the parameters A, B, and C, which reflect the various physico-chemical factors responsible for micellization discussed in the Molecular Model Section, can be computed *analytically*. Using this simple model for g_{mic}, one can evaluate *analytically* various mixed micellar solution properties. Specifically, below we present some quantitative predictions as well as discuss qualitative trends of the CMC and the phase separation observed in aqueous solutions of various binary surfactant mixtures.

In Ref. 15, we derived an expression for the mixture CMC which is identical to the well-known expression derived in the context of the pseudo-phase separation model using the regular-solution theory with an empirical interaction parameter ϵ *(9,10)*. Specifically, in the context of the molecular-thermodynamic theory with g_{mic} computed using the simple analytical expression given above, we showed that the CMC of a binary surfactant mixture can be expressed as a function of the CMC's of the constituent single surfactants as follows

$$\frac{1}{CMC} = \frac{\alpha_1}{f_A CMC_A} + \frac{1-\alpha_1}{f_B CMC_B} , \tag{6}$$

where $\ln f_A = -\beta C(1-\alpha^*)^2$, $\ln f_B = -\beta C(\alpha^*)^2$, $CMC_A = e^{\beta(A+C)-1}$ is the CMC of pure surfactant A, $CMC_B = e^{\beta B - 1}$ is the CMC of pure surfactant B, and α^* is the optimum micellar composition. The variables f_A and f_B are equivalent to the micellar activity coefficients of each surfactant, and the parameter $-\beta C$ is equal to the empirical interaction parameter ϵ used in the pseudo-phase separation model *(9,10)* for mixed micelles. Our approach, therefore, enables us to rationalize the physical basis behind Eq. (6), an expression which has been utilized extensively to analyze and predict CMC's of mixed surfactant solutions *(1,2)*. Furthermore, our approach allows us to *quantitatively predict* mixture CMC's form a knowledge of the parameters A, B, and C, which can be estimated using the simple model for g_{mic} or computed more accurately utilizing the detailed molecular model for g_{mic} presented in the previous section and Ref. 16.

In particular, through f_A and f_B, the parameter C determines the magnitude of the synergistic (attractive) or antagonistic (repulsive) interactions that lead respectively to negative or positive deviations of the mixture CMC (and other micellar properties) from ideality. Indeed, positive C values correspond to synergistic interactions while negative C values correspond to antagonistic interactions. More specifically, we have also shown (15) that in mixtures of hydrocarbon-based surfactants, the parameter C, to leading order in α and l_c, can be obtained from g_{elec}. Indeed, by computing g_{elec}, we have estimated the value of C for various surfactant mixtures. Preliminary results of these calculations are presented in Table I, where we compare the predicted C values with experimentally reported values (2), and as can be seen there is good agreement between theory and experiment.

Regarding the phase separation phenomena, we have investigated the effects of adding small quantities of surfactant A to an aqueous solution of surfactant B which exhibits liquid-liquid phase separation on (i) the variation of the critical temperature of the mixed micellar solution with total surfactant composition, and (ii) the distribution of the two surfactant species between the coexisting micellar-rich and micellar-poor phases. Specifically, in aqueous solutions of nonionic-monovalent ionic surfactant mixtures, we find that the closed-loop coexistence curve, associated with the nonionic surfactant (B), shrinks upon adding small quantities of a monovalent ionic surfactant (A). This prediction is consistent with measured cloud-point curves in aqueous solutions of $C_{12}E_6$-SDS (26). We also find that the ionic surfactant partitions preferentially into the micellar-poor phase, where the overall electrostatic repulsions are minimized. On the other hand, when small quantities of a nonionic surfactant are added to another aqueous solution of a nonionic surfactant, the two-phase region remains practically unaltered. In addition, the two nonionic surfactants partition evenly between the two coexisting phases (for details, see below). Qualitative predictions along the same lines have also been made (15) for other aqueous solutions of binary surfactant mixtures including nonionic-zwitterionic, zwitterionic-zwitterionic, zwitterionic-ionic, and anionic-cationic surfactants, as well as hydrocarbon-fluorocarbon based surfactants.

Quantitative Predictions for Binary Mixtures of C_iE_j Nonionic Surfactants. In the previous sub-section, we presented predictions of micellar solution properties using a simple analytical phenomenological model for g_{mic}. In this section, we utilize the detailed molecular model of mixed micellization to compute more accurately g_{mic} as a function of micellar composition. Subsequently, we use the calculated values of g_{mic}, in the context of the thermodynamic theory, to predict various mixed micellar properties of aqueous solutions containing binary mixtures of C_iE_j nonionic surfactants.

Figure 2 shows predictions of the CMC for aqueous solutions of $C_{12}E_6$-$C_{12}E_8$ (full line) and $C_{12}E_6$-$C_{10}E_4$ (dashed line) at 25°C as a function of α_{soln}, the fraction of $C_{12}E_6$ in each mixture. The various symbols represent experimental CMC values obtained using the surface-tension method. For both systems, pure $C_{12}E_6$ has the lowest CMC, and as the fraction of $C_{12}E_6$ in the mixture is decreased, the CMC increases monotonically and approaches the CMC of the second surfactant ($C_{12}E_8$ or $C_{10}E_4$). The agreement between the theoretical predictions and the experimental data is quite good.

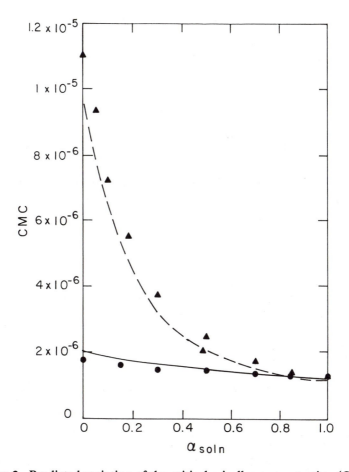

Figure 2. Predicted variation of the critical micellar concentration (CMC) of aqueous solutions of $C_{12}E_6$-$C_{12}E_8$ (full line) and $C_{12}E_6$-$C_{10}E_4$ (dashed line) at 25 °C as a function of α_{soln}, the fraction of $C_{12}E_6$ in each mixture. The two symbols represent experimentally measured CMC values using the surface-tension method: ($\bullet$) $C_{12}E_6$-$C_{12}E_8$, and ($\blacktriangle$) $C_{12}E_6$-$C_{10}E_4$.

Table I. Theoretical and experimental (2) values of C for various surfactant mixtures

Surfactant Mixture	Theoretical C (kT)	Experimental C (kT)
$C_{12}SO_4Na + C_9\phi E_{15}$	5.0	4.6-5.9
$C_{12}SO_4Na + C_9\phi E_{15} + 0.65M$ NaCl	1.8	1.5
$C_{10}SO_4Na + C_8E_4 + 0.05M$ NaBr	3.2	4.1
$C_{10}N(CH_3)_2Br + C_{10}OSO_3Na + 0.05M$ NaBr	12.3	13.2
$C_{12}SO_4Na + C_{12}N(CH_3)_3Br$	19.0	25.5

Figure 3 shows the predicted average mixed micellar composition α_{mic}, at the CMC, for aqueous solutions of $C_{12}E_6$-$C_{12}E_8$ (full line) and $C_{12}E_6$-$C_{10}E_4$ (dashed line) at 25°C as a function of α_{soln}, the fraction of $C_{12}E_6$ in each mixture. The figure reveals that the mixed micelles are enriched with $C_{12}E_6$ ($\alpha_{mic} > \alpha_{soln}$) for both systems. This reflects the fact that, as stated earlier, $C_{12}E_6$ has the lowest CMC, and therefore the highest propensity to form micelles. Furthermore, since $C_{10}E_4$ has a higher CMC than $C_{12}E_8$, micelles are considerably more enriched with $C_{12}E_6$ in the $C_{12}E_6$-$C_{10}E_4$ mixture than in the $C_{12}E_6$-$C_{12}E_8$ mixture.

Figure 4 shows predictions of the weight-average mixed micelle aggregation number $<n>_w$ as a function of temperature for aqueous solutions of $C_{12}E_6$-$C_{12}E_8$ (at $X = 10^{-3}$) having various surfactant compositions α_{soln}. Curve a corresponds to a solution of $C_{12}E_6$-H_2O ($\alpha_{soln} = 1$), and, as we proceed form curve a to curve f, the value of α_{soln} decreases to zero, corresponding to a solution of $C_{12}E_8$-H_2O. For all compositions, Figure 4 clearly shows that the mixed micelles remain small ($<n>_w \approx 50$) below a certain temperature T* which depends on α_{soln}. For temperatures higher than T*, $<n>_w$ increases quite dramatically. In this respect, we have recently shown (17) that pure C_iE_j nonionic surfactants in aqueous solutions can exhibit varying degrees of one-dimensional growth, from spheroidal to cylindrical structures, as a function of temperature. In particular, the transition temperature T* increases (17) from about 15°C for pure $C_{12}E_6$ to about 50°C for pure $C_{12}E_8$ (at a total surfactant concentration of 1 wt%). In this spirit, it is reasonable to suggest that the predictions in Figure 4 are consistent with a "sphere-to-cylinder" shape transition of the mixed micelles which occurs over a narrow temperature range (captured

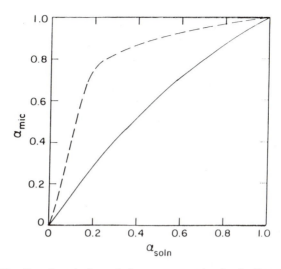

Figure 3. Predicted variation of the average mixed micellar composition α_{mic}, at the CMC, for aqueous solutions of $C_{12}E_6$-$C_{12}E_8$ (full line), and $C_{12}E_6$-$C_{10}E_4$ (dashed line) at 25°C as a function of α_{soln}, the fraction of $C_{12}E_6$ in each mixture.

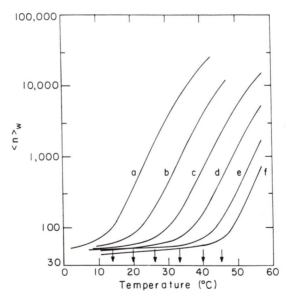

Figure 4. Predicted variation of the weight-average mixed micelle aggregation number $<n>_w$ as a function of temperature for aqueous solutions of $C_{12}E_6$-$C_{12}E_8$ having the following compositions: a ($\alpha_{soln}=1.0$, pure $C_{12}E_6$), b ($\alpha_{soln}=0.8$), c ($\alpha_{soln}=0.6$), d ($\alpha_{soln}=0.4$), e ($\alpha_{soln}=0.2$), and f ($\alpha_{soln}=0.0$, pure $C_{12}E_8$). The predictions were made at a total surfactant mole fraction $X=10^{-3}$. The various arrows denote shape transition temperatures deduced from dynamic light scattering measurements.

by T^*) which increases as α_{soln} decreases from 1 (corresponding to pure $C_{12}E_6$, $T^* \approx 13°C$) to 0 (corresponding to pure $C_{12}E_8$, $T^* \approx 47°C$). The predicted transition temperatures T^* compare very favorably with experimentally deduced transition temperatures, indicated by the arrows in Figure 4, obtained using dynamic light scattering measurements (Puvvada,S.; Chung, D.S.; Thomas, H.; Blankschtein, D.; Benedek, G.B., *to be published*).

Figure 5 shows the measured cloud-point curves of aqueous solutions of $C_{12}E_6$-$C_{10}E_6$ for various values of α_{soln}, the fraction of $C_{12}E_6$ in the mixture, indicated by the number next to each curve. As expected, the cloud-point curves lie between those corresponding to the constituent pure surfactant solutions. Thus, as the fraction of $C_{12}E_6$ in the mixture is increased, the cloud-point curves move to lower temperatures (recall that $T_c(C_{10}E_6)$ = 335.33K, and $T_c(C_{12}E_6) = 324.29K$). The solid lines are theoretical predictions of the coexistence curves for various α_Z values (indicated by the number next to each curve), where α_Z is the composition of the micellar-rich phase. We have also evaluated the composition of the micellar-poor phase α_Y, and find that $\alpha_Y \approx \alpha_Z$. This supports the equivalence of the cloud-point curves and the coexistence curves. The predicted equality of α_Y and α_Z has also been verified by physically partitioning the two micellar phases and determining the values of α_Y and α_Z using high performance liquid chromatography (HPLC) analysis. Similar predictions and measurements have been made for aqueous solutions of $C_{12}E_6$-$C_{12}E_8$ and $C_{12}E_6$-$C_{10}E_4$ (*16*).

Concluding Remarks

Mixed micellar solutions are currently a subject of considerable practical importance because they can exhibit properties which are superior to those of solutions containing the constituent single surfactants. Consequently, detailed information is becoming available on the critical micellar concentration, synergistic and antagonistic interactions between the surfactant species, and micellar solution phase behavior including phase separation. In view of these experimental developments, it becomes increasingly necessary to construct a theoretical approach capable of unifying the rich variety of seemingly unrelated experimental findings into a single coherent computational framework. It has been the purpose of this paper to review our recent contributions to this much needed theoretical unification.

Accordingly, in this paper, we have reviewed a recently proposed molecular-thermodynamic theory of mixed micellar solutions which can be utilized to make *quantitative self-consistent predictions* of a broad spectrum of mixed micellar solution properties as a function of surfactant molecular architecture, surfactant composition, and solution conditions such as temperature. Our approach represents a generalization of a recently developed molecular-thermodynamic theory for single-surfactant solutions, which has been successfully utilized to predict a wide range of micellar properties of aqueous solutions of pure nonionic surfactants and zwitterionic surfactants with and without added solution modifiers such as salts and urea (*17-19,27-32*).

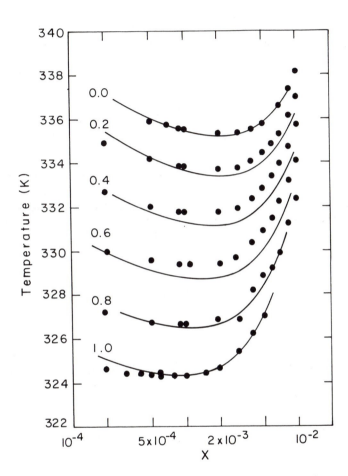

Figure 5. Predicted coexistence curves (lines) and experimentally measured cloud-point temperatures (●), corresponding to various total surfactant mole fractions X, of aqueous solutions of $C_{12}E_6$-$C_{10}E_6$ for various values of α_{soln}, the fraction of $C_{12}E_6$ in the mixture. The number next to each theoretical line denotes both the value of α_{soln}, corresponding to the measured cloud-point temperatures, as well as of the composition of the micellar-rich phase α_Z used to predict each coexistence curve.

The theoretical approach presented in this paper can be extended to treat similar phenomena in other, more complex, self-assembling surfactant systems. It is clear that by explicitly including electrostatic interactions our analysis can be extended to predict quantitatively micellar properties of surfactant solutions containing ionic and zwitterionic surfactants, with or without added salts. The presence of other solution modifiers, both organic and inorganic, can also be incorporated into the theoretical framework. The theory can also be extended to describe more complex self-assembling microstructures, such as, spontaneously forming vesicles, recently discovered in aqueous solutions of some anionic-cationic surfactant mixtures (33).

To conclude, we would like to reiterate that the molecular-thermodynamic theory presented in this paper is *predictive* in nature. As such, we believe that beyond its fundamental value, this theory can be utilized effectively by the surfactant technologist to identify, select, and possibly even tailor new surfactants for a particular application without the need of performing time consuming and often expensive measurements of a large number of equilibrium micellar solution properties. These possibilities appear particularly relevant in times when the search for new biodegradable and environmentally-safer surfactants is being vigorously pursued.

Acknowledgements

We would like to thank Leah Abraham for her invaluable assistance with some of the coexistence-curve and cloud-point curve measurements reported in this paper. We would also like to thank Claudia Sarmoria for her assistance in the calculations presented in Table I. This research was supported in part by the National Science Foundation (NSF) Presidential Young Investigator (PYI) Award to Daniel Blankschtein, and an NSF Grant No. DMR-84-18718 administered by the Center for Materials Science and Engineering at MIT. Daniel Blankschtein is grateful for the support of the Texaco-Mangelsdorf Career Development Professorship at MIT. He is also grateful to the following companies for providing PYI matching funds: BASF, British Petroleum America, Exxon, Kodak, and Unilever.

Literature Cited

1.   Scamehorn, J.F. *Phenomena in Mixed Surfactant Systems*, ACS Symposium Series 311, ACS:Washington, **1986**; and references cited therein.
2.   Rosen, M.J. *Surfactants and Interfacial Phenomena* second ed. Wiley: New York, **1989**; and references cited therein.
3.   Rosen, M.J. Hua, X.Y., *JAOCS*, **1982**, *59*, 582.
4.   Mukerjee, P.; Handa, T. *J. Phys. Chem.* **1981**, *85*, 2298.
5.   Handa, T.; Mukerjee, P. *J. Phys. Chem.* **1981**, *85*, 3916.
6.   Lange, V.H. *Kolloid Z.* **1953**, *96*, 131.
7.   Clint, J.H. *J. Chem. Soc. Faraday Trans. 1*, **1975**, *71*, 1327.
8.   Shinoda, K. *J. Phys. Chem.* **1954**, *58*, 541.

9.   Rubingh, D.N. In *Solution Chemistry of Surfactants*, Vol. 1, Mittal, K.L., Ed.; Plenum: New York, **1979**, pp. 337.

10. Holland, P.M.; Rubingh, D.N. *J. Phys. Chem.* **1983**, *87*, 1984.

11. Nagarajan, R. *Langmuir*, **1985**, *1*, 331.

12. Ben-Shaul, A.; Szleifer, I.; Gelbart, W.M. *J. Chem. Phys.* **1987**, *86*, 1044.

13. Stecker, M.M.; Benedek, G.M. *J. Phys. Chem.* **1984**, *88*, 6519.

14. Meroni, A.; Pimpinelli, A.; Reato, L. *Chem. Phys. Lett.*, **1987**, *135*, 137.

15. Puvvada, S.; Blankschtein, D. *J. Phys. Chem.* **1991**, *submitted*.

16. Puvvada, S.; Blankschtein, D. *J. Phys. Chem.* **1991**, *submitted*.

17. Puvvada, S.; Blankschtein, D. *J. Chem. Phys.* **1990**, *92*, 3710; and references cited therein.

18. Blankschtein, D.; Puvvada, S. *MRS Symposium Proceedings*, **1990**, *177*, 129.

19. Puvvada, S.; Blankschtein, D. *Proceedings of the 8th International Symposium on Surfactants in Solution*; Mittal, K.L., Shah, D.O., Eds.; Plenum: New York, in press.

20. Abraham, M.H. *J. Chem. Soc. Faraday Trans. 1* **1984**, *80*, 153.

21. Fredenslund, A.; Jones, R.L.; Prausnitz, J.M. *AIChE. J.* **1975**, *21*, 1087.

22. Jorgensen, W.L. *Acc. Chem. Res.* **1989**, *22*, 184.

23. Tolman, R.C. *J. Chem. Phys.* **1948**, *16*, 758.

24. Szleifer, I.; Ben-Shaul, A.; Gelbart, W.M. *J. Chem. Phys.* **1985**, *83*, 3612.

25. Sarmoria, C.; Blankschtein, D. *J. Phys. Chem.* **1992**, *96*, 1978.

26. Souza, L.D.S.; Corti, M.; Cantu, L.; Degiorgio, V. *Chem. Phys. Lett.* **1986**, *131*, 1260.

27. Blankschtein, D.; Thurston, G.M.; Benedek, G.B. *Phys. Rev. Lett.* **1985**, *54*, 955.

28. Blankschtein, D.; Thurston, G.M.; Benedek, G.B. *J. Chem. Phys.* **1986**, *85*, 7268; and references cited therein.

29. Puvvada, S.; Blankschtein, D. *J. Phys. Chem.* **1989**, *93*, 7753.

30. Briganti, G.; Puvvada, S.; Blankschtein, D. *J. Phys. Chem.* **1991**, *95*, 8989.

31. Carvalho, B.L.; Briganti, G.; Chen, S-H. *J. Phys. Chem.* **1989**, *93*, 4282.

32. Carale, T.R.; Blankschtein, D. *J. Phys. Chem.* **1992**, *96*, 459.

33. Kaler, E.W.; Murthy, A.H.; Zasadzinski, J.A.N. *Science*, **1989**, *245*, 1371.

RECEIVED March 19, 1992

Chapter 6

Modeling Polydispersity in Multicomponent Nonideal Mixed Surfactant Systems

Paul M. Holland

General Research Corporation, Santa Barbara, CA 93111

A method for treating polydispersity in nonideal mixed surfactant systems is presented. This simplifies the more general regular solution approach for treating multicomponent nonideal mixtures for the special case of homologous groups of surfactant components which behave ideally within the group, and nonideally with other components in the mixture. This results in a tractable approach for treating nonideal surfactant mixtures with many components, a situation which often arises when polydisperse commercial surfactants are used in nonideal mixtures. Application of this method is used to demonstrate the effects of polydispersity on the properties of nonideal surfactant mixtures and predicted trends are shown to be consistent with experimental results.

Polydisperse surfactant mixtures are used in nearly all practical applications of surfactants. This polydispersity usually arises from the presence of different isomers and the various chain length surfactant molecules that are produced when synthesizing typical commercial surfactants. Each of the different single-specie surfactants in the polydisperse mixture may exhibit different properties such as critical micelle concentration (CMC), limiting surface tension, etc. Because of this, measurement of the gross properties of a polydisperse surfactant mixture may not be adequate to properly describe the behavior of the system.

In addition to the unavoidable polydispersity that arises from impurities in starting materials and the natural variability of reaction products during synthesis, mixtures of different types of surfactants are often deliberately formulated. These formulations can be used to improve surfactant system performance by exploiting synergism based on nonideal interactions between different surfactant types (1,2). As a result, mixed surfactant systems used in practical applications can often be highly complex when considered at the level of single-specie components. This situation highlights the need for a tractable approach to polydispersity in nonideal mixed surfactant systems.

A number of studies on ideal and nonideal mixtures with polydisperse surfactants are reported in the literature (3-8). These include experimental mixed CMC (4-6) and

0097–6156/92/0501–0114$06.00/0

ultrafiltration (*5,6*) measurements on "binary" nonideal surfactant mixtures where one of the "surfactants" is a polydisperse mixture. These studies show CMCs to be well described using a "binary" nonideal model with a single interaction parameter (*4-6*) whereas micellar compositions and monomer concentrations (*5,6*) show deviations similar to those observed in ideal mixtures of polydisperse nonionic surfactants (*3*).

Multicomponent Nonideal Surfactant Mixtures

A generalized approach has been previously developed for treating multi-component nonideal mixed surfactant systems and successfully demonstrated in modeling ternary mixed systems (*9-11*). Key results of this pseudophase separation model will be summarized here to provide a starting point for examining the effects of polydispersity (see also Chapter 2, this volume).

The CMC of a system of any number of components is given by the generalized expression

$$\frac{1}{C^*_{Mix}} = \sum_{i=1}^{n} \frac{\alpha_i}{f_i C_i^*} \tag{1}$$

where C^*_{Mix} is the mixed CMC, α_i the mole fraction of component i in the total surfactant mixture, f_i the activity coefficient, and C_i^* the CMC of pure component i (see Legend of Symbols). This allows the mixed CMC to be readily calculated from the composition of the system if the activity coefficients and CMCs of the single-specie components are known.

For monomer concentrations and micellar mole fractions, the generalized expression

$$\sum_{i=1}^{n} \frac{\alpha_i C}{C + f_i C_i^* - M} = 1 \tag{2}$$

applies, where

$$M = \sum_{i=1}^{n} C_i^m \tag{3}$$

It is straight forward to solve this for any number of single-specie components given the activity coefficients. This allows the individual monomer concentrations to be obtained by

$$C_i^m = \frac{\alpha_i f_i C_i^* C}{C + f_i C_i^* - M} \tag{4}$$

and the mole fractions in the micelle to be obtained from

$$x_i = \frac{\alpha_i C}{C + f_i C_i^* - M} \tag{5}$$

In the case of ideal mixing, with unit activity coefficients, the multicomponent model can be readily solved for an arbitrary number of different surfactant components. For nonideal systems, it is first necessary to determine the activity coefficients.

A fully generalized expression for the activity coefficients based on pairwise (binary) interaction parameters has been developed for any number of components (*9,10*). In this expression

$$\ln f_i = \sum_{j=1}^{n} \beta_{ij} x_j^2 + \sum_{j-1}^{n} \sum_{k=1}^{j-1} (\beta_{ij} + \beta_{ik} - \beta_{jk}) x_j x_k \tag{6}$$

β_{ij} represents the net interaction between any two components i and j, and x the mole fractions in the mixed micelle (where $i \neq j \neq k$). It can be seen that there are both direct interaction terms, similar to form for binary systems and cross terms in the expression. Since this approach assumes that the binary interaction parameters β_{ij} have been independently determined, activity coefficients in the the multicomponent system can be determined without adjustable parameters.

The pairwise interaction parameters for this model can be obtained by measurements in binary mixed surfactant systems, from tabulated values in the literature, or by estimates based on results for similar systems. For n different components in the system there will be a total of

$$C_2^n = \frac{n!}{(n-2)!\ 2!} \tag{7}$$

different pairwise parameters needed, with n - 1 direct terms and C_2^n - n + 1 cross terms in appearing in Equation 6.

A severe tractability problem with the generalized approach becomes obvious as the number of components (and pairwise interactions) grows. This situation is demonstrated in Table I.

Table I. Number of Pairwise Interactions versus Number of Surfactant Components in Generalized Multicomponent Model	
Components	Pairwise Interactions
3	3
4	6
5	10
10	45
20	190
30	435
50	1225
100	4950

It can be readily seen that as the number of surfactant components becomes large, the number of pairwise interactions grows geometrically. While realistic situations of interest are unlikely for the largest numbers of components shown in the table, polydispersity in commercial surfactants types such as alkyl ethoxylates may routinely reach 30 or so single-specie components (e.g. ethoxylation from one to ten units, with three different hydrocarbon chain lengths, $C_{10\text{-}14}E_{1\text{-}10}$). At such levels of polydispersity, it is clearly not tractable to solve for the properties of nonideal mixtures of these with other surfactant types using the generalized approach of Equation 6.

Group Activity Coefficients

Inspection of Equation 6, suggests that in many cases a significant simplification of the problem might be realized. In particular this may occur for mixtures which contain groups of components of the same structural type (e.g. alkyl ethoxylates, C_xE_y) where components within the group might be expected to behave ideally with respect to each other, and have equivalent or very similar nonideal interactions with surfactant types outside the group.

It can be readily demonstrated for situations where $\beta_{ik} = \beta_{jk}$ and $\beta_{ij} = 0$, then f_i must equal f_j. The net result of this is that the effective number of pairwise interaction parameters in equation 6 can be decreased substantially each time two components in the mixture can be identified which behave ideally with respect to each other and have approximately equivalent pairwise interaction with other components in the mixed system (i.e. by $C_2^n - C_2^{n\text{-}1}$, where n is the equivalent number of components or groups. This can lead to the definition of group activity coefficients, as shown in the simple example below for two groups A and B, of n and N-n components respectively (total of N single-specie components). Here, the micellar mole fraction for each group is simply defined as the sum of the mole fractions of components in that group

$$x_A = \sum_{i=1}^{n} x_i \tag{8}$$

$$x_B = \sum_{i=n+1}^{N} x_i \tag{9}$$

and group activity coefficients for the two groups in this case become analogous to those for a binary mixed system, mainly

$$f_A = \exp[\beta(1-x_A)^2] \tag{10}$$

$$f_B = \exp[\beta x_A^2] \tag{11}$$

Generally, group activity coefficients will take a form analogous to those given by Equation 6, based on the number of groups defined and using sums of mole fractions of the components in each group in place of the x_i.

Polydispersity Effects on CMCs

The effects of polydispersity on CMCs in multicomponent systems can be readily examined using the group activity coefficient method. For example, the CMC of a system of N components

$$\frac{1}{C_{Mix}^*} = \sum_{i=1}^{N} \frac{\alpha_i}{f_i C_i^*} \tag{12}$$

might be broken into two groups and can be rewritten as

$$\frac{1}{C_{Mix}^*} = \sum_{i=1}^{n} \frac{\alpha_i}{f_A C_i^*} + \sum_{i=n+1}^{N} \frac{\alpha_i}{f_B C_i^*} \tag{13}$$

using group activity coefficients. Group mole fractions as total surfactant, α_A and α_B, can be defined in the usual way and pulled out of the summations along with group activity coefficients.

$$\frac{1}{C_{mix}^*} = \frac{\alpha_A}{f_A} \sum_{i=1}^{n} \frac{\alpha_i/\alpha_A}{C_i^*} + \frac{\alpha_B}{f_B} \sum_{i=n+1}^{N} \frac{\alpha_i/\alpha_B}{C_i^*} \tag{14}$$

What remains in the summations are seen by Equation 1 to be identical to the ideal mixed CMCs of each group. This demonstrates that the group activity coefficient method can be used to rewrite expressions for the mixed CMC of multicomponent systems entirely in terms of group parameters

$$\frac{1}{C_{Mix}^*} = \frac{\alpha_A}{f_A C_A^*} + \frac{\alpha_B}{f_B C_B^*} \tag{15}$$

It can also be seen that the forms of the expression for calculating the mixed CMC (Equations 1 and 15) are identical whether written with single component or group parameters, making it clear that results for the mixed CMC of the system will be identical whether single component (C_i^*) or group CMCs (C_A^*) are used. This demonstrates that as long as conditions for the use of group activity coefficients hold, it makes no difference whether one uses a single experimentally measured CMC value for each polydisperse group of ideal components in a mixture, or individual CMC values for each component in the mixture. This conclusion is consistent with experimentally measured results which show that mixed CMCs in such systems are well described using a strictly binary nonideal model with a single interaction parameter (4,5).

Polydispersity Effects on Monomer Concentrations

Although results show that polydispersity effects do not significantly impact mixed CMCs in multicomponent systems, there are substantial effects on monomer concentrations and micellar compositions. These can be demonstrated by example calculations using the group activity coefficient method. For example, if one considers N surfactant components divided into two groups (A and B) consisting of 1 and N-1 single-specie surfactants, respectively, then group parameters can be defined as in Equations 8-11 with n = 1. Given the interaction parameter β between the two groups and CMCs of the pure single-specie components, one can then solve the following relationship for any composition of the mixture

$$x_A + \sum_{i=2}^{N} \frac{\alpha_i C}{f_B C_i^* - f_A C_A^* + (\alpha_A C/x_A)} = 1 \qquad (16)$$

Example Calculations. To illustrate the effects of polydispersity on monomer concentrations, example calculations have been carried out on four different equimolar ternary surfactant mixtures. These systems consist of mixtures of a single component (A) which can behave nonideally with respect to a second group of two components (B) which behave ideally with respect to each other. These example mixtures are designed such that the ideal mixed CMC of group B is always equal to that of the single component A. This means (based on the results of the previous section) that the mixed CMC of the system, for any given degree of nonideality, will always be the same regardless of how different the CMCs of the group B components are. The relative CMCs of the components in the four different example mixtures are listed in Table II and show increasing differences in the CMCs of the two group B components to demonstrate polydispersity effects.

Table II. Relative CMCs of Surfactant Components in Example Mixtures			
Mixture	Component A	Group B Components	
1	1.00	1.00	1.000
2	1.00	2.00	0.666
3	1.00	4.00	0.571
4	1.00	8.00	0.533

Calculated results for the total monomer concentration versus concentration for the four different equal CMC mixtures are shown in Figure 1 for ideal mixing. It can be readily seen that as the CMCs of the two components in the group B diverge (polydispersity effectively increases), the total monomer concentration above the CMC in these ideal mixtures increases with increasing surfactant concentration. Figure 2 shows the relatively modest effect on the same mixtures for nonideal mixing between component A and group B with an interaction parameter β of -3 (typical for anionic-nonionic mixtures). In Figure 3 results for a more nonideal β of -9 show the total monomer concentration (scaled to the mixed CMC) to increase much more rapidly.

The effects of nonideality are examined directly in Figures 4 and 5, which show the total monomer relative to the mixed CMC as β changes from zero to -12, for the second and fourth mixtures, respectively. It is seen that as differences in the CMCs of group B components increase (in going from mixture 2 to 4), overall effects on the monomer concentrations become larger, but the relative magnitude of the nonideality effect diminishes. It should also be noted that as nonideality increases, the mixed CMC (to which results are scaled) decreases.

Finally, effects of nonideality on individual monomer concentrations are examined in Figures 6-8 for the third mixture. These show results for the individual monomer concentrations of each component (relative to the mixed CMC) for β equals zero, -3, and -12, respectively. It is interesting to note that individual monomer concentrations above the CMC can either increase or decrease with increasing concentration, and that with increasing nonideality the relative monomer concentrations of components A and B2 are reversed.

Such polydispersity effects on monomer concentrations can be quite important in the behavior of mixed surfactant systems since the monomer concentration (or activity) establishes the chemical potential of each component in solution. The bulk solution chemical potentials, in turn, effectively control all of the equilibrium properties of the mixed system, including those at solution interfaces.

Comparison with Experiment. There have been a number of experimental studies on mixed surfactant systems in which a polydisperse nonionic surfactant has been used as one of the components of a "binary" surfactant mixture (4-8). It is useful to consider results of these studies in light of the polydispersity effects that would be predicted by the group activity coefficient method.

Perhaps the most relevant experimental measurements in the literature for this purpose are an extensive series of CMC and ultrafiltration results on five different binary mixtures of anionic alkyl benzene sulfonate surfactants and polydisperse nonionic nonyl phenol ethoxylates by Osbourne-Lee et al. (5,6). The polydisperse nonyl phenol ethoxylate (NPE) surfactants used in their measurements had average ethylene oxide chain lengths of 10, 20 and 50. While the compositions and CMCs of individual surfactant components in these NPE surfactants are not available, useful comparisons of this study with the predicted trends for polydispersity effects can be made.

First, experimental CMC results (5) for each of these five different mixtures across a wide range of compositions show excellent agreement with the strictly binary regular solution model using a single interaction parameter (values ranging from -1.5 to -2.6). These results appear to confirm the group activity coefficient results in the previous section which indicate that polydispersity should have no effect on the mixed CMCs of the system.

Second, when experimental ultrafiltration results (5) for micellar mole fractions of the anionic surfactant are compared with predictions using the same strictly binary regular solution model and interaction parameters, the model predictions are

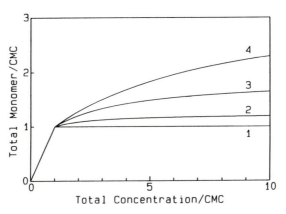

Figure 1. Total monomer concentration versus surfactant concentration for ideal mixing ($\beta = 0$) as polydispersity increases from mixtures 1 to 4 (see text).

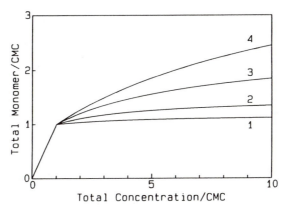

Figure 2. Total monomer concentration versus surfactant concentration for non-ideal mixing ($\beta = -3$) as polydispersity increases from mixtures 1 to 4 (see text).

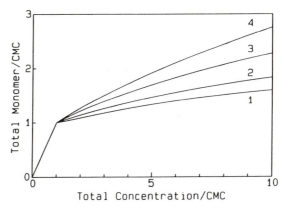

Figure 3. Total monomer concentration versus surfactant concentration for non-ideal mixing ($\beta = -9$) as polydispersity increases from mixtures 1 to 4 (see text).

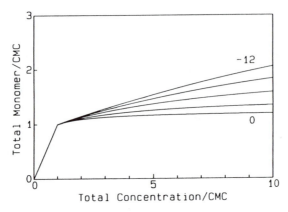

Figure 4. Total monomer concentration versus surfactant concentration showing nonideality effects as β changes from 0 to -12 in second mixture (see text).

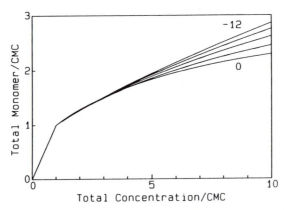

Figure 5. Total monomer concentration versus surfactant concentration showing nonideality effects as β changes from 0 to -12 in fourth mixture (see text).

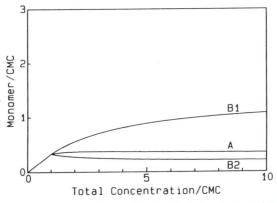

Figure 6. Monomer concentrations of individual group A and B components versus total surfactant concentration ($\beta = 0$) in third mixture (see text).

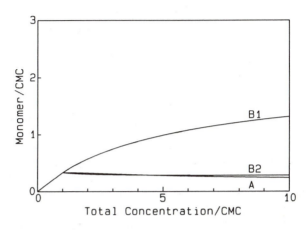

Figure 7. Monomer concentrations of individual group A and B components versus total surfactant concentration ($\beta = -3$) in third mixture (see text).

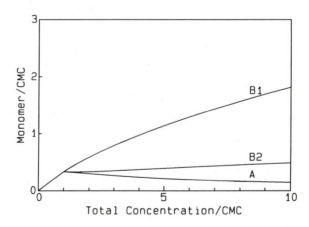

Figure 8. Monomer concentrations of individual group A and B components versus total surfactant concentration ($\beta = -12$) in third mixture (see text).

consistently low, with increasingly larger deviations in going from NPE_{10} to NPE_{50}. This observation was considered to be evidence that the regular solution model is inadequate for such predictions, and lead directly to the development of an alternative model accounting for the excess entropy of mixing (5,6). However, in the present case it is interesting to consider these same results in light of predicted polydispersity effects. Here, the observation that experimental results for the alkyl benzene sulfonate micellar mole fractions are always higher than the predictions of a strictly binary model also means that experimental total monomer concentrations of the polydisperse NPE surfactant are higher than model predictions. This trend agrees well with the increase in total monomer concentration due to polydispersity predicted by the group activity coefficient method (Figures 1 and 2). Furthermore, the observation of increasingly larger deviations as NPE becomes more polydisperse (going from NPE_{10} to NPE_{50}) is in qualitative agreement with the predicted effects of polydispersity using the group activity coefficient method.

While the simple calculations shown here can be used to demonstrate the effects of polydispersity on surfactant behavior in nonideal mixed systems, the primary utility of the group activity coefficient method is to make calculations for complex nonideal systems tractable. This feature has been demonstrated on a personal computer using example group activity coefficient calculations on nonideal mixtures with more than one hundred surfactant components.

Conclusions

An approach for treating polydisperse surfactants in nonideal multicomponent mixtures has been developed based on the use of group activity coefficients. This is demonstrated to provide a tractable method for calculating the properties of multicomponent systems consisting of polydisperse groups of components that behave ideally within the group and have similar nonideal interaction parameters with other surfactants or groups. Results obtained when applying this to simple model polydisperse systems shows that surfactant polydispersity will have significant effects on both micellar compositions and monomer concentrations and that these effects will increase with the degree of nonideality in the system. However, it is also demonstrated that surfactant polydispersity does **not** directly effect mixed CMCs. That is, one can utilize a single (ideal) mixed CMC for treating each polydisperse surfactant type in calculating CMCs in nonideal mixed systems. The group activity coefficient method predictions for both of these effects are shown to be consistent with available experimental results on polydisperse systems in the literature.

Legend of Symbols

C	total surfactant concentration
C_i^*	CMC of pure surfactant i
C_{mix}^*	CMC of mixed surfactant system
C_i^m	monomer concentration of surfactant i
f_i	activity coefficient of surfactant i in mixed micelles
f_i^s	activity coefficient of surfactant i in mixed monolayer
R	gas constant

T   absolute temperature
x_i mole fraction of surfactant i in mixed micelles
α_i mole fraction of surfactant i in total surfactant
β kdimensionless interaction parameter in mixed micelle

Literature Cited

1. Hua, X. Y.; Rosen, M. J. *J. Colloid Interface Sci.* **1982**, *90*, 212.
2. Rosen, M. J. In *Phenomena in Mixed Surfactant Systems;* Scamehorn, J. F, Ed.; ACS Symposium Series 311, American Chemical Society: Washington, DC, 1986; pp 144-162.
3. Warr, G. G.; Griese, F.; Healy, T. W. *J. Phys. Chem.* **1983**, *87*, 1220.
4. Scamehorn, J. F.; Schechter, R. S.; Wade, W. H. *J. Dispersion Sci. Technol.* **1982**, *3*, 261.
5. Osborne-Lee, I. W.; Schechter, R. S.; Wade, W. H.; Barakat, Y. J. *J. Colloid Interface Sci.* **1985**, *108*, 60.
6. Osborne-Lee, I. W.; Schechter, R. S. In *Phenomena in Mixed Surfactant Systems;* Scamehorn, J. F, Ed.; ACS Symposium Series 311, American Chemical Society: Washington, DC, 1986; pp 30-43.
7. Graciaa, A.; Lachaise, J.; Bourrel, M.; Osborne-Lee, I. ; Schechter, R. S.; Wade, W. H. *Soc. Pet. Eng. J.* **1987**, *2*, 305.
8. Yoesting, O. E.; Scamehorn, J. F. *Colloid Polym. Sci.* **1986**, *264*, 148.
9. Holland, P. M.; Rubingh, D. N. *J. Phys. Chem.* **1983**, *87*, 1984.
10. Holland, P. M., *Adv. Colloid Interface Sci.* **1986**, *26*, 111.
11. Graciaa, A.; Ben Ghoulam, M.; Marion, G.; Lachaise, J. *J. Phys. Chem.* **1989**, *93*, 4167.

RECEIVED January 6, 1992

PHENOMENA IN MIXED MICELLAR SOLUTIONS

Chapter 7

Thermodynamics of Mixed Micelle Formation
Solutions of a Cationic Surfactant, a Sulphobetaine, and Electrolyte

D. G. Hall[1], P. Meares[2], C. Davidson[2], E. Wyn-Jones[3], and J. Taylor[3]

[1]Unilever Research Port Sunlight Laboratory, Quarry Road East,
Bebington, Wirral, Merseyside, L63 3JW, United Kingdom
[2]Department of Chemistry, University of Aberdeen, Scotland
[3]Department of Chemistry and Applied Chemistry, University of Salford,
Salford, M5 4WT, United Kingdom

Thermodynamic arguments are developed which lead to a micellar-Gibbs-Duhem equation for ionic surfactants. For mixed micelles of ionic and nonionic surfactants this equation can be tested by measuring the ionic surfactant monomer concentration as a function of micelle composition and supporting electrolyte using EMF methods.

For the system $C_{12}TAB$, C_{14} sulphobetaine in 0.2 M NaBr, the mixed micelles conform within experimental error to regular solution theory. Studies at a series of electrolyte concentrations show that micellar degrees of dissociation obtained from counterion activities agree with estimates found from the behaviour of the monomer concentration.

The dependence of the micellar degree of dissociation α on micelle composition is well described by the simple expression $(1+n_s)/\alpha = 1 + n_s/\alpha_o$, where n_s is the ratio of the ionic to nonionic surfactant and α_o is the degree of dissociation of the pure surfactant.

Thermodynamic treatments of micellar surfactant solutions have two key objectives. The first is to derive expressions which enable explicit information about micelles such as aggregation number, polydispersity, micelle composition and degree of counterion binding to be obtained from thermodynamic measurements such as turbidity, surfactant activity, surface tension and the behaviour of the CMC. The second objective is to provide a framework which enables the properties of micellar solutions to be predicted from a limited number of well chosen measured parameters. For dilute solutions of non interacting uncharged micelles whose contents are well defined the first objective is met by the essentially exact multiple equilibrium/small systems thermodynamics approach[(1-4)]. Unfortunately this approach cannot be applied as it stands to solutions of ionic surfactants. Two reasons for this are that there is no satisfactory way of defining the counterion content of the micelles and that interactions between micelles cannot be ignored.

0097–6156/92/0501–0128$06.00/0

A popular way of dealing with these difficulties has been to treat interactions between micelles and small ions by supposing that each micelle occupies the centre of a cell and solving the Poisson-Boltzman equation[5-7]. However the predictive capability of this approach is limited because it requires some prior knowledge of micelle geometry and structure which is not easily obtained.

This paper is concerned with an alternative approach as outlined below[8-10]. The application of this approach to mixed micelles in the presence of different amounts of supporting electrolyte leads to expressions that are testable experimentally. Appropriate experiments are described and the results are compared with theory. The agreement is good.

Thermodynamic Theory

Noninteracting Micelles. The main results of the multiple equilibrium/small systems thermodynamics approach as applied to polydisperse multicomponent micelles can be summarised in the following two equations

$$kT \, dlnC_m = \bar{S} \, dT - \bar{V} \, dp + \sum_i \bar{N}_i \, d\mu_i \tag{1}$$

$$kT \left[\frac{\partial \bar{N}_k}{\partial \mu_i} \right]_{T,p,\mu_j} = \overline{N_i N_k} - \bar{N}_i \, \bar{N}_k \tag{2}$$

where $\bar{S}$ and $\bar{V}$ respectively denote the average entropy and volume of a micelle, C_m is the concentration of micelles, i and k refer to different micellar components and μ_i is the chemical potential of component i. $\bar{N}$ and $\overline{N_i N_k}$ are given by

$$\bar{N}_i = \sum_r N_i^r C_r / \sum_r C_r$$

$$\overline{N_i N_k} = \sum_r N_i^r N_k^r C_r / \sum_r C_r \tag{3a,b}$$

C_r is the concentration of micellar species r which contains N_i^r molecules of i.

Equation 1 is the micellar Gibbs-Duhem equation. Equation 2 relates the dependence of the micellar chemical potentials on micelle composition to fluctuations in micelle composition. For single component micelles this equation relates the dependence of aggregation number on surfactant monomer concentration to the second moment of the micelle size distribution.

A major feature of equation 1 is that it enables the approximations inherent in regarding the micelles as a separate phase[11,12] to be clearly stated. For large micelles this approximation turns out to be very good and this is probably the most important conclusion to emerge from the multiple equilibrium/small systems thermodynamics approach.

Combining the multiple equilibrium/small systems thermodynamics approach with Kirkwood Buff solution theory[13,14] leads to the expressions[9,15]

$$kT \, d \ln C_m \quad = \quad (\bar{S} + \sum_i N_i^+ s_i) \, dT \; - \; (\bar{V} + N_i^+ v_i) \, dp$$

$$+ \; \sum_i (\bar{N}_i + N_i^+) \, d\theta_i \qquad\qquad (4)$$

$$kT \left(\frac{\partial \bar{N}_k}{\partial \theta_i} \right)_{T,p,\theta_j} \quad = \quad \overline{N_i N_k} + \overline{N_k N_i^+} - \bar{N}_i \bar{N}_k - \bar{N}_k N_i^+ \qquad (5)$$

where

$$\theta_i \; = \; (\mu_i - \frac{v_i}{v_c} \mu_c) \qquad\qquad (6)$$

and is the chemical potential of the electrically neutral combination of ions formed by one i ion and sufficient ions of species c to neutralise its charge.

Also

$$N_i^+ \; = \; \sum_r C_r N_i^{+r} / \sum_r C_r$$

$$\overline{N_k N_i^+} \; = \; \sum_r C_r N_k^r N_i^{+r} / \sum_r C_r \qquad\qquad (7a,b)$$

where N_i^{+r} is the average amount of non micellar i adsorbed by a micelle of type r.

Other notation is given in references 9 and 15.

Equations 4 and 5 are essentially exact for solutions of non interacting micelles. These include solutions of ionic surfactants at their CMC and more concentrated solutions when a large excess of supporting electrolyte is present. An advantage of equations 4 and 5 is that micelle-counterion interactions can be allowed for without invoking specific counterion binding by including the entire excess of counterions in the appropriate terms. Indeed this strategy is applicable to any species for which there is significant ambiguity in defining the amount contained by the micelles.

Equation 4 is of considerable value in interpreting the behaviour of the CMC. In particular for ionic surfactants it shows that the micellar degree of dissociation obtained from the effect of electrolyte on the CMC is equal to twice the negative adsorption of surfactant ion and coions per micellar surfactant ion provided that activities rather than concentrations are plotted.

The methods used to derive equations 4 and 5 can also be applied to solutions of interacting micelles but the equations which result are not easy to apply. The main reason for this is that the interactions are coupled with fluctuations in composition[9,15].

Interacting Micelles of Ionic Surfactants. A notable feature of ionic surfactant solutions is that the effect of electrolyte on the CMC and the behaviour of colligative properties above the CMC are simply related[8,9]. The corresponding relationship for solutions of polyelectrolytes is that between the osmotic coefficient in the absence of added salt and the Donnan exclusion parameter in the presence of excess salt[16,17]. This relationship does not follow from thermodynamic arguments alone. Consequently any theory which reproduces it must have some extra-thermodynamic content. A theory of this kind has been developed and shown to work well for solutions of polyelectrolytes and ionic surfactants in the presence and absence of added salt with a common counterion[8,9]. An advantage of this theory is that it results in equations of the same form as equations 1 and 2. As originally stated some of the assumptions underlying this theory are somewhat arbitrary and hard to assess. Recently, however, an alternative derivation has been forwarded in which the assumptions involved are more readily evaluated[10].

The new version of the theory, which can incorporate mixtures of counterions having different valencies, runs as follows. We consider a solution of a charged colloid in Donnan osmotic equilibrium with a colloid free solution of electrolyte and let ψ_m denote the Donnan potential. Suppose now that we have two charged plates immersed in the colloid free solution that the mid plane potential is ψ_m and that the concentrations at the mid plane are m_i. We identify the m_i with the free concentrations in the colloidal solution and write

$$(C_i - m_i) = (\bar{N}_i + \Gamma_i) C_m \qquad (8)$$

where $\bar{N}_i$ is the average amount of i specifically bound to a colloidal unit and Γ_i is the average amount non specifically bound. Also, when p and p_b respectively are the pressures of the colloid solution and the colloid free solution and $(p' - p_b)$ is the repulsive force per unit area between the charged plates we suppose that $(p' - p_b)$ is the electrostatic contribution to the osmotic pressure and that $(p - p') = nkT$, where n is the number density of colloidal particles. This set of assumptions lead to the expression

$$kT \, d \ln C_m = \bar{S}_i \, dT - \bar{V} \, dp + \sum_i (\bar{N}_i + \Gamma_i) \, d\mu_i \qquad (9)$$

where the μ_i denote single ion chemical potentials in the colloidal solution.
These quantities are given by

$$\mu_i = \mu_i^{\theta} (T,p) + kT \ln m_i \gamma_i \tag{10}$$

where

$$\ln \gamma_i = - \frac{z_i^2 A \sqrt{I}}{1 + \sqrt{I}} \tag{11}$$

$$I = \frac{1}{2} \sum_i m_i z_i^2 \tag{12}$$

and A is the appropriate constant given by Debye-Huckel theory.
The summation in equation 12 does not include the charged colloidal units.
Finally we assume that for indifferent coions, species 3,

$$\bar{N}_3 = \Gamma_3 = O \tag{13}$$

For large micelles the LHS of equation 9 is much smaller than the individual terms on the RHS and it is a good approximation to set it to zero. We refer to this approximation as "the phase approximation".

Mixed Micelles of an Ionic and a Nonionic Surfactant. Let 1, 2 and s respectively denote the ionic surfactant the counterion and the nonionic surfactant and let a denote activity. Equations 9 and 13 together with the phase approximation enables us to write that constant T and p

$$O \approx \bar{N}_1 d \ln a_1 + \bar{N}_s d \ln a_s + \bar{N}_1 (1 - \alpha) d \ln a_2 \tag{14}$$

where α defined by $\bar{N}_1 (1 - \alpha) = $ may be regarded as an effective degree of dissociation.
On rearrangement equation 14 gives

$$d [\ln a_s + n_s \ln a_1] = \ln a_1 dn_s - n_s (1 - \alpha) d \ln a_2 \tag{15}$$

where $n_s = \bar{N}_1/\bar{N}_s$

on cross differentiation equation 15 gives

$$1 + \left(\frac{\partial \ln a_1}{\partial \ln a_2} \right)_{n_s} = \left(\frac{\partial \alpha n_s}{\partial n_s} \right)_{a_2} \tag{16}$$

Also if counterions interact with mixed micelles of constant composition in the same way as with single component micelles then the RHS of equation 16 can be expected to be independent of a_2 within experimental error.
    Two predictions of the theory are as follows
1. a_1 should not depend on surfactant concentration at constant n_s and a_2.
2. α given by integration of equation 16, i.e.

$$\alpha = \frac{1}{n_s} \int_0^{n_s} (1 + \left(\frac{\partial \ln a_1}{\partial \ln a_2}\right)_{n_s}) \, dn_s \tag{17}$$

should agree with α given by

$$\alpha = \frac{m_2 - (m_1 + C_3)}{(C_1 - m_1)} \tag{18}$$

Experimental Tests of the Theory. The experimental information required to test the above theory has been obtained from EMF measurements using surfactant ion selective electrodes and commercially available electrodes for the other ionic species present.

The surfactant ion selective electrodes are described in detail elsewhere(18,19) and the design currently in use is shown in figure 1. The surfactants used were Dodecyl trimethyl ammonium bromide (DTAB) and Tetradecyl dimethyl amino propion sulphonate ($C_{14}SB$). The latter material was chosen because its CMC is sufficiently small that the error introduced by assuming it all to be micellar is negligible at the concentrations used in the experiments.

The experimental procedure consisted of checking the surfactant electrodes response with a dilute solution of DTAB + NaBr then adding then adding a solution of the sulphobetaine + NaBr followed by aliquots of a concentrated solution of DTAB + NaBr.

For each different solution the EMFs of either the first or both of the following two cells were measured.

Surfactant Electrode		test solution		Coion Electrode
				Cell 1

Surfactant Electrode		test solution		Counterion Electrode
				Cell 2

From cell 1 changes in the ratio a_1/a_3 are obtained. When $\gamma_1 = \gamma_3$ this gives changes in the ratio m_1/m_3 which together with the calibration in the absence of sulphobetaine leads to values of m_1 and the composition of the mixed micelles.

EMF measurements of cell 2 lead to estimates of the quantity $m_1 m_2 \gamma^2$ from which m_2 is readily obtained by iteration using equations 11 and 12. α then follows immediately from equation 18.

Results

Experiments in 0.2 M NaBr. Under these circumstances interactions between micelles are insignificant. Also a_{Br^-} and γ are approximately constant and the EMF of cell 2 is not required. The data obtained at 35°C are presented graphically in figs 2-4. It is clear from figure 2 that the mixed micelles show negative deviations from

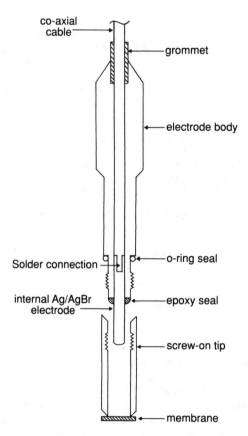

Figure 1. Construction of surfactant ion selective electrode currently in use at Salford University.

Raoult's law. If these deviations can be described by regular solution theory we should have

$$m_1 = m_1^o x_1 \exp - \frac{\beta(1 - x_1)^2}{RT} \tag{19}$$

which leads to the conclusion that $\ln(m_1/x_1)$ vs $(1 - x_1)^2$, where x_1 is the micellar mole fraction of 1 and m_1^o is the cmc of pure 1, should be linear. Figure 3 shows that for the present system this is indeed the case.

A similar result was obtained at 45°C. The respective values of β/RT are - 0.600 at 35°C and - 0.585 at 45°C. The respective values of β are - 1.543 and - 1.546 kJ mole^{-1}. This indicates that β is independent of temperature. Hence, the deviations from ideal mixing are almost entirely enthalpic in origin and the entropy of mixing is ideal. The thermodynamics functions of mixing at 35°C are presented in figure 4. Overall, the agreement with regular solution theory is remarkably good.

Figure 5 shows the effect of micellar surfactant concentration on the relationship between a_1 and x_1. The agreement between the two sets of data is excellent and shows that the approximation of regarding the micelles as a separate phase is valid with experimental error for this system.

Effect of NaBr Concentration. Experiments were done at several different NaBr concentrations ranging from 10^{-4} molar to 10^{-1} molar. For each run the data was plotted as $\ln a_1$ vs n_s and $\ln a_2$ vs n_s. A series of n_s values was chosen and for each of these $\ln a_1$ vs $\ln a_2$ was plotted. A typical graph for $n_s = 0.5$ is shown in figure 6 and it is clear that, as expected, the graph is linear. According to equation 16 the slope gives $(\partial \alpha n_s / \partial n_s)$. The dependence of this quantity on n_s is shown in figure 7. From this graph αn_s was obtained by numerical integration. Together with equations 18 11 and 12 this information enables a_{Br^-} to be calculated for any solution and compared with the experimental value obtained from cell 2. The results of this comparison for three runs at different salt concentrations are shown in figure 8. The agreement at the two lower salt concentrations is very good indeed. The agreement at the higher salt concentration is less good, but even in this case the difference is much less than the overall change in a_{Br^-}. Also the errors in a_{Br^-} based on measurements of cell 2 are greater at higher salt concentrations. We conclude that overall the agreement between the theory summarised in equations 9-14 and experiments is really quite good.

We now turn to the dependence of α on n_s. Let σ denote the charge per unit area at the micelle surface. On electrostatic grounds we expect the following limiting behaviour $\alpha \rightarrow 1$ as $\sigma \rightarrow o$ and $\alpha \rightarrow o$ as $\sigma \rightarrow \infty$. These conditions are met by the simple expression

$$\alpha = \frac{\omega}{\omega + \sigma} \tag{20}$$

where ω is constant. Moreover a noteworthy feature of equation 20 is that $\alpha\sigma \rightarrow \omega$ as $\sigma \rightarrow \infty$. This feature of equation 20 is in good accord with the prediction of the Poisson-Boltzmann equation that the negative adsorption of coions becomes insensitive to changes in charge density when the charge density is large[20] and is

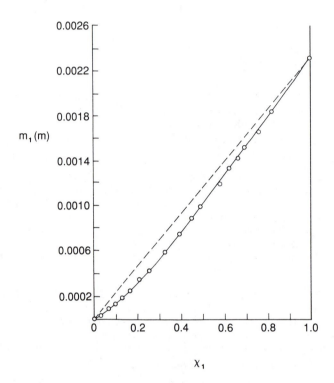

Figure 2. m_1 vs x_1 for mixed micelles of C12TAB and C14S/B at 35°C.

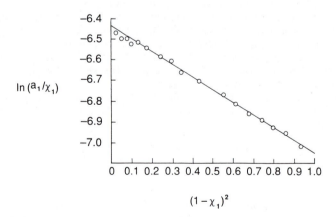

Figure 3. $\ln(m_1/x_1)$ vs $(1 - x_1)^2$ for mixed micelles of C_{12}TAB and C_{14}S/B at 35°C.

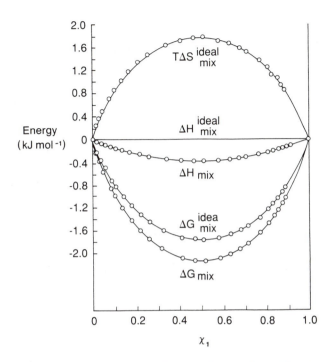

Figure 4. Thermodynamic functions of mixing for mixed micelles of $C_{12}TAB$ and $C_{14}S/B$ at 35°C.

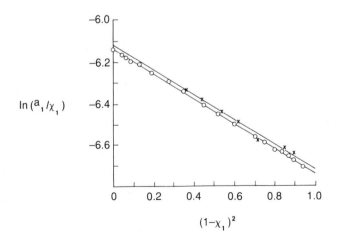

Figure 5. Effect of surfactant concentration on the thermodynamics of mixing of $C_{12}TAB$ and $C_{14}S/B$ micelles at 45°C.   x Concentration range of micellar surfactant 0.00638 - 0.01049 moles dm^{-3}. o Concentration range of micellar surfactant 0.044 - 0.104 moles dm^{-3}.

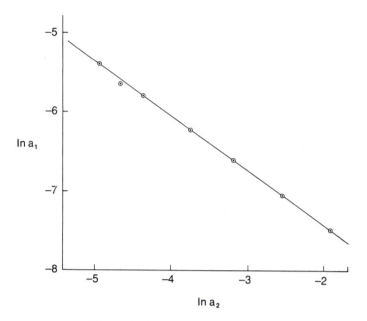

Figure 6. $\ln a_1$ vs $\ln a_2$ for n_s = 0.5.

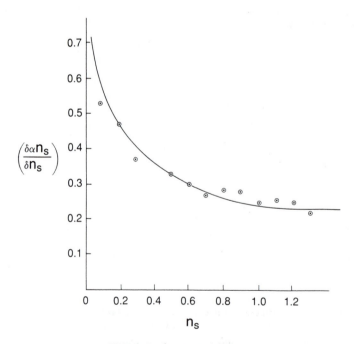

Figure 7. $(\partial \alpha n_s / \partial n_s)$ vs n_s.

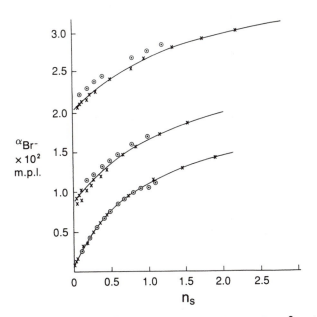

Figure 8. a_{Br^-} vs n_s. - values from experimental estimates of $m_2\gamma^2$ + denotes experimental points. o values calculated from equations 17 and 18.

reminiscent of the counterion condensation that features in theories of linear polyelectrolytes[17,21].

If we now suppose that σ is proportional to the micellar mole fraction of the nonionic surfactant it is straight forward to obtain from equation 20 the following result

$$\frac{1 + n_s}{\alpha} = 1 + \frac{n_s}{\alpha_o} \tag{21}$$

where α_o is the value of α for the pure ionic surfactant. According to equation 21 the LHS should be a linear function of n_s with slope $1/\alpha_o$ and intercept 1.

The appropriate plot is given in figure 9. The slope gives $\alpha_o = 0.23$ which is perfect agreement with estimates of α_o for DTAB found from the effect of salt on the CMC[22]. Clearly equation 21 is obeyed remarkably well for this system. The same equation stated in a different form also gives a good account of other data[23].

Concluding Remarks

The above experimental study confirms two general features of micellar thermodynamics, firstly that the phase approximation is valid for dilute solutions of noninteracting micelles and secondly that the theory based on equations 8-14 gives a reasonable account of the thermodynamic behaviour of dilute solutions of interacting mixed micelles.

Numerous studies of mixed micelle formation in solutions of ionic and nonionic surfactants have shown that for many systems the CMCs of mixtures can be fitted by

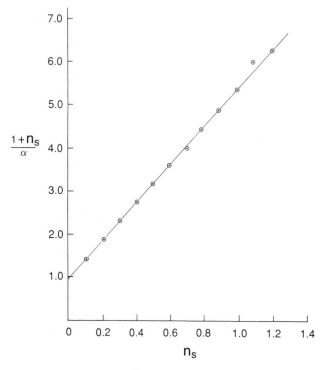

Figure 9. $1 + \dfrac{n_s}{\alpha}$ vs α. slope = 4.28

regular solution theory[24]. However this is not a sensitive test that regular solution theory is obeyed for two reasons. Firstly the micelle composition is inferred from the CMC measurements. Secondly the theory used to treat the data is demonstrably wrong in some cases because it ignores changes in counterion activity. The methods described above allow micelle compositions to be obtained directly from the experimental measurements and for this reason provide a more stringent test of mixing theories than CMC data. Although the system described above conforms well to regular solution theory this is not generally the case, for example application of the same method to mixed micelles of Sodium Dodecyl Sulphate (SDS) and $C_{12}E_6$[25] show significant deviations from regular solution theory.

Finally we note that although equation 21 appears to work quite well for the system described above and for other systems too it has no firm theoretical basis and need not be generally valid.

References

1.   T. L. Hill, Thermodynamics of Small Systems, Volumes 1 & 2 Benjamin, N.Y. 1963, 1964.

2.   D. G. Hall and B A Pethica, Chapter 16 in Nonionic Surfactants Ed. M. Schick, Marcel Dekker, N.Y. 1967.

3.   J. M. Corkill, J. F. Goodman, T. Walker and J. Wyer, Proc. Roy. Soc. A 1969, 312, 243.

4.   D. G. Hall, Trans Faraday Soc. 1970, 66, 1350 1359.

5.   G. M. Bell and A. J. Dunning, Trans Faraday Soc 1970, 66, 500.

6.   M. Mille and G. Vanderkooi. J. Colloid and Interface Sci. 1977, 59, 211.

7.   G. Gunnarson, B Jonsson and H Wennerstrom. J. Phys. Chem. 1980, 84, 3114.

8.   D. G. Hall, J. Chem. Soc. Faraday Trans II 1981, 77, 1121.

9.   D. G. Hall, Chapter 2 of Aggregation Processes in Solution Ed. E. Wyn-Jones and J. Gormally, Elsevier, Amsterdam 1983.

10.   D. G. Hall, J. Chem. Soc. Faraday Transactions, 1991, 87, 3529.

11.   E. Hutchinson, A. Inaba and L. G. Baily, Z. Phys. Chem. F 1955, 5, 344.

12.   K. Shinoda and E. Hutchinson, J. Phys. Chem 1962, 66, 577.

13.   J. G. Kirkwood and F. P. Buff, J. Chem. Phys, 1951, 19, 774.

14.   D. G. Hall, Trans Faraday Soc, 1971, 67, 2516.

15.   D. G. Hall, J. Chem. Soc. Faraday Trans 2, 1972, 68, 1439, 1977, 72, 897.

16.   A. Katchalsky, Pure and Applied Chem, 1972, 26, 327.

17.   F. Oosawa, Polyelectrolytes Marcel Dekker, N.Y. 1971.

18.   C. Davidson, P. Meares and D. G. Hall, J. Membrane Sci. 1988, 38, 511.

19.   D. Painter, PhD Thesis, University of Salford, 1988.

20.   H. J. van den Hol and J. Lyklema, J. Colloid and Interface Sci, 1967, 23, 500.

21.   G. S. Manning, J. Chem. Phys. 1969, 51, 524, 3249.

22.   E. M. P. Cathrae, PhD Thesis, Aberdeen University 1982.

23.   D. G. Hall and T. J. Price, J. Chem. Soc. Faraday 1 1984, 80, 1193.

24.   D. N. Rubingh, Solution Chemistry of Surfactants volume 1, p. 337, K. Mittal Ed. Plenum Press N.Y. 1979.

25.   C. Davidson, PhD Thesis, Aberdeen University 1983.

RECEIVED January 6, 1992

Chapter 8

Physicochemical Properties of Mixed Surfactant Systems

Keizo Ogino and Masahiko Abe

Faculty of Science and Technology, Science University of Tokyo, 2641, Yamazaki, Noda, Chiba, 278, Japan

Within the past several years, a great number of solution properties for mixed surfactant systems have been published. Our research group has also investigated the solution properties of various mixed surfactant systems, based on surface tension and electrical measurements over the past fifteen years. This paper will discuss recent developments of mixed surfactant systems, especially, of anionic-nonionic surfactant systems. Included in the set of examples will be (1) differences in properties of mixed micelles forming with different alkyl chain lengths and / or polyoxyethylene chain lengths in nonionic surfactants, and (2) protonation and fading phenomena of azo oil dyes occurring in aqueous solutions containing mixtures of surfactants.

Recently, it has been reported that the surface activity of mixed surfactant systems is superior to that of single surfactant systems (*1-3*). In fact, surfactants used in practical applications almost always consist of a mixture of surface-active compounds. Mixed surfactant systems are also of great theoretical interest. It can be assumed that the tendency to form aggregated structures in solutions containing mixtures of surfactants is substantially different from that in solutions involving only the pure surfactants. Therefore, many publications have reported solution properties of mixed surfactant systems, including those cited in references (*4-7*) and papers from our laboratory (*8-13*).

Micelle Formation of Anionic-Nonionic Mixed Surfactant Systems

ECL-$C_m POE_n$ systems
Surface Tension of Aqueous Solution of Binary Surfactants Systems. We have obtained surface tension data for aqueous mixed surfactant system: sodium 3,6,9-trioxaicosanoate (ECL, Ethercarboxylate, an anionic surfactant which has both nonionic and anionic properties) - alkyl polyoxyethylene ethers ($C_m POE_n$, a nonionic

0097–6156/92/0501–0142$06.75/0

surfactant: m=12, 14, 16 and 18, n=10, 20, 30 and 40) (*12, 13*). Figure 1 shows the dependence of surface tension on the mole fraction of ECL in mixed surfactant solutions above the CMC. In the case of the ECL-$C_{12}POE_{20}$ system, the surface tension decreases monotonically with increasing mole fraction of ECL (X_{ECL}). However, these surface tension vs. mole fraction curves are shifted from a monotonic curve to curves with an inflection point with increasing alkyl chain length of the nonionic surfactant. In the case of the ECL-$C_{18}POE_{20}$ system, the surface tension decreases with increasing X_{ECL} up to 0.3, remains almost constant until X_{ECL}=0.7, then decreases again.

Figure 2 shows the surface tension of ECL-$C_{12}POE_{20}$ mixed surfactant solutions containing various mole fractions of ECL, plotted against the total concentration of surfactant. In the case of the ECL-alone and the $C_{12}POE_{20}$-alone systems, each surface tension value decreases with increasing concentration, but remains constant above the CMC. On the other hand, in the case of the ECL-$C_{12}POE_{20}$ system (Fig. 2), the surface tension values of aqueous solutions at various mole fractions also decrease with increasing total concentration. However, the line breaks at two points, in the vicinity of the CMC for $C_{12}POE_{20}$ alone, and of that for ECL alone. The intervals between the two breakpoints decrease with increasing alkyl chain length in nonionic surfactant. Finally, in the case of the ECL-$C_{18}POE_{20}$ system the surface tension decreases with increasing total concentration, and breaks at only one point, as shown in Fig. 3.

Figure 4 shows the relation between surface tension and the concentration in bulk for the ECL-$C_{12}POE_{20}$ system. The concentration in bulk (a) and the surface tension (b) are plotted against the concentration in solution. Here, points A and B represent the CMC values of aqueous solutions for $C_{12}POE_{20}$ alone and for ECL alone, respectively. The concentration in bulk phase for ECL and $C_{12}POE_{20}$ increases in direct proportion with increasing total concentration up to point A. Then, in the case of $C_{12}POE_{20}$, the concentration in the bulk phase remains constant at point A by forming the micelle. On the other hand, in the case of ECL, this concentration increases with increasing total concentration up to point B.

The surface tension is closely related to the concentration of the bulk phase. The additivity of surface tension values occurs because interactions between ECL and $C_{12}POE_{20}$ are entirely absent. Thus, a plot of surface tension against total concentration is broken at the two points of CMC for ECL alone and the $C_{12}POE_{20}$ alone, as shown by the dashed line in Fig. 4 (b). The solid line in Fig. 4 (b) is the observed value. As can be seen from Fig. 4 (b), there is good agreement between the observed values (solid line) and calculated values (dashed line). This indicates that there are two kinds of micelles are coexisting (one rich-in ECL and the other rich in $C_{12}POE_{20}$).

Results for the ECL-$C_{18}POE_{20}$ system can be treated in a similar way. Figure 5 shows (a) the concentration in bulk phase and (b) the surface tension, plotted against the total concentration in the mixed surfactant solution. As seen in Fig. 5 (b), the observed value (solid line) is different from the calculated value (dashed line). This is attributed to the fact that a mixed micelle is formed in mixtures of ECL and $C_{18}POE_{20}$. Thus, the mixed micelle is formed more easily by a nonionic surfactant including a long alkyl chain than by one having shorter alkyl chains.

Similar results were obtained on the effect of oxyethylene chain length in nonionic

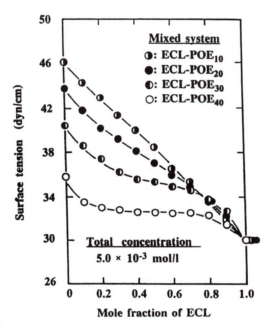

Figure 1. Relation between the surface tension and the mole fraction of ECL in mixed surfactant solutions above CMC. (Reproduced with permission from ref. 13. Copyright 1985 Academic Press.)

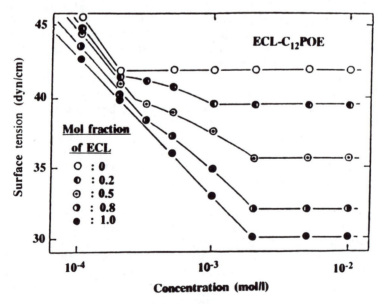

Figure 2. Relation between the surface tension and the total concentration in mixed surfactant solutions of ECL/C_{12}POE system. (Reproduced with permission from ref. 12. Copyright 1985 Academic Press.)

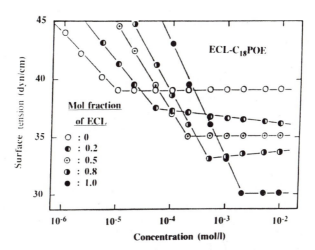

Figure 3. Relation between the surface tension against the total concentration in mixed surfactant solutions of ECL/C_{18}POE system. (Reproduced with permission from ref. 12. Copyright 1985 Academic Press.)

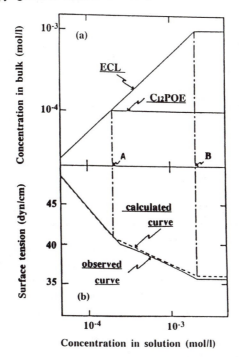

Figure 4. The surface tension against the concentration in bulk of ECL/C_{12}POE system. (a) The concentration in bulk and (b) the surface tension against the concentration in solution. (Reproduced with permission from ref. 12. Copyright 1985 Academic Press.)

surfactants on surface tensions of mixed surfactant systems (*13*). That is, the surface tension vs. mole fraction curves are shifted from the specific "S" shaped curves to a monotonic curve with increasing polyoxyethylene chain length in the nonionic surfactant. Moreover, in the case of the mixed systems including a nonionic surfactant which has a shorter polyoxyethylene chain length, the surface tension vs. total concentration curve shows only one breakpoint. In the case of systems including the nonionic surfactant with longer polyoxyethylene chain lengths, the curve shows two breakpoints in the vicinity of the CMC for ECL alone and that for POE_n alone. This may be attributed to the fact that the mixed micelle is formed more easily by a nonionic surfactant having a shorter polyoxyethylene chain length than by one with a longer chain. Thus, in the mixed surfactant systems, there are two kinds of micelles coexisting (one rich in anionic surfactant and the other rich in nonionic). We suggest that the mixed micelle does not always form in the hydrocarbon-hydrocarbon mixed surfactant solutions. This may be reflect the extent of hydrophobic interaction between an anionic surfactant molecule and a nonionic one. The above suggestion is illustrated by the micellization models for $ECL-C_mPOE_n$ mixed surfactant system in Fig. 6.

Rosen and Hua (*14*) have reported that the mixed micelles should show more nearly ideal behavior as the number of oxyethylene units is increased in the nonionic surfactant. However, in their study, the number of oxyethylene units in the nonionic surfactant was varied from 3 to 12, a range much smaller than ours (n=10-40). Thus, we infer that a value of n near 10 is optimum for the formation of ideal mixed micelles.

Electrical Properties of Binary Surfactant System. Electrical properties of mixed anionic-nonionic surfactants in aqueous solutions above the CMC have been studied in relation to the pNa value, electrical conductivities, and dielectric constants. Electrical conductivity and dielectric constant values were measured in the frequency range 220 Hz to 1M Hz (*10*). Table I shows the degree of ionic dissociation of mixed micelles, calculated from the pNa values. The value of α increases with increasing concentration of POE, and becomes nearly 1.0 at a mole ratio of POE/ECL=1.

Figure 7 shows the temperature dependence of electrical conduction for a given concentration of ECL (5.0×10^{-3} mole/liter) in mixed surfactant solutions. The electrical conductivities increase with increasing temperature, and when plotted as a function of inverse temperature, these plots are linear. From these results, we have obtained the activation energy for conduction in the mixed surfactant systems.

Table II shows the activation energies of ECL alone and of ECL mixed with POE, obtained from the slopes of the lines in Fig. 7; these values increase with increasing concentrations of POE. As in the case of ECL-POE mixed surfactant systems discussed above, the degree of ionic dissociation (α) of mixed micelles increases with increasing concentration of POE. However, as can be seen in Table II, the activation energy for conduction increases with increasing concentration of POE, in spite of the increase in the degree of ionic dissociation (α) for mixed micelles.

In a previous paper (*15*), we have reported that the activation energy for conduction of anionic surfactants above the CMC is governed by the surface charge density of the counterion of the micelle. That is, the activation energy for conduction begins to decrease with increasing surface charge density of the counterion of the micelle. Here, the surface charge density of a mixed micelle depends not only on the

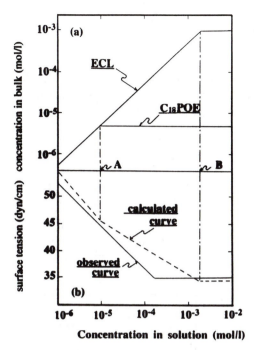

Figure 5. The surface tension against the concentration in bulk of ECL/C_{18}POE system. (a) The concentration in bulk and (b) the surface tension against the concentration in solution. (Reproduced with permission from ref. 12. Copyright 1985 Academic Press.)

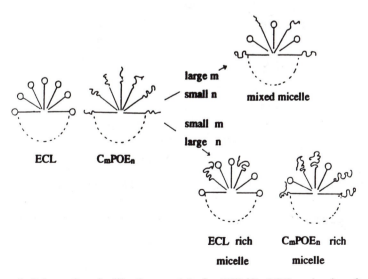

Figure 6. Schematic micellization models for ECL/C_mPOE$_n$ mixed surfactant systems. (Reproduced with permission from refs. 12 and 13. Copyright 1985 Academic Press.)

Table I The Degree of Ion Dissociation of Mixed Micelle (α) in Mixed Surfactant Solutions at 30 °C (Reproduced with permission from ref. 10. Copyright 1984 Academic Press.)

Concentration of POE (mol/l)	α
0	0.46
1.3×10^{-3}	0.66
3.3×10^{-3}	0.95
5.0×10^{-3}	0.99

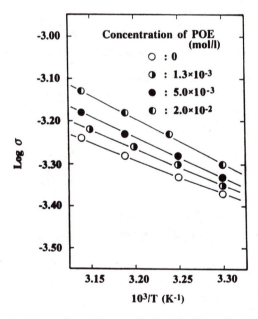

Figure 7. Temperature dependence of electrical conduction for a given concentration of ECL (5.0×10^{-3} mol/l) in mixed surfactant solutions. (Reproduced with permission from ref. 10. Copyright 1984 Academic Press.)

degree of ionic dissociation but also on the micellar size of the mixed micelle. So, we attempted to determine the radius of the mixed micelle including the electric double layer.

As can be seen in Table III, the radius (R) of the mixed micelle including the electric double layer increases with increasing concentration of POE. Thus, the surface charge density seems to decrease with increasing concentration of POE. Consequently, the degree of ionic dissociation (α) of the mixed micelle increases with increasing concentration of POE. However, the activation energy of conduction also increases with increasing concentration of POE. This discrepancy can be accounted for by changes in the degree of surface charge density of counterion (Na^+) with increasing size of the mixed micelle.

Next, we determined how (a) the alkyl chain length (11) and (b) the oxyethylene chain length (9) in nonionic surfactants affect electric properties. From these results, we have also determined that mixed micelles are formed more easily by a nonionic surfactant including a long alkyl chain length (and/or shorter oxyethylene chain) than by one having a shorter alkyl chain (or longer oxyethylene chain).

SDS-C_mPOE_n Systems.   Next, we obtained properties of mixed systems containing the most popular anionic surfactant, SDS, and C_mPOE_n (16, 17). The observed trends in these systems are similar to those for ECL-C_mPOE_n systems. Table IV shows the degree of ionic dissociation of the micelle (α) in SDS-C_mPOE_n mixed surfactant solutions. For pure SDS, the value is 0.21, which is only one-half the value for ECL obtained previously. When nonionic surfactants are added to the solution, the values of α increase. Moreover, the values of α increase with a decrease in the number of oxyethylene groups in the nonionic surfactant. The effect of nonionic surfactant on the rate of increase of α is larger for SDS mixed systems (maximum 3.5 times) than for ECL mixed ones (maximum 2 times). Table V shows activation energies of SDS alone and of SDS mixed with $C_{16}POE_n$; these values decrease continuously as the number of oxyethylene groups in the nonionic surfactant increases, in spite of the increase in the degree of ionic dissociation. Values of activation energy for the SDS-$C_{16}POE_{40}$ mixed surfactant system are equal to the values in the pure SDS system. The radius of the mixed micelle including the electric double layer is larger for a nonionic surfactant having a shorter polyoxyethylene chain length than for one having a long polyoxyethylene chain (Table VI).

The degree of ionic dissociation of mixed micelle shows a minimum with an increase in the number of alkyl chains in the nonionic surfactant (C_mPOE_{20}, m=6, 8, 10, 12, 14, 16, and 18) (17). As chain length increases, the electrical conductivities of the mixed surfactant solution decrease, in spite of decreased activation energy for conduction. The radius of mixed micelle will the electric double layer is larger for a nonionic surfactant with longer alkyl chain length than one with short chain length.

We have also considered the properties of mixed micellar systems interred from surface tension measurements (16). Rubingh (18), Rosen and Hua (19) and others have proposed theoretical equations relating to the miscibility of micelles. Rubingh calculated values of a parameter (β) related to the degree of molecular interaction between the two surfactants in the mixed micelle. We have values of β from cmc data, and these are given in Table VII. β becomes more negative as the number of oxyethylene groups in the nonionic surfactant decreases. The large negative values of

Table II The Activation Energies for Conduction for a Given Concentration of ECL in Mixed Surfactant Solutions (Reproduced with permission from ref. 10. Copyright 1984 Academic Press.)

ECL: 5.0×10^{-3} mol/l		ECL: 1.0×10^{-2} mol/l	
Concentration of POE (mol/l)	ΔE (kJ/mol)	Concentration of POE (mol/l)	ΔE (kJ/mol)
0	15.6	0	15.3
1.30×10^{-3}	16.5	2.5×10^{-3}	16.3
5.0×10^{-3}	17.7	1.0×10^{-3}	17.5
2.0×10^{-2}	20.3	4.0×10^{-2}	20.2

Table III Dielectric Parameters and the Radius of Mixed Micelle for a Given Concentration of ECL (5.0×10^{-3} mol/l) in Mixed Surfactant Solutions at 30 °C (Reproduced with permission from ref. 10. Copyright 1984 Academic Press.)

Concentration of POE (mol/l)	τ (sec)	σ (mho/cm)	U (cm2/V x sec)	R (Å)
0	8.60×10^{-4}	4.30×10^{-4}	8.92×10^{-4}	66.2
1.3×10^{-3}	9.25×10^{-4}	4.43×10^{-4}	8.67×10^{-4}	67.3
5.0×10^{-3}	9.89×10^{-4}	4.65×10^{-4}	8.61×10^{-4}	69.8
2.0×10^{-2}	1.05×10^{-3}	5.02×10^{-4}	8.40×10^{-4}	71.0

Table IV The Degree of Ion Dissociation of Micelle (α) in $C_{16}POE_n$/SDS Mixed Surfactant Solutions[a] (Reproduced with permission from ref. 16. Copyright 1988 Am. Oil Chem. Soc.)

Mixed systems	α
$C_{16}POE_{10}$/SDS	0.74
$C_{16}POE_{20}$/SDS	0.64
$C_{16}POE_{30}$/SDS	0.54
$C_{16}POE_{40}$/SDS	0.49
SDS	0.21

[a]Temperature, 30 °C; molar ratio, SDS/$C_{16}POE_n$=1/1.

Table V   The Activation Energies for Conduction in $C_{16}POE_n$/SDS Mixed Surfactant Solutions[a]   (Reproduced with permission from ref. 16. Copyright 1988 Am. Oil Chem. Soc.)

Mixed systems	ΔE (kJ/mol)
$C_{16}POE_{10}$/SDS	17.6
$C_{16}POE_{20}$/SDS	14.6
$C_{16}POE_{30}$/SDS	14.0
$C_{16}POE_{40}$/SDS	13.8
SDS	13.2

[a]Molar ratio, SDS/$C_{16}POE_n$=1/1.

Table VI   Dielectric Parameters and the Radius of Mixed Micelle for a Given Concentration of SDS (2.0 x 10^{-2} mol/l) and C16POEn (2.0 x 10^{-2} mol/l) in Mixed Surfactant Solutions at 30 °C   (Reproduced with permission from ref. 16. Copyright 1988 Am. Oil Chem. Soc.)

Mixed systems	τ (sec) x 10^{-4}	σ (mho/cm) x 10^{-3}	U (cm^2/V sec) x 10^{-4}	R (Å)
$C_{16}POE_{10}$/SDS	7.14	1.91	8.32	58.3
$C_{16}POE_{20}$/SDS	6.42	1.61	8.25	55.1
$C_{16}POE_{30}$/SDS	6.05	1.58	8.20	53.3
$C_{16}POE_{40}$/SDS	5.79	1.41	8.28	52.4
SDS	5.62	1.35	8.09	51.0

Table VII   Molecular Interaction Parameters for Mixed Surfactant Systems. (Reproduced with permission from ref. 16. Copyright 1988 Am. Oil Chem. Soc.)

Mixed systems	$-\beta$
$C_{16}POE_{10}$/SDS	6.58
$C_{16}POE_{20}$/SDS	6.18
$C_{16}POE_{30}$/SDS	4.29
$C_{16}POE_{40}$/SDS	–
SDS	0.21

β in Table VII show that the mixed micelle form more easily by a nonionic surfactant including shorter polyoxyethylene chains than by one having long polyoxyethylene chains. A similar trend was recognized in results for alkyl chain length in the nonionic surfactant systems. Consequently, we propose the micellization model for the SDS-C_mPOE_n mixed surfactant system, which is similar to the model in Fig. 6, but with ECL replaced by SDS.

Fading Phenomena of Azo Oil Dyes in Anionic-Nonionic Mixed Surfactant Solutions

The fading phenomenon is observed when 4-phenylazo-1-naphthol (4-OH) is added to an aqueous solution of a mixed surfactant system, although it does not occur in single surfactant solutions. We have interpreted the fading behavior of an azo oil dye to infer molecular characteristics of surfactants and the optimum condition for phase phenomena. Scheme 1 shows the tautomer (an azo form and a hydrazo form) and the maximal absorption wavelength of 4-OH. One hydrogen atom of the hydroxide group at the 4-position of the naphthalene ring transfers to the β-position of the azo group, and the tautomerism is established between the hydrazo form of the keto type and the azo form of the enol type.

Figure 8 shows the absorption spectra of 4-OH in pure SDS, in pure $C_{16}POE_{20}$, and in a mixture of these surfactant solutions just after the addition of dye at 30°C (19). The maximal absorption wavelength (λ_{max}) of 4-OH appears at 480 nm in the solution of SDS, but at 415 nm in the $C_{16}POE_{20}$ solution, where a shoulder emerges at 480 nm. Tautomerism of 4-OH is also observed in these mixed solutions.

Figure 9 shows the time dependence of absorption spectra of 4-OH in an SDS/$C_{16}POE_{20}$ (1/1) mixed solution at 30°C (19). The absorbance at both 415 and 480 nm decrease with time, and the complementary color of 4-OH in the solution vanishes completely after 48 h.

Figure 10 represents the changes in the absorbance at 480 nm with time. The absorbance (at 480 nm) in each pure surfactant solution is independent of time. In mixtures of these surfactants, however, the absorbance decreases rapidly with the time. A similar tendency is recognized in results for anionic surfactants other than SDS (for example, sodium dodecyl benzene sulfonate and/or sodium dodecyl sulfonate). However, in carboxylate-$C_{16}POE_{20}$ systems, a long time is required for the absorbance to vanish.

As mentioned above, no fading occurs in pure SDS or in pure $C_{16}POE_{20}$, but it does occur in mixtures of these surfactants. As is evident from Fig. 11, the effect becomes a maximum at mole ratios in the vicinity of 0.5. To increase the hydrophilic-hydrophilic interaction between two surfactants, it is necessary to make the molar ratios nearly equal and to make adjacent molecules heterogeneous. Figure 12 shows the time dependence of the 4-OH absorbance in CTAB/$C_{18}POE_{20}$ mixed solutions at 30°C. The absorbance at 480 nm is independent of time and is constant; no fading phenomenon occurs.

Next, we observed the effect of alkyl chain length and number of oxyethylene groups in C_mPOE_n on the fading rate (20). As can be seen from Fig. 13 for longer alkyl chain lengths, the fading rate accelerates in the mixed surfactant systems. The opposite effect of the number of ethylene oxide groups is shown in Fig. 14. As the polyoxyethylene chain length decreases, the fading rate accelerates in the mixed systems. From these results, this phenomenon is dependent on the molecular characteristics of C_mPOE_n added to the mixed surfactant systems. As mentioned above, we found that for nonionic surfactants including long alkyl chain length, ACL (or shorter polyoxyethylene chain length, PCL), the mixed micelle is formed more easily than for other systems. In systems containing a surfactant with a shorter ACL (or long PCL), there are two kinds of micelles coexisting (one rich in anionic surfactant and the other rich in nonionic surfactant) as can be seen in Fig. 6. Therefore, the fading phenomenon of 4-OH in the mixed surfactant systems appears to be related to mixed micelle formation between SDS and C_mPOE_n.

We also consider why the fading phenomenon of 4-OH occurs only in anionic-nonionic mixed surfactant solutions. Because many organic substances are oxidized and decomposed with oxygen, especially singlet oxygen, we have performed a number of experiments on surfactant mixtures. Figure 15 shows that bubbling O_2 gas into the mixed surfactant solutions accelerates the fading rate, while bubbling N_2 gas decelerates the rate. We have speculated that the long lifetime of singlet oxygen can make the fading rate of 4-OH faster.

We have found that the fading rate is increased under the following conditions:
(1) the anionic surfactant has a strongly polar group;
(2) the nonionic surfactant incorporates a long alkyl chain (or short PCL chain);
(3) the molar ratio is approximately 0.5.
The above effects are included in the hydration model for SDS-C_mPOE_n mixed surfactant system shown in Fig.16. When the mixed micelle at a molar ratio of 0.5 is formed, the water molecule would be mechanically trapped in regions between the oxygen atoms of the hydrophilic group of SDS and ethyleneoxide in the C_mPOE_n molecule to depress the vibrational motion. It can be speculated that, when the transition energy of oxygen is not exchanged to the internal energy of the water molecule, the singlet oxygen lifetime becomes quite long. As mentioned above, the effect on the fading behavior of 4-OH is larger for a system which can easily form a mixed micelle than for a system in which two kinds of micelles coexist.

Next, we have studied the effects of solubilizates on the fading behavior of 4-OH (*21*). The effect of solubilizates on the fading behavior of 4-OH is dependent on the differences in their solubilization sites in mixed micelles. A solubilizate with a strong polar group supported a hydrophilic-hydrophilic interaction, whereas one including a weak polar group does not.

We have also studied the amount of oxygen both in the surfactant solutions and in the mixed micelles and the effect of some quenchers and sensitizers for the active oxygen species on the fading behavior of 4-OH. We have reconfirmed in detail the possibility of singlet oxygen formation in the anionic-nonionic mixed micelles (*22*).

The oxygen molecule usually exists as triplet oxygen in the ground state. This triplet oxygen can be converted into an active oxygen species such as the hydroxyl radical, hydrogen peroxide, superoxide anion radical, and singlet oxygen, which are more reactive than triplet oxygen.

We have also studied the possible contribution of each active oxygen species which may be formed to the fading phenomena in the mixed micelle. As can be seen Fig. 17, the dissolved oxygen concentration in the surfactant solution is decreased, and the

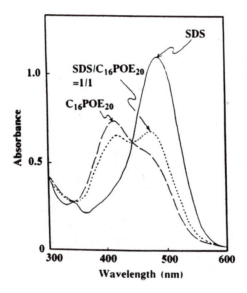

Scheme 1. Tautomerism equilibrium of 4-phenylazo-1-naphthol and each maximal absorption wavelength.

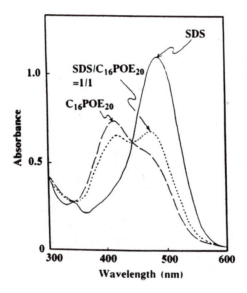

Figure 8. Absorption spectra of 4-phenylazo-1-naphthol in SDS, $C_{16}POE_{20}$, and a mixture of both solutions at 30 °C. (Reproduced with permission from ref. 19. Copyright 1987 Academic Press.)

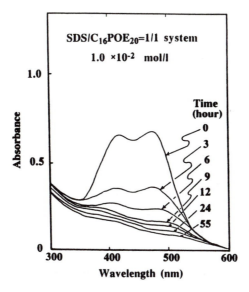

Figure 9. Time dependence of absorption spectra of 4-phenylazo-1-naphthol in SDS/$C_{16}POE_{20}$ (molar ratio, 1/1) solutions at 30 °C. (Reproduced with permission from ref. 19. Copyright 1987 Academic Press and from ref. 20. Copyright 1987 Steinkopff Verlag Darmstadt.)

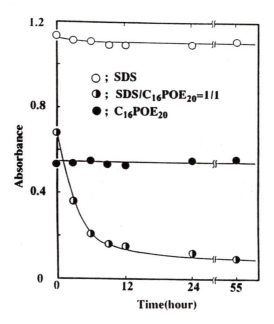

Figure 10. Time dependence of absorbance at 480 nm in SDS, $C_{16}POE_{20}$, and a mixture of both systems at 30 °C. (Reproduced with permission from ref. 19. Copyright 1987 Academic Press.)

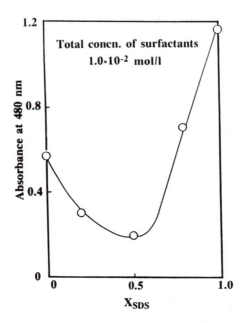

Figure 11. Relation between absorbance at 480 nm and mole fraction of SDS at 30 °C. (Reproduced with permission from ref. 19. Copyright 1987 Academic Press.)

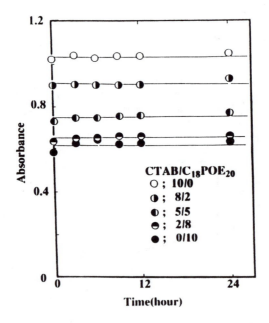

Figure 12. Time dependence of absorbance of 4-OH in various $CTAB/C_{16}POE_{20}$ solutions at 30 °C. (Reproduced with permission from ref. 19. Copyright 1987 Academic Press.)

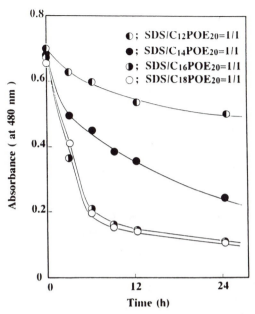

Figure 13. Effect of alkyl chain length in C_mPOE_{20} on the fading rate in SDS/C_mPOE=1/1 (total concentration; 1.0×10^{-2} mol/l) mixed systems at 30 °C. (Reproduced with permission from ref. 20. Copyright 1987 Steikopff Verlag Darmstadt.)

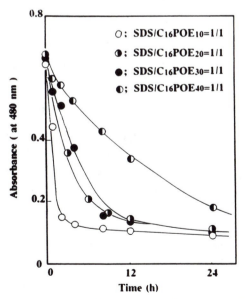

Figure 14. Effect of the number of ethylene oxide groups in $C_{16}POE_n$ on the fading rate in SDS/$C_{16}POE_n$=1/1 (total concentration; 1.0×10^{-2} mol/l) mixed systems at 30 °C. (Reproduced with permission from ref. 20. Copyright 1987 Steikopff Verlag Darmstadt.)

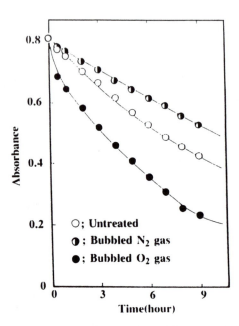

Figure 15. Changes in absorbance of 4-OH with or without bubbling O_2 gas (or N_2 gas) versus time at 30 °C in SDS/$C_{18}POE_{20}$ (1/1) solutions. (Reproduced with permission from ref. 19. Copyright 1987 Academic Press and from ref. 20. Copyright 1987 Steikopff Verlag Darmstadt.)

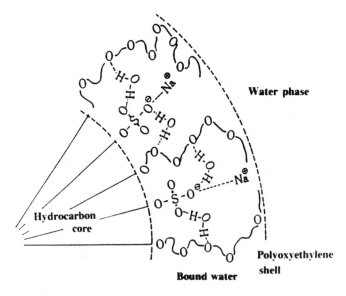

Figure 16. Hydration model for SDS/C_mPOE_n systems. (Reproduced with permission from ref 19. Copyright 1987 Academic Press and from ref. 20. Copyright 1987 Steikopff Verlag Darmstadt.)

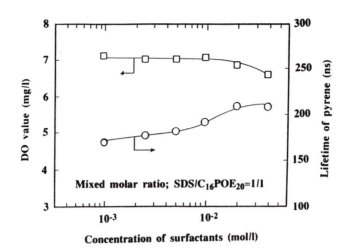

Figure 17. Dissolved oxygen (DO) value and fluorescence lifetime of pyrene monomer against total concentration of surfactants in $SDS/C_{16}POE_n$ mixed systems at 30 °C.

pyrene fluorescence lifetime is increased at surfactant concentrations above 1.0×10^{-2} mol/L. The fact that the amount of oxygen in the mixed micelle is decreased by increasing the total concentration of surfactants above 1.0×10^{-2} mol/L shows that the fading phenomenon of 4-OH is not related to the amount of oxygen in the mixed micelle.

The dissolved oxygen concentration in the mixed surfactant solution and/or in the mixed micelle is almost constant and independent of the alkyl chain lengths and poly(oxyethylene) chain length. Thus, although the fading rate is dependent on the alkyl and poly(oxyethylene) chain length, the concentration of dissolved oxygen in the micelle is independent of either chain length in the nonionic surfactant.

We have studied the possible effects of formation of each active oxygen species on the fading phenomena in the mixed micelle. First of all, the possibilities of hydrogen peroxide and hydroxyl radical forming in the mixed micelle were studied.

The hydroxyl radical is the most reactive of the active oxygen species. In general, irradiation with light is required to form the hydroxyl radical from hydrogen peroxide or water molecules. Figure 18 shows that the fading rate of 4-OH in the dark is as fast as that under the natural light. The fading phenomenon of 4-OH is therefore independent of the formation of hydroxyl radicals in the mixed surfactant solution.

Secondly, the possibility of superoxide formation in the mixed surfactant solution was investigated. Superoxide dismutase (SOD) is a typical quencher for superoxides. Figure 19 shows that the fading behavior of 4-OH in $SDS-C_{16}POE_{10}$ mixed surfactant systems is the same as that without SOD and it is independent of SOD concentration. It can be concluded that the superoxide does not form in the mixed micelles of SDS/C_mPOE_n.

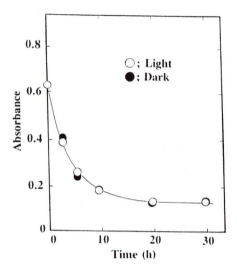

Figure 18. Time dependence of absorbance at 480 nm in SDS/$C_{16}POE_{10}$ mixed surfactant solutions at 30 °C.

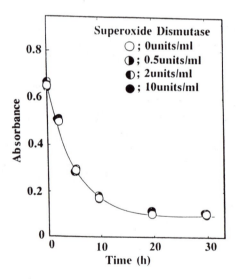

Figure 19. Effect of Superoxide Dismutase (SOD) on the fading rate of 4-phenylazo-1-naphthol in SDS/$C_{16}POE_{10}$ mixed surfactant solutions at 30 °C.

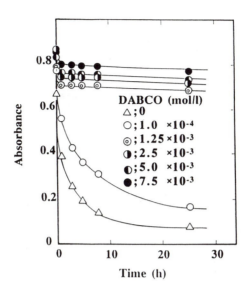

Figure 20. Effect of 1,4-diazobicyclo[2,2,2]octane (DABCO) on the fading rate of 4-phenylazo-1-naphthol in SDS/$C_{16}POE_{10}$ mixed surfactant solutions at 30 °C.

Next, the possibility of a fading reaction attributable to singlet oxygen was investigated. DABCO is a well-known quencher for singlet oxygen. Figure 20 depicts the time dependence of absorbance at 480 nm in SDS/$C_{16}POE_{20}$ containing DABCO. The fading rate of 4-OH is decreased with increasing the concentration of DABCO solubilized in the mixed surfactant solution. Therefore, the fading phenomenon of 4-OH may be caused by the singlet oxygen.

The following experiment was performed in order to confirm these results. The lifetime of the singlet oxygen in H_2O is approximately 2-4.2 μs (*23*). However, in D_2O (*24*), the lifetime of the singlet oxygen is about 55-68 μs, 10 times longer than in H_2O. The effect of replacement of H_2O by deuterium oxide on the fading rate of 4-OH was studied. As can be seen from the Fig. 21, the decrease in the optical densities for the solutions containing D_2O is faster than that in solutions without D_2O. The fading rate of 4-OH is increased as the mixed molar ratio of D_2O increased. This may reflect the fact that the lifetime of the singlet oxygen is lengthened in the D_2O surfactant solution.

Additionally, the effect of the singlet oxygen sensitizer on the fading behavior of 4-OH has been investigated in the mixed surfactant solution. The singlet oxygen sensitizers used in this study were TPP and TPPS, which are water insoluble and soluble, respectively. As is shown in both Fig. 22 and 23, the fading rate of 4-OH is accelerated on addition of the singlet oxygen sensitizer and increases with an increase in the concentration of sensitizer. Consequently, even though the amount of dissolved oxygen in the micelle is almost the same in each surfactant solution, the activated oxygen species, singlet oxygen, forms in the mixed micelle, causing the fading phenomenon of 4-OH.

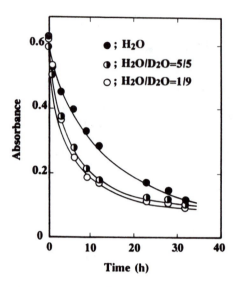

Figure 21. Effect of Deuteriumoxide (D_2O) on the fading rate of 4-phenylazo-1-naphthol in SDS/$C_{16}POE_{10}$ mixed surfactant solutions at 30 °C.

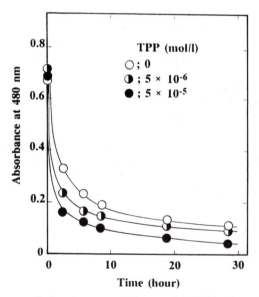

Figure 22. Changes of absorbance at 480 nm of 4-OH in the mixed surfactant solutions solubilized with TPP at 30 °C.

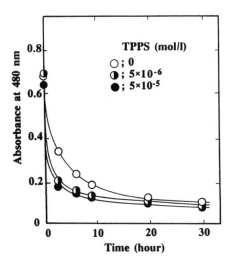

Figure 23. Effect of TPPS on the fading rate of 4-OH in the mixed surfactant solutions.

Literature Cited

1. Lange, H., and Beck, K. H., *Colloid Polym. Sci.* **1973**, *251*, 424.
2. Moroi, Y., Nishikido, N., and Matsuura, R., *J. Colloid Interface Sci.* **1977**, *50*, 344.
3. Moroi, Y., Akisada, H.,Sato, M., and Matsuura, R., *J. Colloid Interface Sci.* **1977**, *61*, 233.
4. Cayias, J. L., Schechter, R. S., and Wade, W. H., *J. Colloid Interface Sci.* **1977**, *59*, 31.
5. Van den Bogaent, R., and Joos, P., *J. Phys. Chem.* **1980**, *84*, 1984.
6. Scamehorn, J. F., Schechter, R. S., and Wade, W. H., *J. Colloid Interface Sci.* **1982**, *85*, 479.
7. Holland, P. M., and Rubingh, D. H., *J. Phys. Chem.* **1983**, *87*, 1984.
8. Ogino, K., Abe, M.,and Tsubaki, N., *Yukagaku*, **1982**, *31*, 953.
9. Abe, M., Tsubaki, N., and Ogino, K., *Yukagaku*, **1983**, *32*, 672.
10. Ogino, K., Tsubaki, N., and Abe, M., *J. Colloid Interface Sci.* **1984**, *98*, 78.
11. Abe, M., Tsubaki, N., and Ogino, K., *Colloid Polym. Sci.* **1984**, *262*, 584.
12. Abe, M., Tsubaki, N., and Ogino, K., *J. Coloid Interface Sci.* **1985**, *107*, 503.
13. Ogino, K. Tsubaki, N., and Abe, M., *J. Colloid Interface Sci.* **1985**, *107*, 509.
14. Rosen, N. J., and X. Y. Hua, *J. Am. Oil Chem. Soc.* **1982**, *59*, 582.
15. Abe, M., and Ogino, K., *J. Colloid Interface Sci.* **1981**, *80*, 56.
16. Ogino, K., Kakihara, T., Uchiyama, H., Abe, M., *J. Am. Oil Chem. Soc.* **1988**, *65*, 405.
17. Abe, M., Kakihara, T., Uchiyama, H., Ogino, K., *J. Jpn. Oil Chem. Soc.* **1987**, *36*, 135.

18. Rubingh, D, N, in solution Chemistry of Surfactants, edited by Mittal, Plenum Press, N. Y., **1979**, Vol. 1; pp. 337-354.
19. Ogino, K. Uchiyama, H. Ohsato, M. Abe, M., *J. colloid Interface Sci.* **1987**, *116*, 81.
20. Ogino, K., Uchiyama, H., Abe, M., *Colloid Polym. Sci.* **1987**, *265*, 52.
21. Uchiyama, H., Abe, M., Ogino, K., *Colloid Polym. Sci.* **1987**, *265*, 838.
22. Uchiyama, H., Akao, S., Abe, M., Ogino, K., *Langmuir,* **1990**, *6,* 1763.
23. Wasserman, H. H., Murray, R. W., *Singlet Oxygen*, Academic Press; N.Y., **1979**.
24. Griffiths, J., Hawkins, C., J. *Chem. Soc. Perkin Trans*, **1977**, *2,* 747.

RECEIVED November 5, 1991

Chapter 9

Ionic–Nonionic Interactions in the Partitioning of Polydistributed Nonionic Surfactants

A. Graciaa[1], M. Ben Ghoulam[2], B. Mendiboure[1], J. Lachaise[1], and Robert S. Schechter[3]

[1]Laboratoire de Thermodynamique des Etats Metastables et de Physique Moleculaire, Centre Universitaire de Recherche Scientifique, Avenue de L'Universite, 6400 Pau, France
[2]University of Casablanca, Morocco
[3]Department of Chemical Engineering, University of Texas, Austin, TX 78712

A phase separation model describing both the composition of the interfacial pseudophase and the partitioning of surfactant molecules into the oil and aqueous phases is proposed. Model parameters include the Critical Micelle Concentrations (CMC) and the partition coefficients of the individual surfactant molecules. These are parameters characteristic of the individual surfactant molecules. A second set of parameters relates to surfactant interactions within the interfacial pseudophase. For the cases studied, including those comprised of ternary surfactant systems, it is found that the binary interaction parameters derived from mixture CMC data are adequate to describe all of the experimental results.

When formulating systems for specific applications, surfactant blends are often preferred over pure monoisomeric surfactants . This is often true simply because the cost to produce suitable monoisomeric surfactants may be excessive. However, for many applications, a distinctly better performance can be achieved using blends rather than pure surfactants (1,2). Mixtures of surfactants have, for example, been proposed for application in enhanced oil recovery (3,4) as well as for detergents (5,6). The benefits derived from using blends are often so profound that the interaction between the different surfactant types is said to be "synergistic" (1,7,8,9). The CMC's or the surface tensions of blends may be smaller than any corresponding property of the individual components.

Because of the practical importance of synergism, a number of studies have reported the mixture CMC's of ionic-nonionic systems (1,2,10-14) and of cationic-anionic mixtures (7,8,15). Models representing these results have been primarily based on the phase separation model that includes a specific interaction between different surfactant types. One successful approach for representing the interactions has been to assume the interfacial pseudophase to be a regular solution (12-14,16-20).

The primary phenomenon modeled has been the mixture CMC and to a lessor extent the surface tension. Another important property relating to both emulsion and microemulsion stability has to do with the partitioning of surfactant into both oil and aqueous phases (21). It has been shown that the partitioning of polyethoxylated octylphenols into an oil phase can be significant, especially for those components containing less than five ethyleneoxides (20) and that this partitioning can be reduced by blending anionic surfactants with the ethoxylates (20,22). Furthermore, the concentrations of the various ethoxylates in the oil phase were found to be predictable using the phase separation model with the interactions between different surfactant types treated as a regular solution (20,21).

In this paper the partitioning of polyethoxylated phenols from blends consisting of anionic-nonionic, cationic-nonionic, and anionic-cationic-nonionic surfactants into an oil phase (in this study, isooctane) is measured. It is shown that a relatively simple model using binary interaction parameters is adequate for predicting the partitioning of ethoxylates into the oil.

Experimental

Systems. Mixtures of water, isooctane, sodium dodecylbenzenesulfonate, tetradecyltrimethylammonium bromide, and polyoxyethylene octylphenols have been studied. The water was deionized and redistilled. The isooctane was a spectrograde MERCK Chemicals product. The sodium dodecylbenzenesulfonate was sodium 6,6'-dodecylbenzene-sulfonate $(C_{12}H_{25}\text{-}C_6H_4SO_3^-,Na^+)$ synthesized at The University of Texas (23) that contains less than 1% impurities It is denoted as SDBS or in some cases with a subscript "a" (for anionic).

The tetradecyltrimethylammonium bromide, $C_{14}H_{29}N(CH_3)_3^+Br^-$ was obtained from Aldrich Chemical at a purity greater than 99%. It will be denoted as TTAB or with a subscript "c" (for cationic).

The polyoxyethylene octylphenols $C_8H_{17}C_6H_4\text{-}(CH_2CH_2O)_{n-1}CH_2CH_2OH$ are commercial products, 100% active, supplied by SEPPIC (France) and were used as received. They will be denoted as $OP(\overline{EO})_n$ where n represents the number of ethylene oxide units; the overbar indicates that n is an average and that surfactant is polydisperse satisfying the Poisson distribution (24).

$$m_i^t = \frac{(n-1)^{i-1}\exp(1-n)}{(i-1)!} \frac{M_i}{M_n} \sum_{j=1}^{N} m_j^t \qquad i=1,2...N \qquad (1)$$

where m_i^t is the mass of the oligomer $OP(EO)_i$ whose molecular weight is M_i. M_n is the molecular weight of the mean oligomer and $N = 2n + 5$ is the maximum oligomer number of the polydisperse surfactant.

Procedures. Aqueous solutions of surfactant were prepared and equilibrated with an equal volume of isooctane by mixing two phases in sealed tubes suspended in a bath that was maintained at a temperature of $25.0 \pm 0.1°C$. The tubes were gently agitated twice a day for one week. They were then kept upright for a period of eight weeks to ensure that equilibrium was attained (25).

At equilibrium, all the systems studied are of the Winsor I type (26) except for those formed with ternary mixtures of surfactants whose concentrations were equimolar in anionic and cationic surfactants. These were of the Winsor III type (26,27).

The partitioning of the ionic surfactants into the excess isooctane phase was generally found to be small and is neglected in all calculations presented here. The concentrations of the various nonionic oligomers in the isooctane phase were measured by gas chromatography. An Intersmat Model 16 with an ionization detector was used. The column was a 50 cm GAW Chromosorb substrate with 5% silicone grease. A temperature increase of 4°C per minute between 150 and 300°C was programmed.

The surfactant compositions in the aqueous phase depend on the mixture CMC and are calculated as shown below.

Model Development

To determine the composition of the interfacial region, a model that takes the composition to be the same within all interfacial regions irrespective of the interfacial curvature is developed. According to this model, the proportions of each surfactant type present at a macroscopic oil-water interface are taken to be the same as those at interfaces separating submicroscopic regions of oil phase from water within a microemulsion phase. This approach is an extension of the phase separation model of micelles (27-31). It is known that, in fact, the adsorption of surfactant at an interface is curvature dependent (26), but this dependence becomes weak for highly swollen micelles (microemulsions). Thus, we model systems exhibiting a variety of phase behavior patterns (i.e., Winsor I, II, or III) all in the same way. Each is composed of three thermodynamically equilibrated phases: an oil phase containing dissolved surfactant molecules, an aqueous phase containing electrolyte and dissolved surfactant, and finally an interfacial pseudophase composed strictly of surfactant (Figure 1).

One of the components of the surfactant mixture, call it j, is, therefore, distributed between the oil and aqueous phase and the pseudophase. The total inventory of component j is found as follows:

$$N_j^t = N_j^W + N_j^O + N_j^P \tag{2}$$

where N_j^W, N_j^O, and N_j^P are, respectively, the moles of component j in the aqueous phase, the oil phase, and the pseudophase.

N_j^P may also be written as

$$N_j^P = x_j \sum_{i=1}^{N+2} N_i^P = x_j N^P.$$

where x_j is the mole fraction of component j in the interfacial pseudophase and the sum N^P includes $N + 2$ different surfactants (the N nonionic components plus the anionic and the cationic ones). Of course, the x_j are linked by the relation

$$\sum_{j=1}^{N+2} x_j = 1 \tag{3}$$

The concentration of each type of surfactant in the aqueous phase is low enough so that the chemical potential of component j can be written as

$$\mu_j^W = \mu_{oj}^W + RT \ln C_j^W \tag{4}$$

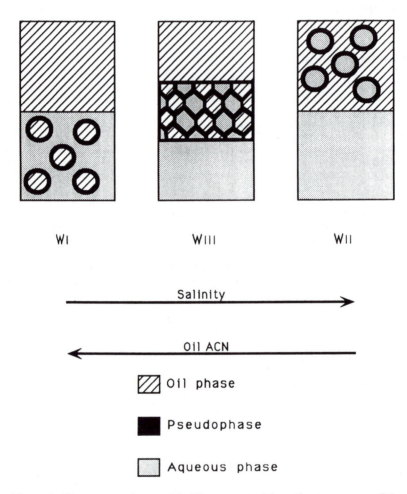

Figure 1. Phase separation model. The same model applies to any one of the three systems depicted.

where $C_j^w = N_j^w / V^w$ is the monomer concentration of the component j in the volume V^w of the aqueous phase and μ_{oj}^w is its standard chemical potential in this phase.

In the pseudophase, the surfactant molecules are closely packed and the interaction energies between the different types (ionic and nonionic, for instance) cannot be neglected. Thus, in the interfacial pseudophase, the chemical potential for component j has the form

$$\mu_j^p = \mu_{oj}^p + RT \ln f_j x_j \tag{5}$$

where f_j is the activity coefficient of the component j in the pseudophase, x_j is the molar fraction, and μ_{oj}^w is the standard chemical potential in this phase.

The equality of chemical potentials in all phases requires that

$$RT \ln C_j^w = \mu_{oj}^p - \mu_{oj}^w + RT \ln f_j x_j \tag{6}$$

When only component j is present, then $x_j = 1$, $f_j = 1$, and $C_j^w = (CMC)_j$. The difference between the two standard chemical potentials then becomes

$$\mu_{oj}^p - \mu_{oj}^w = RT \ln (CMC)_j \tag{7}$$

Based on equations (6) and (7), we find a relation between C_j^w, f_j, x_j and $(CMC)_j$ which allows us to write N_j^w as follows:

$$N_j^w = f_j (CMC)_j V^w x_j \tag{8}$$

The concentration of surfactant in the oil phase can be obtained from its concentration in the aqueous phase by introducing the partition coefficient of surfactant monomer between water and oil,

$$K_j = \frac{C_j^w}{C_j^o} \tag{9}$$

which gives

$$N_j^o = \frac{N_j^w}{K_j z} = f_j x_j (CMC)_j \frac{V^w}{K_j z} \tag{10}$$

where $z = V^w / V^o$ is the water-to-oil volume ratio.

Introducing equations 3, 8, and 10 into equation 2, the following expression for N_j^t is found:

$$N_j^t = \left[N^p + (CMC)_j V^w \left(1 + \frac{1}{z K_j} \right) f_j \right] x_j \tag{11}$$

In equation (11), the subscripts do not refer to any particular type of surfactant. It is now convenient to make the notation explicit for the particular surfactant type under consideration. For the nonionic surfactants, the subscript n is used where n varies from 1 to N and for the anionic and the cationic surfactants, the subscripts a and c, respectively, are used.

The ionic surfactants used in this study are hydrophilic and therefore their presence in the oil phase can be neglected. We can, therefore, consider their partition coefficients between water and oil (respectively, Ka and Kc) to be infinite. Thus, the inventories of each of the surfactant types can be written as follows:

$$N_n^t = \left[N^p + (CMC)_n V^w \left(1 + \frac{1}{zK_n} \right) f_n \right] x_n \tag{11a}$$

$$N_a^t = \left[N^p + (CMC)_a V^w f_a \right] x_a \tag{11b}$$

$$N_c^t = \left[N^p + (CMC)_c V^w f_c \right] x_c \tag{11c}$$

To complete the analysis, the activity coefficients must be stated as a function of the mole fraction of surfactant within the interfacial pseudophase. For binary surfactant mixtures, it has been shown that a regular solution model for the activity coefficients adequately represents the mixture CMC and also the partitioning of surfactant between the phases. Therefore, the activity coefficient for component j given by Rubingh (16) and Holland (17) for multicomponent regular solutions are adopted here. If the β_{ij} are the pair interaction parameters of the surfactant components in the pseudophase, then

$$\ln f_j = \sum_{j=1}^{N+2} \beta_{ij} x_i^2 + \sum_{i=1}^{N+2} \sum_{k=1}^{i-1} (\beta_{ij} + \beta_{jk} - \beta_{ki}) x_i x_j \tag{12}$$

Experimental evidence shows that the interaction between the nonionic components of the polydisperse ethoxylates can be neglected (20), that is, $\beta_{ij} = 0$ for $1 \le i, j \le N$. Then equation 12 becomes

$$\ln f_n = \beta_{an} x_a^2 + \beta_{cn} x_c^2 + \sum_{k=1}^{N} (\beta_{an} - \beta_{ak}) x_a x_k$$

$$+ (\beta_{an} + \beta_{cn} - \beta_{ac}) x_a x_c \qquad \text{for } n=1,2,...,N \tag{12a}$$

For the anionic surfactant monomers we find

$$\ln f_a = \beta_{ac} x_c^2 + \sum_{n=1}^{N} \beta_{an} x_n^2 + \sum_{n=1}^{N} \sum_{k=1}^{n-1} (\beta_{an} - \beta_{ak}) x_n x_k$$

$$+ \sum_{n=1}^{N} (\beta_{ac} + \beta_{cn} - \beta_{an}) x_c x_n \tag{12b}$$

and for the cationic surfactant monomer

$$\ln f_c = \beta_{ac} x_a^2 + \sum_{n=1}^{N} \beta_{cn} x_n^2 + \sum_{n=1}^{N} \sum_{k=1}^{n-1} (\beta_{cn} - \beta_{ck}) x_n x_k$$

$$+ \sum_{n=1}^{N} (\beta_{ac} + \beta_{cn} - \beta_{an}) x_a x_n \tag{12c}$$

These three expressions for activity coefficients can be simplified by using equation 3. They, respectively, become

$$\ln f_n = \beta_{an} x_a + \beta_{cn} x_c - \beta_{ac} x_a x_c - x_a \sum_{n=1}^{N} \beta_{an} x_n - x_c \sum_{n=1}^{N} \beta_{cn} x_n$$

$$\text{for } n=1,2...N \tag{13a}$$

$$\ln f_a = \beta_{ac}(1 - x_a) x_c + (1 - x_a) \sum_{n=1}^{N} \beta_{an} x_n - x_c \sum_{n=1}^{N} \beta_{cn} x_n \tag{13b}$$

$$\ln f_c = \beta_{ac} x_a(1 - x_c) + (1 - x_c) \sum_{n=1}^{N} \beta_{cn} x_n - x_a \sum_{n=1}^{N} \beta_{an} x_n \tag{13c}$$

Substituting equations 13a, 13b, and 13c into equations 11a, 11b, and 11c, we obtain the final expression for the partitioning. To solve these equations, the partition coefficients of the surfactant between the isooctane and water phases and the pairwise interaction parameters must all be known.

The partition coefficients of the nonionic monomer between water and isooctane have been measured (32) (see also Table I) and can be approximated by

$$\ln K_j = 1.02 \, j - 8.83 \tag{14}$$

The interaction parameters represent pairwise interactions and these can be calculated from mixture CMC data for binary surfactant systems. Such measurements have been reported and the following correlations can be inferred from the results (33). For nonionic/anionic interactions, the correlation is given by

$$\beta_{an} = -0.65 \, n - 3 \tag{15}$$

For nonionic/cationic interaction, the correlation is given by

$$\beta_{cn} = -0.65 \, n - 2.1 \tag{16}$$

Table I. Physical Properties of the Surfactants Studied

Surfactant	CMC in Water 25°C, mmole/l	Partition Coefficient Water/Isooctane 25°C
$OP(EO)_1$	0.049	$(1.84 \pm 0.14) \times 10^{-4}$
$OP(EO)_2$	0.076	$(7.17 \pm 0.90) \times 10^{-4}$
$OP(EO)_3$	0.103	$(3.13 \pm 0.52) \times 10^{-3}$
$OP(EO)_4$	0.129	$(9.83 \pm 0.62) \times 10^{-3}$
$OP(EO)_5$	0.172	$(2.46 \pm 0.09) \times 10^{-2}$
$OP(EO)_6$	0.250	$(5.92 \pm 0.33) \times 10^{-2}$
$OP(EO)_7$	0.268	$(1.82 \pm 0.16) \times 10^{-1}$
$OP(EO)_8$	0.283	$(5.04 \pm 0.47) \times 10^{-1}$
$OP(EO)_9$	0.304	1.42 ± 0.13
$OP(EO)_{10}$	0.323	3.85 ± 0.14
$OP(EO)_{16}$	0.430	
SDBS	2.4	
TTAB	3.2	

Finally, for the case of anionic/cationic interactions,

$$\beta_{ac} = -24.1 \qquad (17)$$

These equations are system specific and only apply for the particular surfactants used in this study. However, it is believed that the trends shown are typical of a wide variety of surfactant molecular structures and the conclusions derived are more widely applicable than just to the particular systems studied here.

The $(CMC)_j$ values of the pure component species are presumed to be known as are the parameters K_n, β_{an}, β_{cn}, and β_{ac}. Most often the water-to-oil ratio is fixed as for example in formulating an emulsion or a microemulsion as are the total quantities of the chemicals to be used. The central questions to be answered are to what extent do the surfactants partition into the oil and water phases and what is the composition of the interfacial pseudophase?

Equation 11 taken together with the appropriate expressions for the activity coefficients, equations 13, and 3 form $N + 3$ equations which are to be solved for the $N + 3$ unknowns, x_n, x_a, x_c, and N^p. These equations are nonlinear, so that they must be solved numerically.

Comparison of Theory With Experiment

Nonionic/Anionic Interactions.   In the absence of ionic surfactant, a commercial ethoxylated octylphenol contains compounds that partition preferentially into the oil phase. An inspection of the partition coefficients given in Table I will reveal the strong tendency for those molecules composed of fewer than 5 or 6 EO units to reside in the oil phase. The pseudophase model suggests that because β_{an} is negative, decreasing linearly with increasing n (see equation 15), adding anionic should therefore reduce the tendency to partition into the oil by increasing the tendency for ethoxylates to remain in the interfacial pseudophase.

To verify these predictions, systems composed of equal volumes of water and isooctane have been mixed with $OP(\overline{EO})_5$ and SDBS. The overall concentration of $OP(\overline{EO})_5$ was maintained at 5 g/l while the $SDBS/OP(\overline{EO})_5$ weight ratio was varied from 0 to 1.5. In the absence of SDBS, molecules with $EO \leq 4$ partition strongly into the oil phase. The results are plotted in Figure 2. Note that the concentration of $OP(\overline{EO})_4$ in the oil and pseudophases are about equal. Surfactant molecules with fewer than 4 EO's are found primarily in the oil phase.

For many applications, the fractionation shown in Figure 2 is considered adverse and suggests that the HLB of the interfacial pseudophase will be substantially greater than the HLB of the nonionic surfactant added to the system. Furthermore, the HLB of the interfacial pseudophase will depend on the oil-to-water ratio as shown by equation 11. Therefore, emulsifiers that are effective at one oil-to-water ratio may not be suitable when this ratio is changed.

The solid curve in Figure 2 is calculated using the data given in Table I. It is seen that the comparison is satisfactory. Since, according to equation 15, there is a strong interaction between nonionic and anionic molecules, it is expected that adding SDBS should reduce this partitioning. The results plotted in Figure 3 show this to be the case. To emphasize the degree to which adding the anionic surfactant SDBS reduces the ethoxylate concentrations in the oil phase, a relative concentration is plotted on some graphs rather than an absolute oil phase concentration. This relative concentration is defined as $100[C_n^0 - C_n^0(0)]/C_n^0(0)$ where $C_n^0(0)$ is the ethoxylate concentration found when no SDBS has been added. $C_n^0(0)$ is therefore read from Figure 2. As expected, the concentration of each ethoxylate monomer in the excess oil phase decreases as SDBS is added to the system. Furthermore, the trends are closely predicted by the pseudophase separation model using binary interaction parameters obtained from mixture CMC data.

Note also that, because the interaction between the anionic surfactant and an ethoxylate becomes stronger as n increases, the retention of longer chain ethoxylates within the interfacial pseudophase for a given concentration of SDBS is higher.

Clearly, adding anionic surfactant to a nonionic emulsifier will reduce partitioning of some nonionic ethoxylates into the oil phase. It would, however, be even more beneficial if molecules that are interfacially active and that bind with those ethoxylates with small n could be found. The authors are not aware of such compounds. A systematic search can, however, be carried out by studying mixture CMC's and applying the pseudophase model.

Cationic/Nonionic Interactions. The experiments carried out with mixtures of anionic and nonionic surfactants have been repeated using TTAB instead of SDBS. The results are shown in Figure 4. The interactions between ethoxylated oligimers and TTAB, defined by equation 16, are not quite as strong as with SDBS so that the relative concentration of an oligimer at a given TTAB concentration is slightly smaller than with SDBS. This small difference can be seen by comparing Figures 3 and 4.

As before, the experimental results are predicted using binary interaction parameters derived from mixture CMC data.

Nonionic/Cationic/Anionic Interactions. When only one type of ionic surfactant (SDBS or TTAB) is present, it has been seen that nonionic partitioning into the oil phase is decreased and furthermore this decrease is predicted by the pseudophase separation model. When the two different types of ionic surfactant are both present, the results shown in Figure 5 are obtained. Systems composed of equal volumes of water and isooctane have been mixed at 25°C with a surfactant

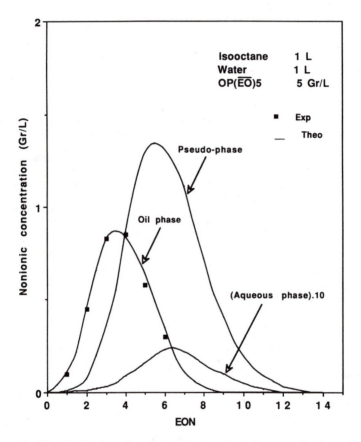

Figure 2. The distribution of octylphenol ethoxylates between the three phases as a function of the number of ethylene oxide units. (Note: The aqueous concentrations have been multiplied by a factor of 100.)

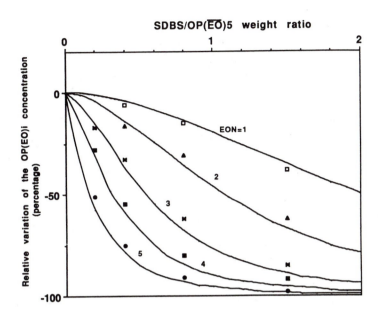

Figure 3. Relative variation of the OP(EO)$_i$ concentration in the oil phases in the function of the SDBS/OP($\overline{EO}$)$_5$ weight ratio for systems whose OP($\overline{EO}$)$_5$ concentration was maintained at 5g/l. (The relative variations are expressed in percentages.)

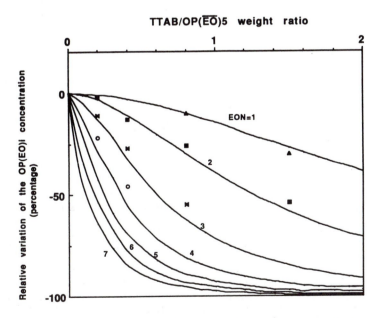

Figure 4. Relative variations of the OP(EO)$_i$ concentrations in the oil phases as a function of the TTAB/OP($\overline{EO}$)$_5$ weight ratio for systems whose OP($\overline{EO}$)$_5$ concentration was maintained at 5 g/l.

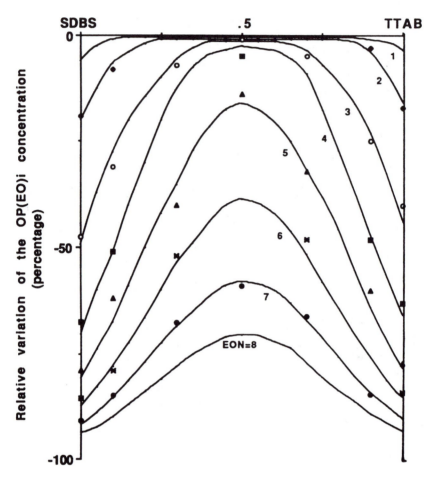

Figure 5. Relative variations of the OP(EO)$_3$ and OP(EO)$_8$ concentrations in the oil phases as a function of the composition of a surfactant ternary mixture. The OP($\overline{EO}$)$_{7.5}$ concentration was maintained at 6.4 g/l.

formulation consisting of 6.4 g/l of OP($\overline{EO}$)$_{7.5}$ and various proportions of SDBS and TTAB. The amount of (SDBS + TTAB) was maintained at 4 g/l.

Figure 5 shows that the partitioning of nonionic monomer into the oil phase is greater when mixtures of the cationic and anionic surfactant are present than when either one is present in the absence of the other. Furthermore, the mixture containing equal moles of cationic and anionic surfactant has the least influence on the partitioning of the nonionic surfactant. While this trend can be rationalized by imagining that some of the cationic surfactant molecules form ion pairs with anionic surfactant molecules and they, therefore, tend to "neutralize" one another producing essentially a new nonionic component that tends to bind loosely with the OP(EO)$_n$ molecules, it is important to recognize that the variations shown by Figure 5 are quite predictable based on the pseudophase separation model taken together with the

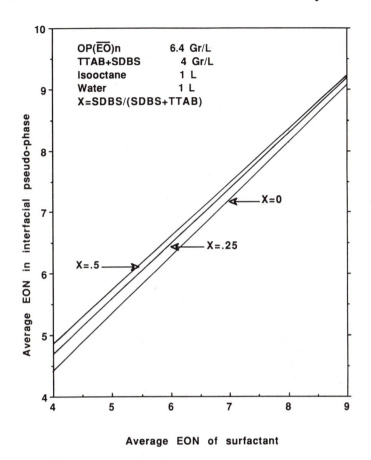

Figure 6. The EON content of the interfacial pseudophase as a function of the EON content of the system. The parameter X defines the ratio of anionic surfactant to the total ionic content.

expressions for the activity coefficients that involve only binary interaction parameters and therefore the "explanation" for the complex behavior shown is in fact embedded within the model. Certainly, the interaction parameter, $\beta_{ac} = -24.1$, indicates a high degree of association between TTAB and SDBS molecules.

The agreement between the predicted concentrations in the oil phase and the experiment shown by Figure 5 is good despite the fact that only binary interaction parameters are used to represent the activity coefficients of the components of a ternary surfactant mixture. This, in some sense, provides further justification for use of the regular solution model as expressed by equation 12 and certainly demonstrates that the model can be used for predicting surfactant partitioning as well as mixture CMC's. Furthermore, in many applications it is the interfacial composition that is important, not the composition of the surfactant used in formulating the system. These can differ substantially as shown by Figure 6. The average ethylene oxide content of the interfacial pseudophase is compared to that of the system. Because of partitioning, these differ. Here we note that this difference is a function of the ratio of SDBS to TTAB. Thus, to formulate a surfactant system composed of all three different surfactant types that yields a certain desired interfacial composition, one must use a value (EO) in the formulation which is smaller than the desired interfacial value. For example, Figure 6 shows that to attain an average of 6 ethylene units in the interfacial pseudophase, the surfactant added to the system must have an average of 5.8 units added together with cationic surfactant ($x_a = 0$). If the two types of ionic surfactant are added in equal quantities, then the surfactant added should have 5.2 units to give a pseudophase having an average of 6 ethylene oxides. The average number of EO to be used in the formulation is given by the model, but it should be noted that its value depends on the water-to-oil ratio. It also depends on the initial distribution of ethoxylates. In the calculations present here, the Poisson distribution expressed by equation 1 has been used. However, commercial polyethoxylate compounds are themselves often blended together and in this case the initial distribution may differ markedly from that expressed by equation 1. Nevertheless, the model can be applied. The only requirement is that the initial distribution be known.

Conclusions

It has been shown that the addition of either SDBS or TTAB can reduce the concentrations of $OP(EO)_i$ in the oil phase. When the two types are both added, they are less effective in reducing the concentration. Furthermore, the reduction of $OP(EO)_j$ in the oil phase increases with increased ionic surfactant concentration and with increased ethylene oxide units.

These results are all predictable using a phase separation model that divides the actual system into an oil, an aqueous, and a surfactant interfacial pseudophase. To calculate the interactions between surfactants within the pseudophase, a regular solution model is assumed. The parameters appearing in this model have been evaluated using mixture CMC data and these then are used successfully to predict surfactant partitioning.

Acknowledgments

We are indebted to the Compagnie Europeenne des Charbons Actifs and to the Societe Nationale Ele Aquitaine (Production) for the financial support of this work.

Literature Cited

1. Scamehorn, J. F. In *Phenomena in Mixed Surfactant Systems*; Scamehorn, J. F., Ed.; ACS Symp. Series 311; American Chemical Society: Washington, DC, 1986, Chapt. 30.
2. Kürzendorfer, C. P.; Schwuger, M. J.; Lange, H.; Bunsenger, B. *J. Phys. Chem.* 1978, *82*, 962.
3. McCoy, D. R.; Naylor, C. G. 1981, U.S. Patent 4288-334.
4. Hinkamp, P. E.; Tomkins, D. C.; Byth, N. J.; Thompson, J. L. 1981, EP Patent 32072.
5. Procter and Gamble 1982, U.S. Patent 4321-165.
6. *Brevet European* 1980, *880*, 537.
7. Lucassen-Reynders, E. H.; Lucassen, J.; Gilles, D. *J. Colloid Interface Sci.* 1981, *81*, 150.
8. Gu, B.; Rosen, J. J. *J. Colloid Interface Sci.* 1989, *129*, 537.
9. Rosen, J. J.; Hua, X. Y. *J. Colloid Interface Sci.* 1982, *86*, 164.
10. Funusaki, N; Hada, S. *J. Phys. Chem.* 1979, *83*, 2471.
11. Schick, M. J.; Manning, D. J. *J. Amer. Chem. Soc.* 1966, *43*, 133.

12. Holland, P. M. In *Structure/Performance Relations in Surfactants*; Rosen, M. J., Ed.; ACS Symp. Series 141; American Chemical Society: Washington, DC, 1984.
13. Scamehorn, J. F.; Schechter, R. S.; Wade, W. H. *J. Disp. Sci. Tech.* 1982, *3*, 261.
14. Hall, D. G.; Huddleston, R. W. *Colloid Surf.* 1985, *13*, 209.
15. Rodakiewicz, J. *J. Colloid Interface Sci.* 1982, *85*, 586.
16. Rubingh, D. N. In *Solution Chemistry of Surfactants*; Mittal, K. L., Ed.; Plenum Press: New York, NY, 1979, I, 337.
17. Holland, P. M.; Rubingh, D. N. *J. Phys. Chem.* 1983, *87*, 1984.
18. Rosen, M. J.; Murphy, D. S. *J. Colloid Interface Sci.* 1986, *110*, 224.
19. Rosen, M. J.; Murphy, D. S.; Zhu, Z. H. *J. Colloid Interface Sci.* 1989, *129*, 468.
20. Graciaa, A.; Lachaise, J.; Bourrel, M.; Osborne-Lee, I.; Schechter, R. S.; Wade, W. H. *Soc. Pet. Eng. J.* 1987, *2*, 305.
21. Graciaa, A.; Lachaise, J.; Marion, G.; Schechter, R. S. *Langmuir* 1989, *5*, 1315.
22. Graciaa, A.; Lachaise, J.; Bourrel, M.; Schechter, R. S.; Wade, W. H. In *Surfactants in Solution*; Mittal, K. L. and Bothorel, P., Eds.; Plenum Press: New York, NY, 1986.
23. Doe, P. H.; El-Emary, M.; Wade, W. H.; Schechter, R. S. *J. Amer. Oil Chem. Soc.* 1977, *54*, 570.
24. Enyeart, C. R. In *Nonionic Surfactants*; Schick, M. J., Ed.; Marcel Dekker: New York, NY, 1967, Chapt. 3.
25. Graciaa, A.; Barakat, Y.; Schechter, R. S.; Wade, W. H.; Yiv, S. *J. Colloid Interface Sci.* 1982, *89*, 1.
26. Bourrel, M.; Schechter, R. S. *Microemulsions and Related Systems*; Marcel Dekker: New York, NY, 1989.
27. Shinoda, K.; Nakagawa, T.; Tamamuschi, B.; Isemura, T. *Colloid Surfactants*; Academic Press: New York, NY, 1963.
28. Lange, H.; Beck, K. M. *Kolloid Z. Z. Polym.* 1973, *251*, 424.
29. Biais, J.; Bothorel, P.; Clin, B.; Lalanne, P. *J. Colloid Interface Sci.* 1981, *80*, 136.
30. Clint, J. H. *J. Chem. Soc.* 1975, *71*, 1327.
31. Biais, J.; Clin, B.; Lalanne, P. *C.R. Acad. Sci. Paris* 1982, *294*, 497.

RECEIVED February 26, 1992

Chapter 10

Interface of Mixed Micelles Formed of Anionic–Cationic and Ionic–Nonionic Surfactants

Electron Spin Studies

Piero Baglioni[1], Luigi Dei[1], Larry Kevan[2], and Elisabeth Rivara-Minten[3]

[1]Department of Chemistry, University of Florence, Florence, Italy
[2]Department of Chemistry, University of Houston,
Houston, TX 77204–5641
[3]Department of Biochemistry, University of Géneve, Ch–1211,
Geneva, Switzerland

Electron spin echo modulation (ESEM) and electron spin resonance of x-doxylstearic acids spin probes (x-DSA, x = 5, 7, 10, 12, and 16) in mixed micellar solutions of sodium dodecylsulfate (SDS) and dodecyltrimethylammonium chloride (DTAC) or mixed micelles of SDS, DTAC and hexakis(ethyleneglycol)monododecyl ether ($C_{12}E_6$), have been studied as a function of the mixed micelle composition in H_2O and D_2O. Analysis of the ESR parameters shows a strong decrease of the polarity and a strong increase of the microviscosity of the surface of SDS/DTAC mixed micelle, as compared to pure micelles. ESEM effects due to x-DSA interaction with water deuteriums give direct evidence that changes in the surface hydration of the mixed micelle are present. It is also shown in the SDS/$C_{12}E_6$ or DTAC/$C_{12}E_6$ mixed micelles that the polar headgroups of the anionic and cationic surfactants are located in the ethylene oxide region of $C_{12}E_6$. The average location of SDS and DTAC polar headgroups in the ethylene oxide region of $C_{12}E_6$ is reported. These results provide an explanation at the molecular level of the different thermodynamical behavior found for mixed micelles composed of anionic-cationic, anionic-nonionic and cationic-nonionic surfactants.

Mixed micellar aggregates are composed of two or more different surfactants in equilibrium with the surfactant monomers. These systems are of theoretical interest and of considerable industrial importance since mixtures of surfactants often exhibit synergism in their physicochemical properties allowing particular applications (*1-9*). Although models for treating mixed micellization in ideal mixed micellar solutions of ionic or nonionic surfactants (*1-5*) have been developed since the early 1950's and have been extended to nonideal mixed micellar solutions of ionic and nonionic surfactants, few descriptions of these systems at the molecular level exist (*1-7,10-13*). Furthermore, thermodynamics fails to explain some aspects of mixed micellar systems. For example, the most difficult problem for mixed ionic-nonionic surfactants is how to treat the electrostatic energy change on mixing charged and uncharged micelles, though regular solution theory has been used to explain the deviation of critical micellar concentration (CMC) from ideality (*3,7,10-16*). This

0097–6156/92/0501–0180$06.00/0

is clearly exemplified by mixtures of SDS/$C_{12}E_7$ and DTAB/$C_{12}E_7$ which show different mean molar volumes of mixing in the mixed micelle *(17)*. In the first system the mixing volume is negative, while in the second it is positive. These results cannot be explained by a simple electrostatic model, since this predicts independence of the sign of the micellar charge. Funasaki et al. *(17)* suggested that a possible explanation could be that polyethylene oxide groups of the nonionic surfactant tend to have a small positive charge. Kwak and co-workers *(18)* showed that for cationic micelles (DTAB) the ethylene oxide groups of nonionic surfactants do not contribute to the hydrophobic interactions in mixed micelle formation, while in anionic micelles (SDS), the ethylene oxide groups impart a contribution to the hydrophobic interactions.

In previous studies, we showed that very detailed information at a molecular level about the micellar interface (water penetration, site of interaction of alcohol solubilized molecules and of crown ethers which can complex the counterions, distribution of toluene, butanol and water in the SDS interfacial film of a five-component microemulsion, microviscosity and polarity of the micellar interface, etc..) can be obtained by application of electron spin echo modulation (ESEM) *(19-25)* and electron spin resonance *(26-29)* via nitroxide spin probes.

In this study, we investigated the interface of mixed micelles of cationic-anionic, cationic-nonionic, and anionic-nonionic surfactants as a function of the mixed micelle composition. The results are mainly analyzed in terms of rotational correlation time (ESR) and deuterium modulation depth (ESEM), which is a direct measure of the strength of the electron-deuterium interaction *(30)*. We followed an approach similar to that of previous studies *(19-25)* where deuteriums were located in specific regions of the mixed micelle: (i) micelle deuterated in the core region, (ii) micelle deuterated in the polar headgroup of the nonionic surfactant, and (iii) nondeuterated micelle in deuterated water. This allows determination of the location of x-doxylstearic acids, used as probes of the mixed micellar interface, and of the hydration of the micellar interface as a function of the mixed micelle composition. It is also shown that the polar headgroups of SDS and DTAC are located in a different region of the polyethylene oxide mantle of the ionic-nonionic mixed micelle. This provides an explanation at a molecular level of the different thermodynamical behavior of the SDS/nonionic and DTAC/nonionic mixed micellar systems.

Experimental Section

Sodium dodecylsulfate (SDS) and dodecyltrimethylammonium chloride (DTAC) were purchased from Eastman Kodak. SDS was recrystallized three times from ethyl alcohol, washed with ethyl ether and dried under moderate vacuum. DTAC was recrystallized three times from acetone and dried under moderate vacuum. Hexakis(ethylene glycol)monododecyl ether ($C_{12}E_6$), hexakis(ethylene-d_4-glycol)monododecyl ether ($C_{12}D_6$) in which all the ethylene oxide groups were deuterated, and hexakis(ethylene glycol)monododecyl-d_{25}-ether (($CD)_{12}E_6$) were gifts from Eniricerche, Milan, Italy. x-Doxylstearic acid spin probes (x-DSA) with x = 5, 7, 10, 12, and 16 were obtained from Molecular Probes, Eugene, Oregon, and were used as received. Stock solutions of 0.05 M surfactants were prepared in triply distilled water and in deuterated water (purity > 99.8%, Aldrich), and were deoxygenated by nitrogen bubbling. A stock solution of x-DSA was prepared in chloroform. Films of the probes generated in vials by evaporating the chloroform were dissolved in the surfactant solutions in a nitrogen atmosphere. The following systems were investigated as a function of the doxyl position, x, along the alkyl chain of the stearic acid spin probe and of the mole ratio between the pure surfactants : SDS/DTAC/D_2O, SDS/$C_{12}E_6$/D_2O, SDS/$C_{12}D_6$/H_2O, SDS/($CD)_{12}E_6$/H_2O, DTAC/$C_{12}E_6$/D_2O, DTAC/$C_{12}D_6$/H_2O, and

DTAC/(CD)$_{12}$E$_6$/H$_2$O. Mixed micelles were obtained by mixing in the appropriate mole ratio the surfactant stock solutions. Only freshly prepared solutions were used. The concentrations were 1×10^{-4} M for x-DSA and 0.05 M for the surfactant (ionic + nonionic or anionic + cationic). The ESR measurements for the SDS/DTAC system were made at 308 K. In the system SDS/DTAC no flocculation was present during the experimental measurements only for the 1.0-0.65 and 0.35-0 SDS mole ratios. The ESEM experiments in the 0.35-0.65 SDS mole ratios were performed by heating the samples overnight at 323 K to obtain homogeneous solutions. The samples were sealed in 2-mm Suprasil quartz tubes and, in the ESEM experiments, frozen rapidly by plunging into liquid nitrogen. Two-pulse electron spin echo signals were recorded at 4.2 K on a home built spectrometer by using 40 ns exciting pulses. ESR measurements were performed with a Bruker 200D spectrometer operating in the X-band, equipped with an Aspect 2000 EPR handling system and an ST100/700 temperature controller. The mean error was ± 0.04 gauss for the nitrogen coupling constant, and about 10% for the rotational correlation time.

Results

Figure 1 shows the correlation time for the probe motion for x-DSA as a function of the composition of SDS/DTAC mixed micelles. The correlation time has been computed by using the approximate equations proposed by Cannon et al. (*31,32*). Correlation times greater than 3 ns (dashed lines) were calculated by computer simulation of the ESR spectra using Freed's program (*33*). Figure 2 and Figure 3 report the normalized deuterium modulation depth as a function of the doxyl position, x, and of the SDS/DTAC mixed micelle composition. The normalized modulation depth has been computed by dividing the depth of the first modulation minimum relative to the unmodulated echo density by the depth to the baseline (*19*). Figure 4 shows the average x-DSA conformation and location in SDS, DTAC, and C$_{12}$E$_6$ micelles computed from the analysis of the deuterium modulation depth as a function of the doxyl position in SDS, DTAC, and C$_{12}$E$_6$, (CD)$_{12}$E$_6$, and C$_{12}$D$_6$ micellar solutions. Figures 5 and 6 show the normalized deuterium modulation depth for SDS/C$_{12}$E$_6$ and DTAC/C$_{12}$E$_6$ equimolar mixed micelles with deuteriums located in different positions of the nonionic surfactant (alkyl chain or polar headgroups) and in the water. Finally, Figure 7 shows the normalized deuterium modulation depth for 10-DSA as a function of the composition of SDS/C$_{12}$E$_6$ and DTAC/C$_{12}$E$_6$ mixed micelles with deuteriums located in different positions of the nonionic surfactant.

Discussion

ESR Analysis of SDS/DTAC Mixed Micelles. Electron spin resonance (ESR) is a magnetic resonance spectroscopy that monitors transitions between magnetic energy levels associated with different orientations taken by an electron spin, generally in an external magnetic field. Analysis of the ESR spectrum can give information about the identity of a molecular species, the geometric structure, the electronic structure and the internal, or overall rotational or translational motion. This technique has been used successfully to investigate the local polarity and microviscosity of the micellar interface, and of a polymer latex interface by using nitroxide spin probes solubilized at the interface of the dispersed systems (*26-29*).

In liquid solution the nitrogen coupling constant, $< A_N >$, is sensitive to the local polarity (*34-37*) and is linearly related to the dielectric constant of the solvent

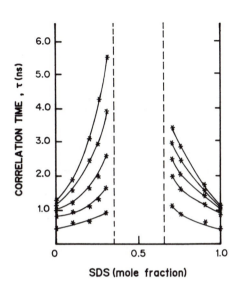

Figure 1. Dependence of the correlation time for x-doxylstearic acid spin probes on the SDS/DTAC mixed micelle composition.

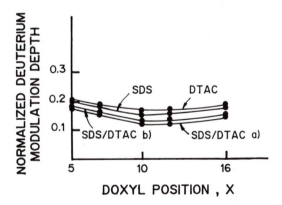

Figure 2. Normalized deuterium modulation depth for x-doxylstearic acid spin probes solubilized in SDS, DTAC, and SDS/DTAC mixed micelles prepared in D_2O with SDS/DTAC mole ratio of a) 0.25 and b) 0.75.

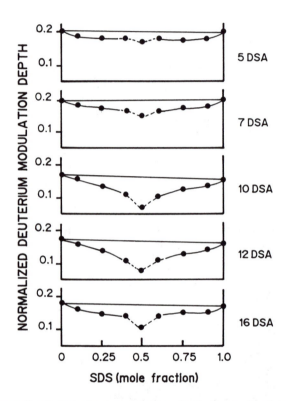

Figure 3. Normalized deuterium modulation depth for x-doxylstearic acid spin probes in SDS/DTAC mixed micellar solutions prepared in D_2O as a function of SDS mole fraction. The dashed lines represent the systems where heating at 50 °C was necessary to obtain homogeneous solutions.

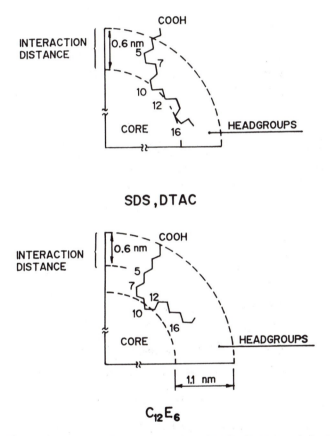

SDS , DTAC

C₁₂E₆

Figure 4. Schematic drawing of a cross section of SDS, DTAC and $C_{12}E_6$ micelle showing a probable conformation of an x-doxylstearic acid molecule and its average location within the micelle. The doxyl group is not shown, and its position along the stearic acid chain is indicated by the corresponding number.

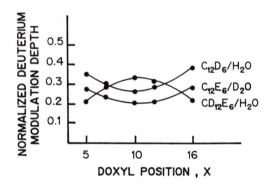

Figure 5. Dependence of the normalized deuterium modulation depths on the doxyl group position (x) in the stearic acid alkyl chain for SDS/$C_{12}E_6$ (1:1 mole ratio) mixed micelles. The nonionic surfactant has been selectively deuterated along the alkyl chain ($(CD)_{12}E_6$) or in the ethylene oxide groups ($C_{12}D_6$).

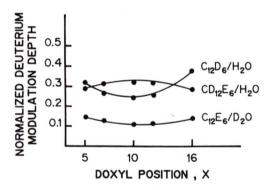

Figure 6. Dependence of the normalized deuterium modulation depths on the doxyl group position (x) in the stearic acid alkyl chain for DTAC/$C_{12}E_6$ (1:1 mole ratio) mixed micelles. The nonionic surfactant has been selectively deuterated along the alkyl chain ($(CD)_{12}E_6$) or in the ethylene oxide groups ($C_{12}D_6$).

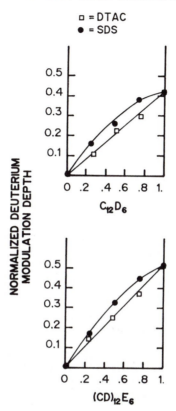

Figure 7. Dependence of the normalized deuterium modulation depth for 10-DSA probe as a function of the composition of DTAC/$C_{12}E_6$ and SDS/$C_{12}E_6$ mixed micelles. The nonionic surfactant has been selectively deuterated along the alkyl chain ((CD)$_{12}E_6$) or in the ethylene oxide groups ($C_{12}D_6$).

(38,39), to Kosover's Z values *(40)*, to Reichardt's E_T parameter *(41)*, the dipolar moment of the solvent *(42)*, and its ability to form hydrogen bonds *(37)*. A decrease of $<A_N>$ corresponds to a decrease of the polarity sensed by the probe. This decrease is usually found at the CMC of the surfactant, corresponding to the transfer of the probe from the continuous phase to the less polar surface of the micelle, or when some water molecules are expelled from the micellar surface, i.e., by the addition of salt to the micellar solution *(26,27)*. Table 1 shows that the nitrogen coupling constant decreases from 15.8 gauss in pure water (pH =10) to 15.3 and 15.1 gauss for SDS and DTAC micelles indicating that the probe is solubilized at the micellar surface, as also reported by using different techniques *(43)*. The difference found for the $<A_N>$ parameter for SDS and DTAC micelles indicates that the probes experience a different surface polarity and, in particular, that the polarity of SDS surface is higher than that of DTAC, in agreement with the hydration numbers obtained with different techniques, 9 and 5 water molecules per polar headgroup of SDS and DTAC respectively *(44)*.

Table I. Nitrogen Coupling Constant, $<A_N>$, for 5-, 7-, 10-, 12-, and 16-DSA in SDS/DTAC Mixed Micelle

SDS/DTAC (mole ratio)	$<A_N>$ (Gauss)
1.0	15.3
0.9	15.1
0.75	14.7
0.25	14.7
0.1	14.9
0.0	15.0
Water (pH=10)	15.8

It is interesting to note that the surface polarity decreases for the mixed micelles, and that this decrease depends on the cationic/anionic surfactant molar ratio of the mixed micelle, suggesting that the electrostatic interaction between the SDS and DTAC polar headgroups occurs with expulsion of water molecules present in the polar headgroup region. This result is also in agreement with the ESEM analysis (see below).

The analysis of Figure 1 shows that the correlation times are dependent both on the composition of the mixed micelle and on the doxyl position. As was stated above, the correlation time is related to the microviscosity of the environment sensed by the probe. In particular, an increase in the correlation time is related mainly to an increase in the microviscosity of the probe environment. The addition of DTAC to SDS to form mixed micelles increases the microviscosity of the micellar interface. Furthermore, the DTAC-rich mixed micelles have a higher surface viscosity compared to SDS rich mixed micelles. These results suggest that the electrostatic interaction between the polar headgroups of SDS and DTAC leads to an increase in the viscosity of the micellar surface. Furthermore the correlation times show that the increase of microviscosity of the mixed micelle is in the sequence 5-DSA > 7-DSA > 10-DSA > 12-DSA > 16-DSA. ESEM results show (see also below) that the 5-DSA probe is located at the mixed micellar surface, while 7-DSA and 10-DSA are located deeper inside the mixed micelle. Therefore the above trend for the correlation time seems to suggest that the microviscosity increase mainly occurs in the polar headgroup region of the mixed micelle while the inner part of the mixed micelle shows minor increases of microviscosity.

ESEM Analysis of SDS/DTAC, SDS/$C_{12}E_6$ and DTAC/$C_{12}E_6$ Mixed Micelles. Electron spin echo (ESE) spectroscopy is a pulsed version of ESR (30). The most common type of pulse sequence consists of a $90°$ focusing pulse followed by a precession time and then a $180°$ spin flip pulse which causes the spin to refocus within another precession time period. The refocusing produces a burst of microwave energy, the "echo". This echo intensity is measured as a function of the time between the two pulses to generate an echo decay envelope. In many cases the echo decay envelope is modulated by weak dipolar hyperfine interactions between the paramagnetic species and neighboring magnetic nuclei (in the present study deuterium) within the range 0.2 to 0.6 nm. This modulation is primarily a function of the number and distance of the closest surrounding magnetic nuclei. Analysis of the electron spin echo modulation (ESEM) provides significant structural information, which is generally not available from electron spin resonance.

In order to apply ESEM to micellar systems, which have a complex structure, the modulation depth is analyzed semiquantitatively. Typically, deuterium modulation is observed from deuterium localized in specific positions in the dispersed system.

In liquid solutions, the anisotropic hyperfine interactions, which are responsible for the modulation, are averaged out by rapid molecular tumbling, and consequently, modulation can only be observed in rapidly frozen solutions. It has already been established that upon fast freezing the micellar structure is retained in normal micellar systems and microemulsions (21-25). In a recent study, using ESEM, we demonstrated that the micellar structure is retained upon fast freezing even in reverse micellar systems (21). In particular we showed that the results obtained in frozen AOT reverse micelles, as a function of the dimensions of the water pool of the reverse micelle, are in excellent agreement with results obtained in liquid solutions via Nuclear Magnetic Resonance, Quasi Elastic Light Scattering, Infrared and Raman spectroscopies. Further evidence that the micellar and microemulsion structure is retained upon fast freezing has been obtained from luminescence quenching, from freeze-fracture electron microscopy, from the method of direct imaging and from SANS (45-50).

The doxylstearic acid spin probes, in ionic micellar solutions, are comicellized with the surfactant molecules with the acidic group at the micellar interface (43), while the doxyl group located further from the micellar surface in passing from the 5-DSA to the 10-DSA position along the stearic acid chain. Since the probe can assume on average two limiting conformations, depending on the structure of the dispersed system (normal or reversed micelles), the doxyl group of 16-DSA can be located in a region of the micellar interface similar to that of either 12-DSA (all-trans or extended conformation) or 5-DSA (bent or U-folded conformation) (19,24). Therefore, in normal micelles with deuterated water as solvent, when the probe presents an all-trans conformation, increasing x moves the doxyl group further away from the surface of the micelle and one expects a decrease in the normalized deuterium modulation depth. On the other hand, a decrease of the deuterium modulation depth from $x=5$ to 10 and an increase at $x = 12$ and 16 indicates that the doxyl probe bends near $x = 10$ and 12, thus having a U-shape introduced by a few gauche links.

ESEM Analysis of SDS/DTAC. Figure 2 shows that the normalized deuterium modulation depth for the SDS, DTAC micellar solutions is maximum for 5-DSA, reaches a minimum for 10-DSA, and increases for 16-DSA indicating that the probe has a bent conformation. It is interesting to note that this conformation does not change in SDS/DTAC mixed micelles, since the mixtures show the same trend of the normalized deuterium modulation depth as that of the pure surfactants. As an example, the trend of the normalized deuterium modulation depth for two SDS/DTAC micellar solutions is reported in Figure 2. The other mixtures exhibit similar trends.

By employing a space-filling spherical model for the micelle and by considering that deuterium modulation is generally only detectable at interaction distances less than 0.6 nm, it is possible to calculate the the average probe conformation and location within the micellar interface. A schematic drawing of a cross section of SDS and DTAC micelle showing the probable conformation and location of a x-doxylstearic acid molecule is shown in Figure 4.

It is well known that mixed micelles formed of anionic-cationic surfactants show strong deviations from ideal behavior (*11-15*). This is largely due to electrostatic interactions among the polar headgroups. These interactions lead to a decrease of the effective micellar surface charge, and for a 1:1 mole ratio of the surfactants, the micelles behave, from a macroscopic point of view, as if they were neutral (*51*). The surface "neutralization" produces a gradual increase in the aggregation number of the mixed micelle (*52*) and, according to the results reported above, a strong increase of the microviscosity of the micellar interface (see also ref. *53*). Figure 3 shows that the normalized deuterium modulation depth decreases and is a minimum for a 1:1 mole ratio. Since the deuterium modulation depth depends on the number of interacting deuteriums and on their distance from the spin probe, a decrease in the deuterium modulation depth reflects a decrease in the number of water molecules present at the micellar interface. It follows that the interaction between the SDS and DTAC polar headgroups occurs with expulsion of water molecules from the micellar surface, in agreement with the ESR results reported above. It is interesting to note that this effect is maximum for 10-DSA and 12-DSA, which are located deeper inside the micellar interface, than for 5-, 7-, and 16-DSA, supporting the view that the water expulsion mainly affects the water molecules at the methylenes α and β to the polar headgroup of the surfactants.

ESEM Analysis of SDS/$C_{12}E_6$ and DTAC/$C_{12}E_6$. In a previous study (*54*) we showed that the x-DSA probes have a bent conformation and are located in the ethylene oxide region of the $C_{12}E_6$ micelle. From a comparative analysis of the deuterium modulation depth arising from deuteriums located in different regions of the $C_{12}E_6$ micellar system (micelle core, micelle polar headgroups, water), we showed that ethylene oxide segments form a region around the hydrophobic core that does not allow much water penetration. Only the outer one or two ethylene oxide groups experience significant contact with water molecules. Furthermore, the deuterium modulation depth as a function of the doxyl position shows symmetrically inverse shapes for deuterated alkyl versus deuterated ethylene oxide curves. This is consistent with the different deuterated portions of the surfactant molecules being separated as expected from SANS experiments (*55, 56*). Figure 4 reports a schematic drawing of a cross section of a $C_{12}E_6$ micelle showing a probable conformation of an x-doxylstearic acid molecule and its average location within the micelle. It should be stressed that this is an overall average probe location and conformation.

One interesting problem in the analysis of SDS/$C_{12}E_6$ or DTAC/$C_{12}E_6$ mixed micelles is the location of the cationic or anionic surfactant in the nonionic micelle. It is generally accepted that, since all the surfactants have the same alkyl chain length, changes in the mixing ratios between the ionic and nonionic surfactants should be essentially considered as the replacement of one polar headgroup with another (*57*). This is true only if the polar headgroups of the ionic surfactant penetrate all the ethylene oxide region of the micelle, which is extended about 11 Å, and locate close to the core of the nonionic micelle. However, we observe that this is not the case and probably this is the explanation for the different behavior of SDS/$C_{12}E_6$ or DTAC/$C_{12}E_6$ mixed micellar systems.

Figures 5 and 6 report the trend of the normalized deuterium modulation depth as a function of the x-doxyl position for SDS/$C_{12}E_6$ and DTAC/$C_{12}E_6$ equimolar mixed micelles and for deuterium located in the water phase, in the ethylene oxide region or in the alkyl region of the nonionic surfactant. The values of the normalized deuterium modulation depth and the shapes of these curves indicate that the x-doxylstearic acids have in these mixed micelles the same location as in the pure nonionic micelle, and are bent as they are in the pure nonionic and ionic micelles. Secondly, the symmetrically inverse shapes for the deuterated alkyl versus the deuterated ethylene oxide curves are also consistent with the different deuterated portions of the surfactant molecules being separated, as observed in the pure nonionic surfactant (*54*).

It is relevant for the analysis of the SDS and DTAC location in the mixed micelle that 10-DSA is located about 1-2 Å inside the alkyl region of the nonionic surfactant (*54*). A rough calculation for a droplet model of the mixed micelle shows, for the deuterium modulation depth of 10-DSA as a function of the molar ratio between ionic and nonionic surfactants, that a linear trend is expected when the polar headgroup of the ionic surfactant is located close to the alkyl region of the nonionic surfactant, i.e. when the deuterated alkyl chain of the nonionic surfactant $((CD)_{12}E_6)$ is completely replaced by the nondeuterated chain of the ionic surfactant. Positive or negative deviations are expected when the polar headgroup of the ionic surfactant is located at the surface of the micelle in the ethylene oxide region or inside the alkyl region of the nonionic surfactant.

Figure 7 shows the normalized deuterium modulation depth for 10-DSA as a function of the $SDS/C_{12}E_6$ and $DTAC/C_{12}E_6$ mixed micelle composition, and for the different locations of the deuteriums in the nonionic surfactant. The analysis of the figure shows that the trend of deuterium modulation depth is linear for the system $DTAC/C_{12}E_6$, while a positive deviation is present for the system $SDS/C_{12}E_6$. Therefore, these results support the view that SDS and DTAC have a different location in the ethylene oxide mantle of the nonionic surfactant in mixed micelles formed of anionic and cationic surfactants. A more thorough analysis of the SDS and DTAC location shows that SDS is located closer to the surface of the mixed micelles at the 2nd-3rd ethylene oxide group while DTAC is located at the 5th-6th ethylene oxide group of the nonionic surfactant (*58*).

Conclusions

From the analysis of the results of electron spin resonance and electron spin echo modulation of doxylstearic acids in SDS/DTAC, $SDS/C_{12}E_6$, and $DTAC/C_{12}E_6$ mixed micelles the following conclusions can be drawn:

1) The doxylstearic acid spin probes have a bent conformation in the pure SDS, DTAC and $C_{12}E_6$ micellar systems and in all the mixed micellar systems studied.

2) In SDS/DTAC mixed micellar systems, the "electrostatic interaction" among the surfactant polar headgroups occurs with the expulsion of water molecules, probably located near the a and b methylenes, and with an increase in the microviscosity of the interface of the mixed micelle. These effects depend on the mixed micelle composition.

3) In $SDS/C_{12}E_6$ and $DTAC/C_{12}E_6$ mixed micellar systems, SDS and DTAC solubilize with their polar headgroups in the ethylene oxide region of the nonionic surfactant. From the trend of the normalized deuterium modulation depth as a function of the doxyl position, and from a comparative analysis of the modulation depths in mixed micelles with the nonionic surfactant selectively deuterated in the alkyl chain or in the ethylene oxide region, it is found that DTAC polar headgroups are located close to the core region of the mixed micelle (5th-6th ethylene oxide of the nonionic surfactant), while SDS polar headgroups are located close to the mixed micellar surface in a more "hydrophilic region" (2nd-3rd ethylene oxide of the nonionic surfactant).

These results provide an explanation at the molecular level of the different thermodynamical behavior found for mixed micelles composed of cationic-nonionic and anionic-nonionic surfactants.

Acknowledgments

This work is supported by grants from the Italian Council of Research (CNR) and the Division of Chemical Sciences, Office of Basic Energy Sciences, Office of Energy Research, U.S. Department of Energy.

Literature Cited

1. Shinoda, K. *J. Phys. Chem.* **1954**, *58*, 541.
2. Shinoda, K. *J. Phys. Chem.* **1954**, *58*, 1136.
3. Lange, H. *Kolloid Z. Z. Polym.* **1953**, *131*, 96.
4. Shedlovsky, L.; Jakob, C.W.; Epstein, M.B. *J. Phys. Chem.* **1963**, *67*, 2075.
5. Corkill, M.; Goodman, J. *Proc. Roy. Soc.* **1963**, *273*, 84.
6. Clint, J. *J. Chem. Soc. Faraday 1* **1975**, *71*, 1327.
7. Lange, H.; Beck, K.H. *Kolloid Z. Z. Polym.* **1973**, *251*, 356; 424.
8. *Phenomena in Mixed Surfactant Systems*; Scamehorn, J. F., Ed.; American Chemical Society: Washington, DC, 1986; A.C.S. Symposium Series 311.
9. Rosen, M.J. In *Mixed Surfactant Systems* ; Scamehorn , J. F., Ed.; American Chemical Society: Washington, DC, 1986; A.C.S. Symposium Series 311, p 144.
10. Nishikido, N.; Morio, Y.; Matuura, R. *Bull. Chem. Soc. Japan*, **1975**, *48*, 1387.
11. Holland, P. M.; Rubingh, D.N. *J. Phys. Chem.* **1983**, *87*, 1983.
12. Rubingh, D. N. In *Solution Chemistry of Surfactants*; Mittal, K. L., Ed.; Plenum: New York, 1979; Vol. 1, p 337.
13. Holland P.M. *Adv. Colloid Interface Sci.* **1986**, *26*, 111.
14. Scamehorn, J. F.; Schechter, R. S.; Wade, W. H. *J. Dispersion Sci. Technol.* **1982**, *3*, 261.
15. Osborne-Lee, I. W.; Schechter, R. S.; Wade, W. H.; Barakat, Y. *J. Colloid Interface Sci.* **1985**, *108*, 60.
16. Hey, M.J.; MacTaggart, J.W.; Rochester, C.H. *J. Chem Soc. Faraday Trans. 1* **1985**, *81*, 207.
17. Funasaki, N.; Hada, S.; Neya, S. *J. Phys. Chem.* **1986**, *90*, 5469.
18. Marangoni, G.D.; Rodenhiser, A.P.; Thomas, J.M.; Kwak J.C.T. This Book.
19. Kevan, L.; Baglioni, P. *Pure Appl. Chem.* **1990**, *62*, 275.
20. Kevan, L. In *Photoinduced Electron Transfer*, Part B ; Fox, M.A.; Chanon, M. Eds.; Elsevier: Amsterdam, 1988; pp 329-394.
21. Baglioni, P.; Nakamura, H.; Kevan, L. *J. Phys. Chem.* **1991**, *95*, 3856.
22. Baglioni, P.; Kevan, L. *J. Phys. Chem.* **1987**, *91*, 1516.
23. Baglioni, P.; Rivara-Minten, E.; Kevan, L. *J. Phys. Chem.* **1988**, *92*, 4726.
24. Baglioni, P. ; Gambi, C.M.C.; Goldfarb, D. *J. Phys. Chem.* **1991**, *95*, 2577.
25. Baglioni, P. ; Gambi, C.M.C.; Goldfarb, D. *Progr. Colloid Polym. Sci.* **1991**, *84*, 55.
26. Baglioni, P.; Rivara-Minten, E.; Dei, L.; Ferroni, E. *J. Phys. Chem.* **1990**, *94*, 8218.
27. Baglioni, P.; Ottaviani, M.F.; Martini, G. *J. Phys. Chem.* **1986**, *90*, 5878.
28. Baglioni, P.; Cocciaro, R.; Dei, L. *J. Phys. Chem.* **1987**, *91*, 4020.

29. Baglioni, P.; Cocciaro, R.; Dei, L. In *Surfactants in Solution*; Mittal, K. L., Ed.; Plenum: New York, 1989; Vol. 10, p 417.
30. Kevan, L.; Bowman, M. K. *Modern and Continuous-Wave Electron Spin Resonance*; Wiley: New York, 1990.
31. Cannon, B.; Polnaszek, C.F.; Butler, K.W.; Eriksson, G.; Smith I.C.P. *Arch. Biochem. Biophys.* **1975**, *167*, 505.
32. Ernandes, J. R.; Schreier, S.; Chaimovich, Y. *Chem. Phys. Lipids*, **1976**, *16*, 14.
33. Freed, J.H. In *Spin Labelling. Theory and Applications*; Berliner, L.J., Ed.; Academic: New York, 1976, p 85.
34. Janzen, E.G. *Top. Stereochem.* **1971**, *6*, 117.
35. Knauer, B.R.; Napier, J.J. *J. Am. Chem. Soc.* **1976**, *98*, 4395.
36. Abe, B.R.; Tero-Kubota, S.; Ikegami, Y. *J. Phys. Chem.* **1982**, *86*, 1358.
37. Stout, G.; Engberts, J.B.F.N. *J. Org. Chem.* **1974**, *39*, 3800.
38. Deguchi, Y. *Bull. Chem. Soc. Japan* **1962**, *35*, 260.
39. Umemoto, K.; Deguchi, Y.; Takaki, H. *Bull. Chem. Soc. Japan* **1963**, *36*, 560.
40. Brière, R.; Lamarie H.; Rassat, A. *Tetrahedron Lett.* **1964**, *20*, 1775.
41. Wajer, Th. A.J.W.; Machor, A.; deBoer, Th.J.T. *Tetrahedron Lett.* **1969**, *25*, 175.
42. Mukai, K.; ishiguchi, K.; Ishizu, Y.; Deguchi, Y.; Takaki, H. *Bull. Chem. Soc. Japan* **1967**, *40*, 2731.
43. Ramachandran, C.; Pyter, R. A.; Mukerjee, P. *J. Phys. Chem.* **1982**, *86*, 3198; 3206.
44. Wennerström, H.; Lindman, B. *J. Phys. Chem.* **1979**, *83*, 2931.
45. Hashimoto, S.; Thomas, J.K. *J. Am. Chem. Soc.* **1983**, *105*, 5230.
46. Talmon, Y. *Colloids Surf.* **1986**, *19*, 237.
47. Dubochet, J.; Adrian, M., Teixeira, J.; Alba, C.M.; Kadiyala, R.K.; MacFarlane, D.R.; Angell, C.A. *J. Phys. Chem.* **1988**, *88*, 6727.
48. Jahn, W.; Strey, R. *J. Phys. Chem.* **1988**, *92*, 2294.
49. Bellare, J.R., Kaneko, T.; Evans, D.F. *Langmuir* **1988**, *4*, 1066.
50. Alba-Simionesco, C.; Teixeira, J.; Angell, C.A. *J. Chem. Phys.* **1989**, *91*, 395.
51. Schwuger, M.J. In *Anionic Surfactants*; Lucassen-Reynders, E.H., Ed.; Marcel Dekker: New York, 1981; p 267.
52. Malliaris, A.; Binana-Limbele, W.; Zana, R. *J.Colloid Interface Sci.* **1986**, *110*, 114.
53. Baglioni, P. In *Surfactants in Solution*; Mittal, K.L., Bothorel, P., Eds.; Plenum: New York, 1986; Vol. 4, p 393.
54. Baglioni, P.; Bongiovanni, R.; Rivara-Minten, E.; Kevan, L. *J. Phys. Chem.* **1989**, *93*, 5574.
55. Triolo, R.; Magid, L.J.; Johnson, J.S.,Jr.; Child, H.R. *J. Phys. Chem.* **1982**, *86*, 3689.
56. Zulauf, M.; Rosenbush, J.P. *J. Phys. Chem.* **1983**, *87*, 856.
57. Nilsson, P.G.; Lindman, B. *J. Phys. Chem.* **1984**, *88*, 5391.
58. Baglioni, P.; Dei, L.; Rivara-Minten, E.; Kevan, L. *J. Phys. Chem.* submitted.

RECEIVED January 6, 1992

Chapter 11

Interaction of Alcohols and Ethoxylated Alcohols with Anionic and Cationic Micelles

D. Gerrard Marangoni, Andrew P. Rodenhiser, Jill M. Thomas, and Jan C. T. Kwak[1]

Department of Chemistry, Dalhousie University, Halifax, Nova Scotia B3H 4J3, Canada

Critical micelle concentrations (CMC values), distribution constants, and the aggregation numbers of the surfactant (N_s) and alcohol (N_a) have been determined for mixed micelles consisting of ionic surfactants with medium chain length alkoxyethanols as the cosurfactant. Distribution coefficients of the solubilizates (cosurfactants) are determined using the recently developed NMR paramagnetic relaxation experiment. The free energy of transfer of the alcohol from the aqueous to the SDS micellar phase decreases as the number of ethylene oxide (EO) groups in the alcohol is increased for a given alkyl chain length. For SDS, at a given cosurfactant concentration, CMC values and the surfactant aggregation numbers, N_s, decrease as the number of EO groups in the alcohol is increased. This suggests that the EO group imparts a contribution to the hydrophobic interactions. However, in DTAB/alkoxyethanol mixed micellar systems, the distribution constants, the CMC values, and N_s are independent of the number of EO groups in the cosurfactant, at a constant alkyl chain length. This implies that for cationic DTAB micelles, the EO group does not contribute to the hydrophobic interactions in mixed micelle formation.

The determination of the interaction of organic solubilizates with ionic and nonionic surfactant micelles has been studied widely since the pioneering work of Shinoda (*1*) and Herzfeld et al. (*2*). Much of this work has focussed on the effect of the solubilizate on a single, specific property of the surfactant micelle (e.g., aggregation number, micelle shape, thermodynamics of micelle formation, and the degree of counterion binding (*3-10*)). It appears that relatively few investigations have appeared where a number of micellar properties have been examined as a function of the cosurfactant concentration (*11*).

Recently, some interest has been shown in examining the interaction of alcohols other than the well studied n-alcohols, e.g., α,ω-alkanediols (*12,13*) with ionic and nonionic micellar solutions. Although alkoxyethanols (in particular, 2-butoxyethanol)

---

[1]Corresponding author

0097–6156/92/0501–0194$06.00/0

are widely used cosurfactants in a number of different systems, including microemulsions (9,10,14), very little information has been found in the literature on the interaction of alcohol ethoxylates with ionic micelles (6,9,12).

Interactions between small solubilizate molecules and micelles have been studied by many different techniques (15). CMC determinations as a function of added solubilizate concentration are used to estimate the change in the free energy of micellization. Yamashita et al. (16,17) have determined the mean activities of sodium decanoate (SD) in a number of alcohol solutions, using a sodium responsive glass electrode and the silver/silver decanoate electrode developed by Vikingstad (18,19). These authors observed a decrease in the CMC values with an increase in both the chain length and the concentration of alcohol. Of particular interest was the observation that the CMC values of SD/2-butoxyethanol mixed micelles are lower than those of SD/1-butanol mixed micelles at the same concentration of alcohol. Manabe et al. (12) have determined the CMC values of sodium dodecyl sulfate (SDS) conductometrically as a function of the alcohol concentration and the number of ethylene oxide (EO) groups in the alcohol, at a constant alkyl chain length. These authors observed a decrease in the CMC values of the SDS/alkoxyethanol mixed micelles as the number of EO groups in the alcohol was increased, at a constant alkyl chain length and alcohol concentration.

A number of techniques (3-8,15) can be used to determine the mole fraction of solubilizate in the micellar phase, p, which is defined as follows

$$p = \frac{n_{a,mic}}{n_{a,t}} \tag{1}$$

where $n_{a,mic}$ is the number of moles of alcohol in the micellar phase and $n_{a,t}$ is the total number of moles of alcohol. The application of the NMR Fourier transform pulsed-gradient spin echo (FT-PGSE) experiment by Stilbs and co-workers to the determination of p values of solubilizates in the micellar phase has greatly enriched our knowledge of surfactant/alcohol mixed micellar systems (20-23). Recently, an alternative NMR experiment for determining the solubilizate distribution in the micellar phase has been devised by Gao et al. (24). The method is based on the changes in the relaxation rate of the alcohol in surfactant and surfactant-free solutions, in the presence of paramagnetic ions. This experiment has been applied successfully to the determination of the micelle-water distribution coefficients for a number of organic solubilizates in some typical ionic surfactant micelles (24,25).

The use of luminescence probing to determine the surfactant aggregation number is a relatively recent development (3,26-34). The main advantage of luminescence probing techniques is that they can be used to determine directly the number of micelles in solution, at any surfactant concentration, and do not require *a priori* a knowledge of the micellar shape. The disadvantage with these techniques is the need for solubilized probes and quenchers, which may influence the micellar concentration.

The static quenching method, proposed originally by Turro and Yekta (30), has been used extensively in the literature for the determination of the aggregation number of the surfactant in the presence of additives. These authors derived a simple Stern-Volmer type relationship between the bulk quencher concentration and the logarithm of the fluorescence intensities as a function of the quencher concentration.

$$\ln\left[\frac{I_o}{I}\right] = \frac{[Q]}{[M]} \tag{2}$$

The main advantage of the static quenching method is in its ease of implementation (it can be used routinely on any emission spectrophotometer). However, the application of the static quenching method depends on compliance with a number of experimental criteria (3,30). Foremost among these criteria are that both the luminescent probe and quencher are immobile and thus associated with the micelle during the luminescent lifetime of the probe, and that the quenching process is efficient so that luminescence is observed only from micelles containing a solubilized probe and no quencher. These criteria for the successful application of the static quenching method to the determination of the surfactant aggregation number have been discussed by a number of authors (3,28-34).

The complete equation relating the emission intensities of the micellar solubilized probe as a function of the quencher concentration is (3,5,30)

$$\frac{I}{I_o} = \exp - <Q> \sum_{q=0}^{\infty} \frac{<Q>^q}{q! \, (1 + k_q \tau_o)} \qquad (3)$$

where q is the number of quencher molecules in a given micelle, $<Q>$ is the average quencher occupancy ($<Q> = [Q]/[M]$); k_q is the rate constant for the quenching of the fluorescence intensity, and τ_o is the lifetime of the fluorescent probe. From equation 3, it is evident that the ratio of the probe intensities in the presence and absence of quencher is critically dependent on the value of the kinetic ratio, $k_q\tau_o$. In Figure 1, computer generated values of ln (I_o/I) are plotted vs. $<Q>$ for $k_q\tau_o$ values of 0.1 to ∞. At the static limit, $k_q\tau_o = \infty$ and equation 3 reduces to the well known Turro-Yekta equation (equation 2). For low kinetic ratios, i.e., $k_q\tau_o < 10$, it can be easily seen in Figure 1 that the micellar concentration, [M], is severely overestimated; hence, the surfactant aggregation number, N_s, is underestimated. However, for $k_q\tau_o \geq 15$, the underestimation of N_s is less critical, and is on the order of 5-7%. It is generally accepted that the use of the either ruthenium tris-bipyridyl chloride/9-methylanthracene (9-MA) or pyrene/cetylpyridinium ion (CP$^+$) as the probe/quencher pair will yield reasonable estimates of the surfactant aggregation numbers by the static quenching method. Applying the known value of the quenching rate constant and probe lifetime for the pyrene/CP$^+$ probe/quencher pair, $k_q\tau_o = 17$-18 (35), we calculate that the static method used here underestimates the real micellar aggregation number in SDS micellar systems by ca. 5%. In DTAC micelles, we have estimated a kinetic ratio of ca. 15 from the lifetime of the pyrene fluorescence in DTAC micelles (36) and the quenching rate constant of pyrene by CP$^+$ in DTAC micelles, estimated from the $k_q\tau_o$ of pyrene/CP$^+$ in CTAC micelles (36) and the known dependence of the quenching rate constants of 1-methylpyrene fluorescence on the chain length of alkylpyridinium chloride quenchers (37). As well, we have determined the surfactant aggregation numbers of DTAC micelles as a function of the surfactant concentration. These results were in very good agreement with the results of DeSchryver et al. (37,38). For SDS, Almgren and Swarup obtain an aggregation number of 70 at an SDS molarity of 0.0346 (31), using the probe/quencher pair ruthenium tris-bipyridyl chloride/9-MA. Under the same conditions, using pyrene/CP$^+$ as the probe/quencher pair, we obtain an aggregation number of 66, in very good agreement with the results of Almgren and Swarup.

A number of studies in the literature have used the static quenching method to determine the effects of solubilizates on N_s. Almgren and Swarup determined the surfactant aggregation number of SDS micelles as a function of the concentration of both the surfactant and solubilizate for a wide variety of solubilizates, including n-alcohols. A number of trends were reported in these studies, including the decrease in N_s brought about by the addition of alcohols. For a given concentration of surfactant, the rate of the decrease in N_s with an increase in the total alcohol concentration was observed to be larger for the more hydrophobic alcohols. It might be expected that this higher rate of decrease is related to the greater distribution of the alcohol in the micelle. This last parameter can be determined by the two NMR methods mentioned earlier. Almgren and Swarup used the distribution constants of the solubilizates, p, obtained by Stilbs (*20,21*) using the FT-PGSE experiment, the concentration of added alcohol, C_a, and the micellar concentration from the static quenching method, to calculate the alcohol aggregation number, N_a.

$$N_a = \frac{p\,C_{a,t}}{[M]} \tag{4}$$

N_a was observed to increase slowly for the less hydrophobic alcohols; for the more hydrophobic alcohols, N_a increased more quickly, so that the total aggregation number, $N_t = N_s + N_a$, remained relatively constant or increased slowly. Similar results for 0.0400 molal SDS/alcohol mixed micelles were found by Malliaris (*32*), using the static quenching method with pyrene/CP^+ ion as the probe/quencher pair.

In this study, we present CMC values, surfactant aggregation and alcohol aggregation numbers in SDS/alkoxyethanol and dodecyltrimethylammonium bromide (DTAB)/alkoxyethanol mixed micellar systems as a function of the concentration of alcohol and its ethylene oxide (EO) chain length, in order to investigate the role of the EO group in the formation of mixed micelles. The surfactant aggregation numbers, N_s, of the mixed micelles are determined using the static fluorescence quenching method, with pyrene/CP^+ as the probe/quencher pair. The alcohol aggregation number, N_a, and the total micelle aggregation number, N_t, are estimated from the micelle concentrations determined from the static quenching method, and the distribution constants of alkoxyethanols obtained from the paramagnetic relaxation experiment. As well, the pyrene I_1/I_3 ratios (indicative of the micropolarity sensed by the luminescent probe) have been determined. All these results will be discussed in terms of the decrease in the free energy of transfer from H_2O to the micellar phase as a function of the number of EO groups in the alcohol, and the significant differences in the interactions of ethoxylated alcohols with anionic and cationic micelles.

Experimental

Sodium dodecylsulfate (SDS) and dodecyltrimethylammonium bromide (DTAB) were obtained from Sigma and purified by repeated recrystallizations from ethanol and an acetone/ethanol mixture, respectively. The deionized water had a resistivity of 2.5×10^6 ohm · cm.

1-butanol (C_4E_0) and ethylene glycol mono-n-butyl ether (C_4E_1) were reagent grade solvents from the Fisher Chemical Company; they were purified by two distillations and stored over molecular sieves. Diethylene glycol mono-n-butyl ether (C_4E_2, Aldrich, Spectroscopic Grade) was used as received. Triethylene glycol mono-n-butyl ether (C_4E_3) was obtained from Tokyo Kasei and used without further purification.

All alcohol/water mixed solvent systems were prepared on a molality basis; the surfactant solutions made up directly in the mixed solvent. The CMC values of SDS/alcohol mixed micellar systems were obtained from conductivity measurements, using an Industrial Instruments resistance bridge, operating at 1000 Hz, and a dip cell ($K_{cell} = 1.0$ cm^{-1}). The CMC values in DTAB/alcohol mixed micellar systems were determined from EMF titrations using a PVC membrane electrode responsive to the cationic amphiphile (*39,40*), in combination with a calomel reference electrode. The EMF titrations were carried out on an automatic titration system consisting of a Dosimat automatic buret (Metrohm) and a Keithley high impedance electrometer (model A614), controlled by an IBM PC.

All spin-lattice relaxation times (T_1's) were measured on freshly prepared solutions. The NMR experiments and the solution preparation have been described previously (*25*).

Solution preparation for the luminescence quenching experiments was as follows. A small amount of a 0.0050 molal pyrene/ethanol solution was placed in a small flask and the solvent allowed to evaporate, usually overnight, depositing the pyrene as a thin film on the bottom of the vessel. The stock solution of the surfactant/mixed solvent system was prepared directly in the flask containing the pyrene, and it was stirred for at least four hours to ensure complete dissolution of the pyrene in the surfactant solution. This method of solution preparation was found previously (3,28,32) to be a very effective means of dissolving hydrophobic fluorescent probes (i.e., pyrene) into a surfactant solution. The quencher solutions were prepared from one-half of the stock surfactant/probe solutions by adding the desired amount of CP^+ to the solution, and stirring for 1-2 hours. Mixtures at specific quencher concentrations were prepared by mixing portions of the two stock solutions. Steady-state pyrene fluorescence emission spectra were recorded at room temperature ($\approx$ 23°C) on a Perkin-Elmer MPF-66 spectrophotometer, using an excitation wavelength of 338 nm and scanning the emission from 350 to 500 nm. The I_1/I_3 ratios were measured directly from the spectra. The intensity of the emission at 373 nm was used in the plots of $\ln(I_o/I)$ versus the quencher concentration.

Results and Discussion

Anionic Surfactant/Alkoxyethanol Mixed Micelles. CMC values for SDS/alkoxyethanol mixed micelles, obtained from the breaks in the conductance vs. $C_{surf,t}$ plots, are presented in Table 1 and plotted in Figure 2 as a function of the total concentration of alcohol. As expected, the CMC values of SDS/alkoxyethanol mixed micelles decrease as the concentration of the alcohol is increased. More importantly, it is evident from Figure 2 that for a given alcohol concentration, there is a steady decrease in the CMC values from SDS/C_4E_0 mixed micelles to SDS/C_4E_3 mixed micelles. This effect of added ethoxylate groups is in agreement with the previous results of Manabe at al. (12) and Yamashita et al. (16).

Shinoda (1) has noted that the decrease in the CMC of alcohol/surfactant mixed micelles is linear with the total concentration of alcohol, c_a. From the data in Table 1, the rates of change of the CMC of SDS/alkoxyethanol mixed micelles with the total concentration of alkoxyethanol, d CMC/d c_a values, are calculated to be 0.020 for C_4E_0, 0.042 for C_4E_1, 0.065 for C_4E_2, and 0.101 for C_4E_3/SDS mixed micelles. The rates of change can be related to the free energy of transfer of the alcohol hydrophobic group from the aqueous phase to the micellar phase (1)

$$\ln \frac{d\,CMC}{d\,c_a} = \frac{mw}{kT} + c_o \qquad (5)$$

where w is the transfer free energy of the alcohol hydrophobic group from the aqueous to the micellar phase, and c_o is a constant related to the free energy of micellization of the pure surfactant micelles. For the series $C_4E_0 \rightarrow C_4E_3$, we can apply equation 5 to the d CMC/d c_a values for the series of alkoxyethanols in SDS micelles and obtain an estimate of the transfer free energy of the EO group, w_{EO}. From the d CMC/d c_a values reported above, we estimate w_{EO} to be 0.53 ± 0.10 kT or 1.3 ± 0.3 kJ mol^{-1}.

The degree of solubilization, p, of the cosurfactant in surfactant micelles, defined by equation 1, is obtained from NMR relaxation experiments as described previously (24,25). These values are presented in Table 2 for alkoxyethanols in SDS and DTAB micelles. From the degree of solubilization, we can calculate the mole fraction based distribution coefficient, K_x, defined as follows

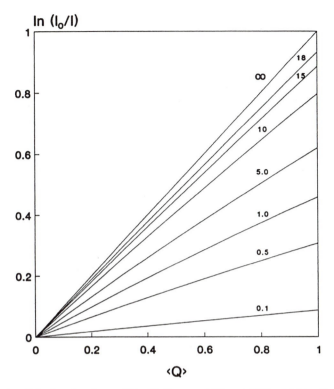

Figure 1. Calculated values of Ln (I_o/I) vs. $<Q>$ at different kinetic ratios, $k_q\tau_o$.

Table 1. CMC values ($\pm$ 0.5 mmolal) for SDS/Alkoxyethanol Mixed Micelles as a Function of the Total Concentration of Added Alcohol

c_a/molal	C_4E_0	C_4E_1	C_4E_2	C_4E_3
0.0000	8.10	8.10	8.10	8.10
0.0025	----	----	----	7.82
0.0050	----	----	----	7.55
0.0075	----	----	----	7.34
0.0100	----	----	7.54	7.08
0.0150	----	----	----	----
0.0200	----	----	6.85	
0.0250	7.58	7.06	----	
0.0300	----	----	6.24	
0.0500	7.18	6.21		
0.0750	6.64	5.09		
0.1000	6.02	4.56		

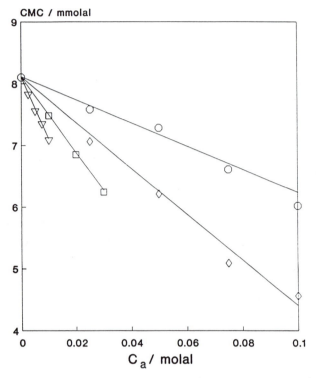

Figure 2. CMC values/mmolal for SDS/alkoxyethanol mixed micelles as a function of c_a (est'd error $\pm$ 0.2 mmolal). $\bigcirc$ C_4E_0, $\diamond$ C_4E_1, $\square$ C_4E_2, ∇ C_4E_3.

Table 2. Distribution Coefficients and Free Energies of Transfer for Several Alcohols in SDS[1] and DTAB[2] Micellar Solutions

Alcohol	p	K_x	ΔG°_t / kJ mol^{-1}
		SDS	
C_4E_0	0.42 ± 0.04	134 ± 24	-12.1 ± 0.4
C_4E_1	0.54 ± 0.03	219 ± 29	-13.4 ± 0.3
C_4E_2	0.66 ± 0.03	364 ± 47	-14.6 ± 0.3
C_4E_3	0.72 ± 0.03	487 ± 71	-15.3 ± 0.4
		DTAB	
C_4E_0	0.32 ± 0.04	128 ± 32	-12.4 ± 0.4
C_4E_1	0.37 ± 0.07	163 ± 33	-13.1 ± 0.7
C_4E_2	0.32 ± 0.07	141 ± 42	-12.7 ± 0.7
C_4E_3	0.31 ± 0.07	131 ± 41	-12.7 ± 0.7

1. $C_{SDS} = 0.243$ m, T = 298 K, $C_{PROXYL} = 0.010$ m, $C_a \approx 0.050$ m.

$$K_x = \frac{X_{mic}}{X_{aq}} \tag{6}$$

X_{mic} and X_{aq} are the mole fractions of solubilizate in the micellar and aqueous phases, respectively, and are calculated as follows

$$X_{mic} = \frac{pC_a}{pC_a + C_{surf,mic}} \; ;$$

$$X_{aq} = \frac{(1-p)C_a}{C_{aq}} \tag{7}$$

where $C_{surf,mic}$ is the surfactant concentration in the micellar phase, and C_{aq} is the heavy water concentration. It should be noted here that the p-value of the alcohol is dependent on the concentration of surfactant; however, K_x is an equilibrium constant and deviates from a thermodynamic equilibrium constant only by neglecting activity coefficients. As an example, Carlfors and Stilbs have determined the distribution equilibria for n-alcohols in SD micellar solutions, at increasing concentrations of surfactant. Their data confirm that, although the p-values increased with increasing surfactant concentration, the distribution coefficients, K_x's, calculated from the p-values in the manner described above, were independent of the amount of surfactant (24). This indicates that in the range of surfactant concentrations investigated, K_x is a true thermodynamic equilibrium constant. The free energy of transfer of the alcohol from the aqueous phase to the micellar phase can then be calculated from the relation

$$\Delta G_t^o = -RT \ln K_x \tag{8}$$

The calculated values of the distribution coefficients, along with the free energy of transfer of the alcohol from the aqueous to the micellar phase, are also presented in Table 2. It can be seen from Table 2 that the p-values of alkoxyethanols in SDS micelles increase with an increase in the number of EO groups. These results are interesting, in that they indicate alkoxyethanols have a greater preference for the micelles than their monohydroxy counterparts, even though the ethoxylated alcohols are more soluble in water than the corresponding n-alcohols (14) The transfer free energy of the EO group can be estimated from the ΔG_t^o data in Table 2. The constant increment in the transfer free energy of ethoxylated alcohols is 1.1 ± 0.4 kJ mol^{-1}. It should be noted that this estimate of the transfer free energy of the EO group compares well with the value $w_{EO} = 1.3 \pm 0.3$ kJ mol^{-1}, calculated above from the dependence of the CMC on the concentration of alcohol.

The emission intensities of micellar solubilized pyrene at 373 nm were plotted against the quencher concentration, [Q], according to equation 2. The slope of these plots is the reciprocal of the micelle concentration, [M]. The slope and the relative error in the slope were calculated by least-squares methods. Good linearity was found for all plots of $\ln(I_o/I)$ versus [Q] (the relative error in the slope was generally around 2-3%). From the micellar concentrations, the aggregation numbers of the surfactant were calculated from the relationship

$$N_s = \frac{[S]_{mic}}{[M]} \approx \frac{C_t - CMC}{[M]} \tag{9}$$

Aggregation numbers of the surfactant, N_s, for 0.0500 molal SDS/alcohol mixed micelles are presented in Table 3 and plotted in Figure 3. The errors reported here are calculated from the relative error in the slope of the plot of $\ln(I_o/I)$ vs. [Q]

Table 3. Aggregation Numbers of Surfactant and Alcohol[1]
for 0.0500 molal SDS/Alkoxyethanol Mixed Micelles as a
Function of the Total Solution Concentration of Alcohol

c_a/molal	C_4E_0			C_4E_1			C_4E_2			C_4E_3		
	N_s	N_a	I_1/I_3	N_s	N_a	I_1/I_3	N_s	N_a	I_1/I_3	N_s	N_a	I_1/I_3
0.000	66	0	1.26	66	0	1.26	66	0	1.26	66	0	1.26
0.025	--	--	----	--	--	----	53	9	1.21	48	10	1.23
0.050	59	9	1.20	53	11	1.18	46	15	1.21	41	16	1.23
0.075	--	--	----	--	--	----	42	21	1.21	37	22	1.24
0.100	55	16	1.19	47	20	1.18	38	26	1.22	36	29	1.23
0.150	48	21	1.14	42	27	1.18	33	32	1.23	30	36	1.23
0.200	46	26	1.12	34	29	1.15	30	38	1.21	27	43	1.23
0.250	43	31	1.11	31	33	1.15	27	44	1.22	--	--	----
0.300	41	34	1.10	29	38	1.14	26	51	1.20	20	51	1.24
0.400	35	41	1.07	23	42	1.14	--	--	----	--	--	----
0.500	30	44	1.05	20	46	1.14	--	--	----	--	--	----

1. $N_s \pm 2$, $N_a \pm 6$.

calculated at the 90% confidence level, regarding all deviations from N_s as random errors. The aggregation number for 0.0500 molal SDS micelles, in the absence of added alcohol, is in excellent agreement with both time-resolved and static experiments involving the same probe/quencher pair (34,35). The alcohol aggregation numbers, N_a, were calculated from equation 4 above, using the micellar concentrations determined via the static fluorescence quenching experiment and the distribution constants obtained from the NMR paramagnetic relaxation experiment in 0.243 m SDS micelles. The reported errors in N_a reflect the sum of the relative errors in the micellar concentration, [M], and the distribution constants (p-values) calculated at the 90% confidence interval, regarding all deviations from the calculated value of N_a as random errors. These numbers are also presented in Table 3. The ratio of the intensities of the first and third peaks of the pyrene emission spectrum, I_1/I_3, are also presented in Table 3. The errors in the I_1/I_3 ratios are estimated to be on the order of 1-2%.

It can be seen that for all alcohols studied in the present paper, N_s decreases when alcohols are added to 0.0500 molal SDS micelles. It is also apparent from Table 3 that N_a increases with an increase in the total alcohol concentration, in agreement with previous investigators (31-34). For a given alcohol concentration, $C_{a,t}$, N_a is greatest for C_4E_2 and C_4E_3, and is smaller for C_4E_0 and C_4E_1. Since these alcohols all possess the same hydrophilic group, i.e., the hydroxyl group, the observed differences in the increase in N_a reflect the differences in the properties of the alkyl or alkyl + EO chain.

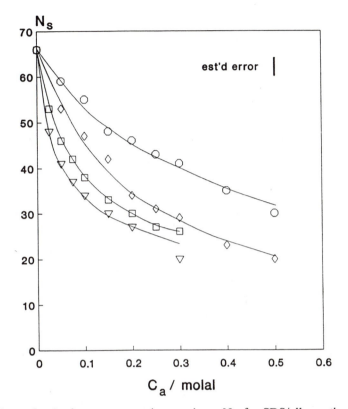

Figure 3. Surfactant aggregation numbers, N_s, for SDS/alkoxyethanol mixed micelles as a function of c_a (est'd error $\pm$ 2).
$\bigcirc$ C_4E_0, $\diamond$ C_4E_1, $\square$ C_4E_2, $\triangledown$ C_4E_3.

For the series SDS/$C_4E_0 \rightarrow C_4E_3$, at the same total concentration of alcohol, the N_s values decrease as the number of EO groups in the alcohol is increased. At the same time, the N_a values tend to increase with an increase in the number of EO groups in the alcohol. For C_4E_0 and C_4E_1, N_t, the total aggregation number, tends to increase slowly with an increase in the alcohol concentration, while for C_4E_2 and C_4E_3, N_t remains relatively constant. This indicates that for all the 0.0500 m SDS/alcohol mixed micelles studied in this paper, the overall core volume of the micelle remains relatively constant, or decreases slightly. At the present range of alcohol concentrations, these results are not unexpected since the micelles would tend to retain their spherical geometry (*38*).

Additional information about the structure of these mixed micelle can be obtained from an analysis of the pyrene I_1/I_3 ratios in Table 3. For mixed micelles of SDS and C_4E_0 or C_4E_1, as the concentration of alcohol in the solution is increased, the pyrene I_1/I_3 ratios decrease indicating a decrease in the micropolarity sensed by the pyrene probe. Zana (*3*) and Thomas (*26*) have interpreted this trend in terms of a looser micelle structure, allowing the luminescent probe (in this case pyrene) to penetrate further into the micelle, thereby sensing a less polar environment. Recently, Gao et al. (*41*) have examined the distribution of aromatic probe molecules, like pyrene, in anionic and cationic micelles. Their findings indicated that SDS solubilized probe molecules e.g., pyrene and naphthalene, are evenly distributed throughout the micelle.

When a short chain alcohol (e.g., C_4E_0) is added to the micelles, this distribution is not expected to be altered, and, hence, the probe molecule would still sample the entire micelle volume, including the region close to the surface of the SDS micelle. Since the I_1/I_3 ratio decreases, this indicates that the presence of the alcohol blocks water penetration into the micelle. For the SDS/C_4E_2 and the SDS/C_4E_3 mixed micellar systems, the I_1/I_3 ratios are invariant with an increase in the concentration of cosurfactant (alcohol). An interesting trend in the I_1/I_3 ratios can be observed upon changing the cosurfactant from C_4E_0 to C_4E_3 in that the I_1/I_3 ratios increase upon addition of EO groups to C_4E_0. This indicates that the pyrene probe senses a more polar environment, possibly due to the penetration of the polar EO groups into the micellar interior.

The decrease in the free energy of micellization, as calculated from the decrease in the CMC values, and the decrease in N_s, is larger for the ethoxylated alcohol/SDS mixed micelles than for the corresponding SDS/C_4E_0 mixed micelles. Two explanations are consistent with these results. The first is that the EO group occupies a larger surface area in the head group region of the mixed micelles; this reduces the electrostatic interactions between the surfactant head groups by increasing the distance between them. Here, w_{EO} would represent an electrostatic contribution to the free energy reduction (11). This interpretation was used by Manabe et al.(12) to account for the observed CMC values of SDS/alkoxyethanol mixed micelles as a function of the EO chain length of the alcohol, and would easily account for the observed decrease in N_s.

Another interpretation of the trends in the CMC and N_s values is that the EO group has a small, but significant contribution to the hydrophobic interactions. In contrast to what was proposed by Manabe et al. (12), this effect would lead to an increase in the solubilization of ethoxylated alcohols into SDS micelles as the number of EO groups is increased. Our data for the distribution constants of alkoxyethanols in SDS micelles are consistent with this explanation. It appears that the trends in the CMC and N_s values of SDS/ethoxylated alcohol mixed micelles may be due, at least in part, to the contribution of the EO group to the hydrophobic interactions. This would lead to an increased solubilization of alkoxyethanols in micelles over the corresponding n-alcohols. Desnoyers et al. (14) stated that C_4E_0 and n-pentanol (C_5E_0) are good cosurfactants for microemulsion systems due to their rather even distribution between oil and water phases. This may explain the acceptance of C_4E_1 as a microemulsion cosurfactant, since its SDS micelle-water distribution coefficient is midway between that of C_4E_0 and C_5E_0 (25).

Cationic Surfactant/Alkoxyethanol Mixed Micelles. CMC values for the DTAB/alkoxyethanol mixed micelles, determined from the break in the EMF_{DTA+} vs. log $C_{surf,t}$ curve, are plotted in Fig. 4 as a function of the total alcohol concentration. It can be seen from Fig. 4 and the tabulated CMC values (Table 4) that, in this case, the addition of an EO group to the alcohol does not result in an additional decrease of the CMC values of the mixed micelles at the same concentration of alcohol. From our data, we calculate d CMC/d c_a values of 0.0194 for DTAB/C_4E_0 mixed micelles (in good agreement with the value reported by Zana (11)), 0.0248 for DTAB/C_4E_1, 0.0237 for DTAB/C_4E_2, and 0.0231 for DTAB/C_4E_3 mixed micelles. Unlike what was observed for SDS/alkoxyethanol mixed micelles above, the d CMC/d c_a values are relatively insensitive to the number of EO groups in the added alcohol, indicating a negligible free energy contribution from the EO group in the formation of cationic DTAB/alkoxyethanol mixed micelles.

The results for the distribution of alkoxyethanols DTAB micelles, again determined using the NMR paramagnetic relaxation method, are presented in Table 4, together with the distribution coefficients (K_x) and free energies of transfer. It is evident that the transfer free energy is independent of the number of EO groups are in the alcohol, in contrast to what we have observed in SDS micelles. This is an indication that the EO group does not contribute to the interaction between ethoxylated alcohols and cationic micelles.

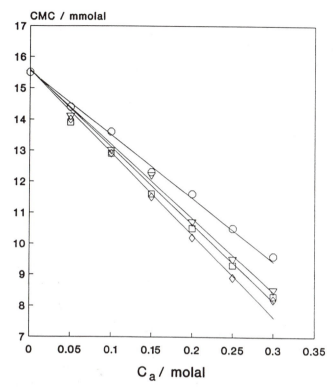

Figure 4. CMC values/mmolal for DTAB/alkoxyethanol mixed micelles as a function of c_a (est'd error $\pm$ 0.3 mmolal). $\bigcirc$ C_4E_0, $\diamond$ C_4E_1, $\square$ C_4E_2, ∇ C_4E_3.

The aggregation numbers of the surfactant, alcohol, the total aggregation number, and the I_1/I_3 ratios of the micellar solubilized pyrene probe for mixed micelles composed of the cationic surfactant DTAB with $C_4E_0 \rightarrow C_4E_3$ are presented in Table 5. The N_s values are plotted in Fig. 5. The N_s value for DTAB in the absence of alcohol is in good agreement with N_s obtained from time-resolved experiments (*42*). As expected, N_s decreases with an increase in the concentration of alcohol for all the DTAB/alkoxyethanol mixed micelles examined here. It is of particular interest to observe that, similar to the trend in the CMC values reported above, changing the alcohol from C_4E_0 to C_4E_1 has a small effect on the aggregation numbers determined from the static quenching experiment. Changing the cosurfactant from C_4E_1 to C_4E_2 or C_4E_3 has a negligible effect on the determined aggregation numbers. Additional information can be obtained from the I_1/I_3 ratios of the pyrene probe in these mixed micellar systems. The pyrene I_1/I_3 ratio is higher in DTAB micelles than in SDS micelles, which can be attributed to the well-known specific interaction between aromatic solubilizates and the head group region of cationic micelles (*3,28*). With the addition of alcohols to DTAB micelles, the I_1/I_3 ratios decrease as noted above. However, the decrease is significantly smaller than was observed for SDS and the difference in the series DTAB/$C_4E_0 \rightarrow C_4E_3$ is only marginally significant, indicating that the EO groups do not penetrate into the micellar interior.

Table 4.   CMC values ($\pm$ 0.5 mmolal) for DTAB/Alkoxyethanol
Mixed Micelles as a Function of the Total Concentration
of Added Alcohol

c_a/molal	C_4E_0	C_4E_1	C_4E_2	C_4E_3
0.000	15.5	15.5	15.5	15.5
0.050	14.4	14.0	13.9	14.1
0.100	13.6	12.9	12.9	13.0
0.150	12.3	11.5	11.6	12.2
0.200	11.6	10.2	10.5	10.7
0.250	10.5	8.9	9.3	9.5
0.300	9.6	8.2	8.3	8.5

In contrast to the results presented above for the SDS/ethoxylated alcohol mixed micelles, it appears that the EO group makes little contribution to the formation of DTAB/alkoxyethanol mixed micelles. The observed difference in the interaction of ethoxylated alcohols with anionic and cationic micelles parallels the interaction of poly(ethylene oxide), or PEO, and other neutral polymers with anionic and cationic micelles, i.e., PEO solubilizes in (or binds to) anionic micelles, but does not interact with cationic micelles (40). As well, Baglioni et al. (43) have probed the location of EO groups in mixed micelles composed of anionic or cationic surfactants with nonionic surfactants containing EO groups. Their conclusions regarding the location and interaction of the EO group in the mixed micelles, obtained from electron spin-echo modulation studies of DOXYL-stearic acid probes in the mixed micelles, are in accord with ours.

Conclusions

From the results of EMF, NMR, and static fluorescence quenching experiments on SDS/alkoxyethanol and DTAB/alkoxyethanol mixed micelles, the following conclusions can be drawn:
1) In anionic surfactant/alkoxyethanol mixed micellar systems, the contribution of the EO group to the hydrophobic interactions results in an decrease of the free energy of transfer of alkoxyethanols from the aqueous to the micellar phase as the number of EO groups in the alcohol are increased. As well, the CMC values and the N_s values of SDS/alkoxyethanol mixed micelles decrease as the number of EO groups in the alcohol are increased, at a given concentration of alcohol, indicating a decrease in the free energy of micellization which may at least in part be due to the contribution of the EO group to the hydrophobic interactions.
2) In cationic surfactant/alkoxyethanol mixed micelles, the free energy of transfer is constant for a series of ethoxylated alcohols and the CMC values of DTAB/alkoxyethanol mixed micelles, within the series $C_4E_1 \rightarrow C_4E_3$, are independent of the number of EO groups in the alcohol. As well, the surfactant and alcohol aggregation numbers are insensitive to the number of EO groups in the alcohol. These results indicate that the EO group has a negligible contribution to the interactions between cationic surfactants and ethoxylated alcohols.

**Table 5. Aggregation Numbers of Surfactant and Alcohol[1]
for 0.0750 molal DTAB/Alkoxyethanol Mixed Micelles as a
Function of the Total Solution Concentration of Alcohol**

c_a/molal	C_4E_0			C_4E_1			C_4E_2			C_4E_3		
	N_s	N_a	I_1/I_3	N_s	N_a	I_1/I_3	N_s	N_a	I_1/I_3	N_s	N_a	I_1/I_3
0.000	52	0	1.40	52	0	1.40	52	0	1.40	52	0	1.40
0.100	47	14	1.39	40	14	1.39	45	13	1.39	44	11	1.41
0.200	40	23	1.39	35	22	1.36	35	19	1.36	34	17	1.39
0.300	38	31	1.38	31	28	1.36	31	22	1.36	28	20	1.39
0.400	34	34	----	28	31	1.31	32	26	1.35	26	24	1.37
0.500	32	45	1.30	28	39	1.30	26	23	1.34	20	22	1.36

1. $N_s \pm 2$, $N_a \pm 6$

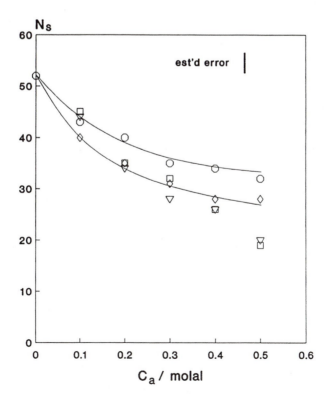

Figure 5. Surfactant aggregation numbers, N_s, for DTAB/alkoxyethanol
mixed micelles as a function of c_a (est'd error $\pm$ 2).
○ C_4E_0, ◇ C_4E_1, □ C_4E_2, ▽ C_4E_3.

Acknowledgements

This work is supported by grants from the Natural Sciences and Engineering Council of Canada (NSERC). DGM is grateful to NSERC and the Killam Foundation for graduate scholarships. The authors thank R. E. Wasylishen, R. Palepu, and Z. Gao for their interest in this project. NMR measurements were carried out at the Atlantic Region Magnetic Resonance Centre (ARMRC) at Dalhousie University.

Literature Cited

1. Shinoda, K. *J. Phys. Chem.* **1954**, *58*, 1136.
2. Herzfeld, S. H.; Corrin, M. L.; Harkins, W. D. *J. Phys. Chem.* **1950**, *54*, 271.
3. *Surfactant Solutions: New Methods of Investigation*; Zana, R., Ed.; Marcel Dekker: New York, 1987.
4. Chachaty, C. *Prog. NMR Spectroscopy* **1987**, *19*, 183.
5. Thomas, J. K. *The Chemistry of Excitation at Interfaces*; A. C. S. Monograph 181; American Chemical Society: Washington, D. C. 1984.
6. Anacker, E. W. in *Cationic Surfactants*; Jungermann, E., Ed.; Marcel Dekker: New York, 1970.
7. Lindman, B.; Wennerström, H. in *Topics in Current Chemistry*; Dewar, M. J. S. et al., Eds.; Springer-Verlag: Berlin, 1980.
8. *Physics of Amphiphiles, Micelles, Vescicles, and Microemulsions*; Degiorgio, V.; Corti, M., Eds.; Italian Physical Society; Elsevier: Amsterdam, 1985.
9. Quirion, F.; Desnoyers, J. E. *J. Colloid Interface Sci.* **1987**, *115*, 176.
10. Desnoyers, J. E.; Hétu, D.; Caron, G. *Colloids and Surfaces* **1989**, *35*, 169.
11. Zana, R.; Yiv, S.; Straizelle, C.; Lianos, P. *J. Colloid Interface Sci.* **1981**, *80*, 208.
12. Manabe, M.; Koda M. *Bull. Chem. Soc. Jpn.* **1978**, *51*, 1599.
13. Boström, G.; Backlund, S.; Blokhus, A. M.; Hoiland, H. *J. Colloid Interface Sci.* **1989**, *128*, 169.
14. Desnoyers, J. E.; Quirion, F.; Hétu, D.; Perron, G. *Can. J. Chem. Eng.* **1983**, *61*, 672.
15. *Phenomena in Mixed Surfactant Systems*; Scamehorn, J. F., Ed.; A. C. S. Symposium Series No. 311; 1986.
16. Yamashita, F.; Perron, G.; Desnoyers, J. E.; Kwak, J. C. T. *J. Colloid Interface Sci.* **1986**, *114*, 548.
17. Yamashita, F.; Perron, G.; Desnoyers, J. E.; Kwak, J. C. T. unpublished results.
18. Vikingstad, E. *J. Colloid Interface Sci.* **1979**, *72*, 68.
19. Vikingstad, E. *J. Colloid Interface Sci.* **1980**, *73*, 260.
20. Stilbs, P. *J. Colloid Interface Sci.* **1982**, *87*, 385.
21. Stilbs, P. *J. Colloid Interface Sci.* **1982**, *89*, 547.
22. Carlfors, J.; Stilbs, P. *J. Colloid Interface Sci.* **1985**, *104*, 489.
23. Stilbs, P. *Prog. NMR Spectroscopy* **1987**, *19*, 1.
24. Gao, Z.; Wasylishen, R. E.; Kwak, J. C. T. *J. Phys. Chem.* **1989**, *93*, 2190.
25. Gao, Z.; Marangoni, D. G.; Labonté, R.; Wasylishen, R. E.; Kwak, J. C. T. *Colloids and Surfaces* **1990**, *45*, 269.
26. Lianos, P.; Zana, R. *Chem. Phys. Lett.* **1980**, *72*, 171.
27. Georges, J. *Spectrochemica Acta Rev.* **1990**, *13*, 27.
28. Thomas, J. K. *The Chemistry of Excitation at Interfaces*; American Chemical Society: Washington, 1984, Chs. 5-7.
29. Thomas, J. K. *Chemical Reviews* **1980**, *80*, 283.

30. Turro, N. J.; Yekta, A. *J. Am. Chem. Soc.* **1978**, *100*, 5951.
31. Almgren, M.; Swarup, S. *J. Colloid Interface Sci.* **1983**, *91*, 256.
32. Malliaris, A. *Adv. Colloid Interface Sci.* **1987**, *27*, 153.
33. Almgren, M.; Swarup, S. *J. Phys. Chem.* **1982**, *86*, 4212.
34. Almgren, M.; Swarup, S. *J. Phys. Chem.* **1983**, *87*, 876.
35. Velazquez, M. M.; De Costa, S. M. B. *J. Chem. Soc., Faraday Trans.* **1990**, *86*, 4043.
36. Malliaris, A.; Lang, J.; Zana, R. *J. Chem. Soc., Faraday Trans. 1* **1986**, *82*, 109. *J. Colloid Interface Sci.* **1986**, *110*, 237.
37. Luo, H.; Boens, N.; Van der Auweraer, M.; De Schryver, F. C.; Malliaris, A. *J. Phys. Chem* **1989**, *93*, 3244.
38. Roelants, E.; DeSchryver, F. C. *Langmuir*, **1987**, *3*, 209.
39. Hayakawa, K.; Kwak, J. C. T. *J. Phys. Chem.* **1982**, *86*, 3866.
40. Hayakawa, K.; Kwak, J. C. T. in *Cationic Surfactants*; Rubingh, D.; Holland, P. M., Eds.; Surfactant Science Series, Vol. 37; Marcel Dekker: New York, 1990, Ch. 5.
41. Wasylishen R. E.; Kwak, J. C. T.; Gao, Z.; Verpoorte, E.; MacDonald, J. B.; Dickson, R. M. *Can. J. Chem.* **1991**, *69*, 822.
42. Lianos, P.; Zana, R. *J. Colloid Interface Sci.* **1981**, *84*, 100.
43. Baglioni, P.; Rivara-Minten, E.; Kevan, L. *A. C. S. Symp. Ser.* **1991**.

RECEIVED February 26, 1992

Chapter 12

Lipase Catalysis of Reactions in Mixed Micelles

Donn N. Rubingh and Mark Bauer

Corporate Research Division, The Procter & Gamble Company, Cincinnati, OH 45239

Enzyme catalyzed reactions on micellized substrates are much less well understood than for soluble substrates. We develop expressions for the lipase catalyzed rate of hydrolysis of surfactant substrates co-micellized with non-hydrolyzable surfactant molecules. The equations contain a term involving inhibition by the non-hydrolyzable surfactant molecule as well as an interfacial activation parameter which may be a function of composition. Under appropriate limiting conditions the well known Michaelis-Menten kinetic expression is obtained except that the concentration of the substrate is that in the micelle. We find a number of surfactant esters and mixtures of surfactant esters show an abrupt change in rate of lipase catalyzed hydrolysis at the cmc. This suggests that both single component and mixed micelles are capable of activating lipase catalysis of ester hydrolysis. The concentration dependence of the rate of hydrolysis of surfactant esters in single component micelles is described by a Michaelis-Menten type expression allowing values of the kinetic parameters, k_{cat} and $K_m{}^*$, to be obtained. In mixtures of hydrolyzable and non-hydrolyzable surfactants both competitive inhibition and exclusion (changes in interfacial activation) can affect the hydrolysis rate depending on surfactant structure.

Enzymes are biological catalysts that greatly accelerate the rate of various chemical reactions. For example, the protease BPN' which catalyzes internal peptide bond hydrolysis in proteins, can accelerate the rate of reactions on certain substrates up to 10^9(*1*). The mechanisms of many enzyme-substrate reactions are well characterized, particularly when the substrate is a water soluble compound(*2*). Many excellent

0097–6156/92/0501–0210$06.00/0

books and reviews are available for the mathematical description of enzyme kinetics for soluble substrates(3)(4).

A wide variety of enzymes, broadly divided into various classes depending on the nature of the reaction they catalyze, are found in nature. Well known examples include proteases for amide bond hydrolysis, esterases for ester bond hydrolysis, and cellulases for cellulose hydrolysis. Some enzymes function on soluble substrates; however, many function on insoluble substrates or, stated differently, at the interface between the insoluble substrate particle and the aqueous solution containing the enzyme. Cellulases and lipases are classes of enzymes where this is true. The mechanisms and mathematical descriptions of enzyme kinetics for insoluble substrate hydrolysis are less well defined than for soluble substrate hydrolysis.

Lipases are active against insoluble esters, particularly insoluble fats or triglycerides(5). Lipases share with phospholipases an interesting feature in that they are relatively inactive until brought in contact with an interface. This phenomenon often called interfacial activation is characterized by an abrupt increase in activity upon crossing a solubility boundary (ie. at the first appearance of an interface) and was first clearly demonstrated by Sarda and Desnuelle(6) for porcine pancreatic lipase.

In the past few years the molecular basis for interfacial activation is beginning to be understood. It has been shown by X-ray diffraction that the catalytic triad of *Mucor mehei* lipase is geometrically identical to serine proteases; however, it is covered by a "flap" which prevents access to the active site(7). Subsequent structural studies on this enzyme have shown that the "flap" is opened when an inhibitor is bound in the active site exposing a hydrophobic patch(8). The implication is that interfacial activation involves a conformational change, upon adsorption to a sufficiently hydrophobic interface, and this change allows entry of the substrate to be hydrolyzed.

Micelle forming materials have been recognized as possible substrates for lipases and phospholipases(5). Most studies appear to have been carried out on phospholipases(9) which share with lipases the ability to be interfacially activated, but differ from lipases in other important respects(10). We have found that a number of surfactant esters, when used as lipase substrates, exhibit a rate discontinuity upon crossing the critical micelle concentration (cmc) (Figure 1). This is similar to the solubility boundary behavior observed by Sarda and Desnuelle. The possibility of using ester linked surfactants as substrates in lipase assays appears not to be widely recognized even though they offer some advantages over emulsions since micellar solutions are thermodynamically stable and reproducibly prepared.

The fact that micelles can activate lipases gave impetus to the idea of using mixed micelles as models to understand the behavior of lipases at oil-water interfaces in systems containing surfactants. The hydrolysis of triglycerides is significantly reduced in the presence of surfactants which limits the efficacy of lipases in certain applications. The mechanism for this reduction in catalytic effectiveness of lipases is not clear. Possibilities include competition between detergent and substrate for the active site, exclusion of the lipase from the substrate by the presence of surfactant at the interface or changed interfacial properties such as increased viscosity due to the surfactant.

We present a theoretical treatment of the enzyme catalyzed hydrolysis of a substrate in a mixed micelle which explicitly recognizes both the competition between the detergent and substrate for the active site as well as the need for interfacial activation. In principle, the two effects can be separated by appropriate application of the model and may shed some light on the more realistic situation where the surfactant and hydrolysis products compete with lipase for the oil-water interface or with substrate for the active site of the enzyme.

Surfactant esters of decanoic acid with pentaoxyethylene and sodium phenol sulfonate were used to verify the proposed description of micellar kinetics. Mixtures of these surfactant substrates with the non-hydrolyzing surfactants $C_{10}E_5$ and $C_{10}E_9$ were studied in an attempt to find the relative importance of competitive inhibition and mechanisms interfering with the activation process to the rate loss in the presence of surfactant .

Experimental

Materials. The substrate $C_{10}OBS$ (Sodium p-Decanoyloxybenzenesulfonate) (Figure 2) was prepared by acylating disodium p-phenol sulfonate with decanoyl chloride in toluene. The product was recrystallized twice from 87:13 methanol/water. The structure and purity were verified by NMR. The substrate $C_{10}E_5$ ester (pentaoxyethylene glycol monodecanoate) (Figure 2) was prepared by acylating pentaoxyethylene glycol (E_5) with decanoyl chloride (C_{10}) in the presence of pyridine. The product was purified by preparative HPLC (Waters Prep500) using 95:5 hexane/isopropanol. Purity was verified by TLC. The detergent $C_{10}E_5$ ether (pentaoxyethylene glycol monodecyl ether) was prepared by NaOH catalyzed alkylation of pentaoxyethylene glycol (excess) with bromodecane. The product was recovered by vacuum distillation and the purity was verified by GC. The detergent $C_{10}E_9$ ether (nonaoxyethylene glycol monodecyl ether) was prepared from $C_{10}E_4Cl$ ($C_{10}E_4$ + $SOCl_2$) and pentaoxyethylene glycol in the presence of NaOH. The product was recrystallized from hexane and the purity was verified by GC. Lipase was purchased from NOVO under the trade name "Lipolase". Lipolase is a 31.5kD single domain glycosylated lipase from *Humicola Lanuginosa*. The commercial sample was further purified by HPLC (Waters 625) over a mono-Q anion exchange column. Purity was verified by SDS-PAGE (sodium dodecyl sulfate polyacrylamide gel electrophoresis).

Methods. Critical Micellar Concentration (cmc) data were obtained by the duNuoy Ring Method of surface tension measurement(*11*). $C_{10}E_5$ ester hydrolysis rates were measured by pH stat(*12*) on a Brinkmann titrimeter system (Metrohm model 665 Dosimat, 614 Impulsomat, and 632 pH meter) at pH 9.0 in 50mM NaCl, 10mM $CaCl_2$ and 25°C. Rates were measured as moles OH^- ions/minute necessary to maintain the pH at 9.0. The number of moles of OH^- added correlates with the number of moles of bonds hydrolyzed by the enzyme. Data acquisition was accomplished by Laboratory Technologies Corporation's "Labtech Acquire" software package. $C_{10}OBS$ hydrolysis rates were also measured on the pH stat

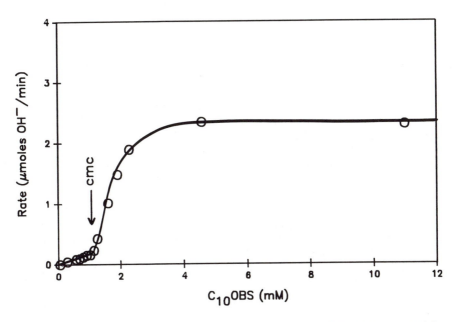

Figure 1. Rate of $C_{10}OBS$ hydrolysis by 8ppm lipolase in 100mM NaCl, pH 9.0 as a function of substrate concentration. The cmc, indicated by arrow, was determined by surface tension measurements.

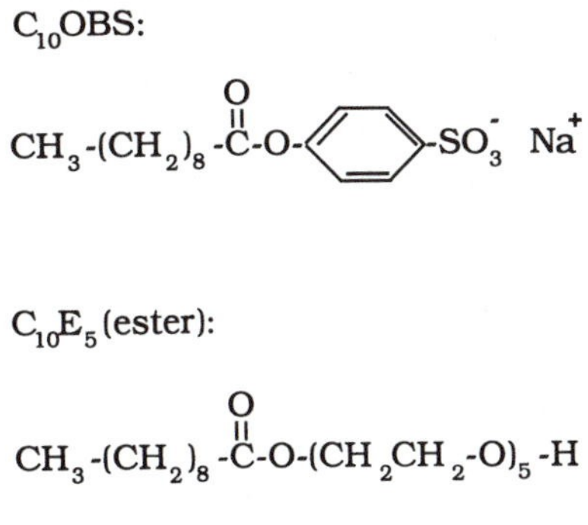

$C_{10}OBS$:

$$CH_3\text{-}(CH_2)_8\text{-}\overset{\displaystyle O}{\overset{\displaystyle \|}{C}}\text{-O-}\langle\!\!\langle\,\,\rangle\!\!\rangle\text{-}SO_3^- \ Na^+$$

$C_{10}E_5$ (ester):

$$CH_3\text{-}(CH_2)_8\text{-}\overset{\displaystyle O}{\overset{\displaystyle \|}{C}}\text{-O-}(CH_2CH_2\text{-}O)_5\text{-H}$$

Figure 2. Structural formulas of the micellizing substrates $C_{10}OBS$ and $C_{10}E_5$ ester.

system described above in 100mM NaCl pH 9.0 and on a Beckmann DU-70 spectrophotometer at 252nm in Tris [Tris(hydroxymethyl)aminomethane] buffer at pH 9.0, 100mM NaCl and 25°C.

Theoretical Development

A few treatments for the kinetics of hydrolysis of substrate incorporated into a mixed micelle exist(13)(14). These have primarily addressed the action of phospholipase on phospholipids incorporated into micelles. One of these differs from the present in that competitive binding between surfactant and substrate (the phospholipid) is not considered(13). The other, although recognizing the possibility, assumes a specific adsorption isotherm (Langmuir) for the enzyme to the surface(14). Neither addresses the solution properties of the substrate or detergent, particularly as it relates to the individual cmcs, and how these influence the composition of the mixed micelle. This is a result of the fact that they were developed to describe insoluble amphiphilic compounds (ie. phospholipids) rather than the soluble substrates and detergents considered here.

Our derivation follows the development of Ransac et. al.(15) which describes the effect of water soluble inhibitors on enzyme action at an oil-water interface. In the present treatment, however, the mixed micelle rather than a mixed interface of substrate and inhibitor (or detergent) provides the activating "phase". The relevant process we need to consider in this case is the adsorption of the lipase to the micelle resulting in enzyme activation as described by equation 1.

$$E + M \underset{k_d}{\overset{k_p}{\rightleftharpoons}} E^*M \tag{1}$$

Here E^*M denotes the activated lipase adsorbed to a micelle. To simplify notation, we will use E^* for E^*M as the activated enzyme at the micelle surface. We specify the rate constant for the formation of activated enzyme as k_p, since it is related to the rate of penetration of the enzyme into a sufficiently hydrophobic environment for activation to occur. The rate constant for desorption from the micelle is k_d.

The activated lipase molecule has the potential to bind in its active site either the surfactant without an ester linkage (hereafter designated detergent) or the surfactant substrate. The binding of detergent is described by equation 2. Since the detergent cannot be hydrolyzed, it acts as a competitive inhibitor for the binding of substrate. The ratio of the forward and reverse binding reaction is K_D^*.

$$E^* + D \underset{k_{-2}}{\overset{k_2}{\rightleftharpoons}} E^*D \tag{2}$$

The reaction described in equation 2 is postulated to occur between a detergent in the micelle which has been responsible for activation as opposed to a monomeric detergent molecule.

The binding of substrate and subsequent reaction to form products is described in equation 3. The species E*S is analogous to the Michaelis-Menten complex for soluble substrate kinetics although here it is postulated to occur between the substrate co-micellized with detergent in the mixed micelle to which the activated lipase is adsorbed. The products of the reaction are fatty acid and alcohol and are indicated by P in equation 3.

$$E^* + S \underset{k_{-1}}{\overset{k_1}{\rightleftarrows}} E^*S \overset{H_2O}{\underset{k_{cat}}{\longrightarrow}} E^* + P \tag{3}$$

To avoid unnecessary complexity, a more detailed mechanistic description, such as the formation of an acyl enzyme with subsequent deacylation by reaction with water, which is known to occur with proteases and quite probably with lipases as well is omitted. However, it is well known that the general form of the resulting equations are similar although the definition of the various constants differs(16).

The equations for the initial rate of reaction for substrates in a mixed micelle are solved in the standard way. The initial rate of substrate disappearance $(\frac{-d[S]}{dt})$ is given by equation 4.

$$v_o = \frac{-d[S]}{dt} = k_{cat}[E^*S] \tag{4}$$

In the steady state approximation the rate of change of concentration of the Michaelis-Menten complex is equal to zero so that:

$$\frac{d[E^*S]}{dt} = k_1[S][E^*] - (k_{-1} + k_{cat})[E^*S] = 0 \tag{5}$$

Additionally, the conservation of enzyme mass implies that:

$$[E_o] = [E] + [E^*S] + [E^*M] + [E^*D] \tag{6}$$

where $[E_o]$ is the total enzyme concentration and $[E]$ is the free enzyme concentration.

Using the steady state approximation for all enzyme intermediates allows one to obtain relationships among the various species and to solve for $[E^*S]$ in terms of the equilibrium constants in equations 1-3 and the substrate concentration $[S]$ and the total enzyme concentration $[E_o]$. After a certain amount of algebra one obtains the equation for the initial rate of substrate disappearance in the mixed micelle of substrate and detergent shown in equation 7. This can be thought of as a micellar Michaelis-Menten equation.

$$v_0 = \frac{x V_{max}}{\frac{K_m^*}{(c-cmc)}[1 + \frac{R(x)}{(c-cmc)}] + \frac{K_m^*}{K_D^*}(1-x) + x} \qquad (7)$$

Here x is the mole fraction of substrate in the mixed micelle, c is the total concentration (substrate + surfactant), V_{max} is the maximal velocity and is equal to $k_{cat}[E_o]$, cmc is the critical micelle concentration of the mixture, and R(x) is equal to $\frac{k_d}{k_p}M$ where M is the molecular weight of the micelle at composition x. It should be noted that both the penetration and desorption rate constants as well as the micelle molecular weight can depend on the composition x. This is why the parameter R is explicitly written as a function of composition. The physical interpretation of this parameter involves the ability of the enzyme to penetrate the micelle and be activated. The other parameters are K_m^* which is the micellar Michaelis-Menten constant and is given by $K_m^* = \frac{k_{-1} + k_{cat}}{k_1}$ similar to its definition for soluble substrates and K_D^*, the micellar detergent inhibition constant, which is given by $K_D^* = \frac{k_2}{k_{-2}}$

A number of limiting cases of equation 7 are of interest. The first is when there is only one surfactant that is also a substrate. In this case x=1 and equation 7 reduces to:

$$v_0 = \frac{V_{max}}{\frac{K_m^*}{(c-cmc)}[1 + \frac{R(x)}{(c-cmc)}] + 1} \qquad (8)$$

This equation is similar to the standard Michaelis-Menten equation except that it contains the interfacial penetration factor R(x) and the concentration scale begins at zero micelle concentration as opposed to zero substrate concentration as is the case for soluble substrates. Equation 8 can be rewritten in a way that demonstrates this by assuming R(x)<<1, and can therefore be dropped, and defining the substrate concentration as [S] = c-cmc. Equation 8 becomes under these conditions:

$$v_0 = \frac{V_{max}[S]}{K_m^* + [S]} \qquad (9)$$

The effect of the interfacial penetration parameter is shown in Figure 3. At low substrate concentrations finite values of R cause significant deviations from the standard Michaelis-Menten rate (R=0). These differences disappear, however, as substrate concentrations increase and the curves approach the same asymptote, namely V_{max}.

A second case of interest is for mixed systems with concentrations of detergent and substrate well above the cmc. In this case the micelle mole fraction (x) and bulk mole fraction (α) are essentially the same since almost all the surfactant is micellized and the term $\dfrac{R(x)}{c\text{-cmc}}$ is necessarily negligible. Under these conditions equation 7 reduces to:

$$v_0 = \frac{\alpha V_{max}}{\dfrac{K_m^*}{(c\text{-cmc})} + \dfrac{K_m^*}{K_D^*}(1-\alpha) + \alpha} \tag{10}$$

A familiar form of equation 10 can be obtained by multiplying both the numerator and denominator by (c-cmc) which is the concentration of micellized substrate and detergent and recognizing that $\alpha(c\text{-cmc}) = [S_m]$ and $(1-\alpha)(c\text{-cmc}) = [D_m]$ where $[S_m]$ and $[D_m]$ are the concentration of substrate and detergent molecules in the mixed micelle environment. Rearranging results in equation 11 which is the well known expression for the rate in the presence of a competitive inhibitor(*3*).

$$v_0 = \frac{V_{max}[S_m]}{K_m^*\left[1 + \dfrac{[D_m]}{K_D^*}\right] + [S_m]} \tag{11}$$

The effect of competitive inhibition by a detergent as a function of micelle composition, assuming constant penetrability, is shown in Figure 4. When the binding of substrate and detergent are equal, $\dfrac{K_m^*}{K_D^*} = 1$, a linear decrease in rate with mole fraction of detergent in the mixed micelle is observed . If detergent binding is very much favored over substrate, $\dfrac{K_m^*}{K_D^*} > 1$, then small amounts of detergent will rapidly decrease the rate while the converse will be true if substrate binds preferentially over the detergent to the active site.

Some of the most interesting cases may occur when the substrate and the detergent are strongly nonideal. In this case the cmc and the composition of the mixed micelle are controlled by the individual cmcs of the materials and their interactions within the micelle. Mixed surfactant theories(*17*)(*18*) provide the required data for micelle mole fraction, x, and the mixed cmc to apply equation 7.

Results

Figures 1 and 5 give typical responses for the rate of micellizing esters as a function of total concentration. A common feature of all the systems studied is a rapid increase in rate as one crosses the cmc. This correspondence between rate discontinuity and cmc is shown explicitly for the $C_{10}E_5$ ester in Figure 5 using the

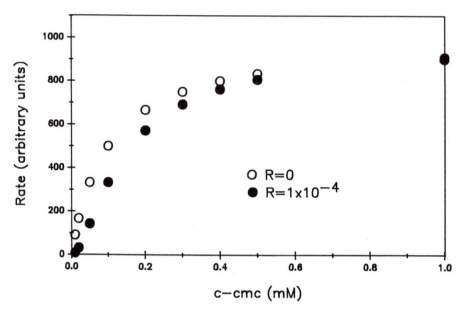

Figure 3. Effect of the exclusion parameter on the concentration dependence of rate. From equation 8 with $V_{max} = 1000$ and $K_m^* = 0.1mM$, R=0 is the normal Michaelis-Menten behavior.

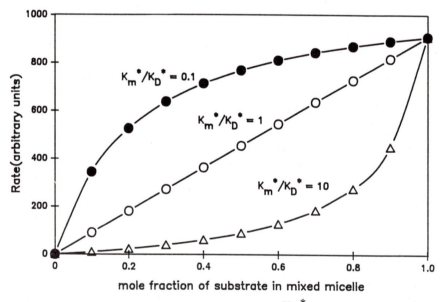

Figure 4. Hydrolysis rates for different values of $\dfrac{K_m^*}{K_D^*}$ from equation 7 with $\dfrac{K_m^*}{(c\text{-}cmc)}[1 - \dfrac{R(x)}{(c\text{-}cmc)}] = 0.1$.

standard surface tension-concentration method to determine the cmc. In addition, we have observed this behavior with PEG-monoleate(a commercial ester detergent from Stepan) and C_9OBS (Data not shown). The maximum in rate at concentrations below the cmc exhibited in Figure 5 for the $C_{10}E_5$ ester was not observed for the other esters; however, detailed studies in this region were not made.

Analysis of the data on pure surfactant systems can be made by rearranging equation 8 in a double-reciprocal or, Lineweaver-Burk, format. This gives the following expression:

$$\frac{1}{v_0} = \frac{1}{V_{max}} + \frac{K_m{}^*}{V_{max}}[\frac{1}{(c-cmc)}] + \frac{K_m{}^*R}{V_{max}}[\frac{1}{(c-cmc)}]^2 \qquad (12)$$

Note that if the last term can be ignored then equation 12 reduces to the normal Lineweaver-Burk expression with the substrate concentration displaced from zero by the cmc. When the linear and quadratic fits were made to the data in Figure 5 a slightly better fit was seen for the quadratic equation; however, the differences were not significant. The results of the two fits in terms of kinetic parameters are shown in Table I.

Table I. The kinetic constants for both pure surfactants including the quadratic fit for $C_{10}E_5$ ester

Substrate	Fit	$K_m{}^*$ (mM)	k_{cat} (sec^{-1})	R
$C_{10}E_5$ ester	linear	3.9	550	----
$C_{10}E_5$ ester	quadratic	2.1	414	1.3×10^{-4}
$C_{10}OBS$	linear	2.3	3	----

The data for $C_{10}OBS$ in Figure 1 was best fit by the linear form of equation 12. The kinetic constants for both pure surfactant esters are shown in Table I. Although the $K_m{}^*$ values are approximately equal, large differences are seen in k_{cat}. One significant difference between the two systems is the presence of Ca^{+2} in the $C_{10}E_5$ ester system. Separate experiments have shown dramatic increases in initial velocity in the $C_{10}E_5$ ester system upon addition of Ca^{+2}. Unfortunately, $C_{10}OBS$ precipitates with added Ca^{+2} making comparison of the two substrates with identical Ca^{+2} concentrations impossible.

We turn our attention now to mixtures where one of the components is a substrate and the second is the detergent component. The first system we examine is a mixture of $C_{10}E_5$ ester and $C_{10}E_5$ ether. The cmc data for pure surfactants and a 1:1 mixture are shown in Table II. The value of the mixed cmc, 1.0×10^{-3} M, is in excellent agreement with the calculated value from the ideal mixing equation(19) and the measured values of the pure components. Thus, this system behaves ideally.

Table II. Comparison of the theoretical to the experimental cmc for a 1:1 mixture of $C_{10}E_5$ ester:$C_{10}E_5$ ether

Ester:Ether Ratio	CMC (mM)	Ideal CMC (mM)
1:0	1.30	----
1:1	1.00	1.01
0:1	0.82	----

Figure 6 shows the rate of hydrolysis as a function of mole fraction $C_{10}E_5$ ester substrate when the total concentration is fixed at 4.23×10^{-2} M. From the pure compound data we can estimate a value of the parameter $\dfrac{K_m{}^*}{(c\text{-}cmc)}[1 + \dfrac{R}{(c\text{-}cmc)}]$ as 0.09 and fit the data using equation 9. The best fit is obtained with a $\dfrac{K_m{}^*}{K_D{}^*}$ of 0.9. The physical meaning of this result is that the substrate is marginally favored over the detergent in the competition for the active site (refer to Figure 4).

The cmcs from surface tension and the breakpoint from hydrolysis rate vs. concentration data for mixtures of $C_{10}OBS$ and $C_{10}E_9$ are shown in Figure 7. Once again excellent agreement between the two discontinuities is observed except in the case of mole fraction $C_{10}OBS = 0.25$ where the breakpoint is most difficult to determine. The implication of the data is that mixed micelles of substrate and detergent are still capable of activating the lipase for hydrolysis.

Figure 7 also gives the best theoretical fit from the regular solution model of mixed micellization to the experimental cmc data(17). The interaction parameter (β) which best describes the data is -0.9. This is considerably less than -4, typical for many mixtures of an anionic with a nonionic surfactant(20). These cmcs were determined in 0.1M NaCl which is expected to reduce electrostatic repulsions and therefore offset some of the favorable enthalpy of mixing from the nonionic surfactant. Even considering the electrolyte effect, the interaction, as measured by β, is smaller than expected and may also reflect a difference between aromatic vs. aliphatic (or ester linked vs non-ester linked) anionic surfactants in mixtures with alkyl ethoxylate nonionic surfactants.

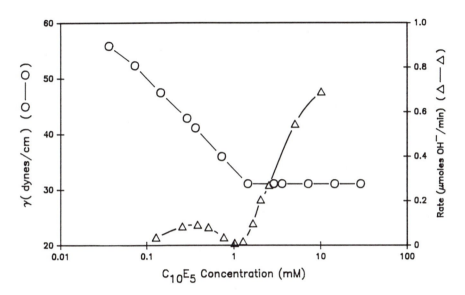

Figure 5. CMC and rate discontinuity comparison for $C_{10}E_5$ ester in 50mM NaCl, 10mM $CaCl_2$ at pH 9.0.   Surface tension data (O), rate data (△).   Enzyme concentration = 0.02ppm.

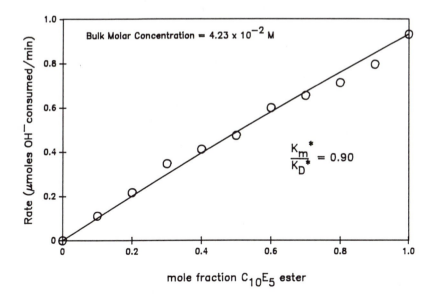

Figure 6. Rate vs. mole fraction for $C_{10}E_5$ ester:$C_{10}E_5$ ether mixtures. Points are experimental data and the line is the fit using equation 10.

In Figure 8 we show data for the rate of $C_{10}OBS$ hydrolysis with added $C_{10}E_9$. As before the experiment is run at a total concentration much higher than the cmc. Contrary to the ideal mixture, however, the theoretical fit is not particularly good. At low mole ratios of $C_{10}E_9$ a value of $\dfrac{K_m^*}{K_D^*} = 4$ fits the data but at higher mole fractions a value of 12 is required. A plausible interpretation of the data within the present model is that the exclusion parameter $R(x)$ is strongly dependent upon composition and penetration becomes increasingly difficult at high mole fraction of $C_{10}E_9$ in the mixed micelle. In fact, this compositional dependence must be very strong to offset the high concentrations which were used to minimize this contribution to the rate loss in these experiments where composition was deliberately varied.

Another way to approach the question of how the activity is lost in a mixed system containing both a detergent inhibitor and substrate is to fix the ratio of detergent and substrate in the mixed micelle and vary the total concentration. This is not generally possible by simply fixing the bulk mole fraction since the micelle mole fraction will change as a function of total concentration near the cmc as is well known from from mixed micelle theories(17). However, at concentrations well above the cmc, particularly in this case where the individual cmcs are nearly equal, this effect becomes negligible. We show the result of such experiments in figure 9 where the concentration is varied in a range well above the cmc so that the micelle mole fraction is reasonably constant.

Inversion of equation 7 would suggest that at fixed x and high total concentrations a linear relationship would exist between $\dfrac{1}{v_o}$ and $\dfrac{1}{c - cmc}$ and the slopes of these lines would vary inversely with mole fraction. Thus at x = 0.5 the slope would be twice that of the pure $C_{10}OBS$. The rate itself could be more strongly affected due to the competitive inhibition, but this effect would be constant at constant x or, stated differently, would exhibit itself in the intercept. Examination of the slope of the lines in Figure 9 shows that they vary much more rapidly than l/mole fraction. The ratio of the slope at mole fraction of 0.5 to mole fraction 1 is 23 rather than 2; while at mole fraction 0.25 this ratio is 105 rather than 4. Both the failure of equation 8 to describe the data in figure 8 and the higher than expected slopes in Figure 9 are consistent with a decreasing ability of the enzyme to be activated by the micelle as the mole fraction of the $C_{10}E_9$ increases. Using the data in Figure 9, we can estimate the exclusion parameter as a function of composition and insert this value into equation 7 to obtain the rate at different compositions. The result of this is plotted in Figure 10. The agreement between theory and experiment is significantly improved although there is still systematic deviation between theory and experiment.

Discussion

We have shown that a theoretical model similar to that presented by Ransac et. al.(15) for interfaces can be applied to mixed micelles containing substrate and detergent molecules. The model describes the single substrate case quite accurately.

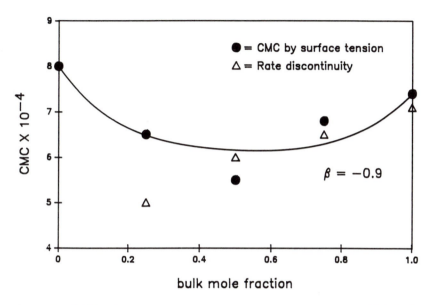

Figure 7. CMC and rate discontinuity vs. mole fraction for $C_{10}OBS:C_{10}E_9$ mixtures. CMC determined from surface tension concentration plot. Rate discontinuities determined from rate vs. concentration plot. The line is theoretical for the regular solution model with $\beta = -0.9$.

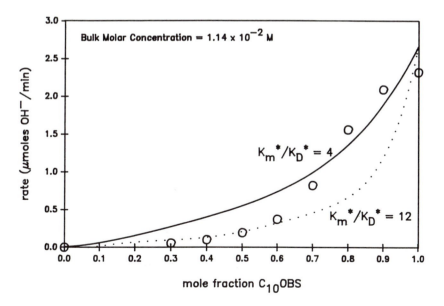

Figure 8. Rate vs. mole fraction for $C_{10}OBS:C_{10}E_9$ mixtures. Experiment vs. theory for the values of $\dfrac{K_m{}^*}{K_D{}^*}$ indicated (using equation 10).

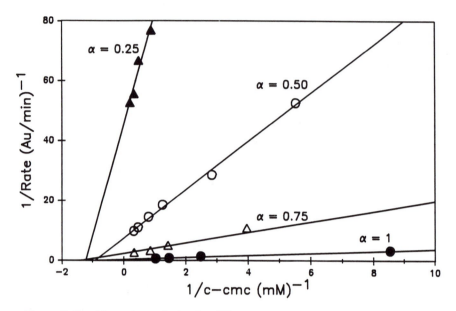

Figure 9. Double reciprocal plot for different (fixed) values of the composition. α represents the mole fraction of substrate.

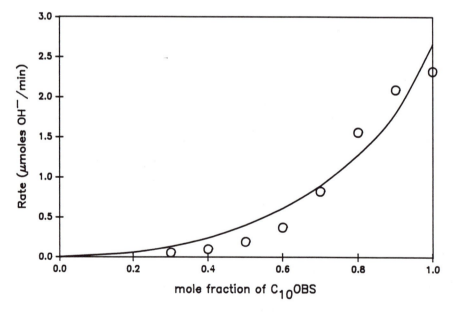

Figure 10. Rate vs. mole fraction for $C_{10}OBS$:$C_{10}E_9$ mixtures. Line is best theoretical fit using equation 7. The points are experimental data with $\frac{K_m^*}{K_D^*} = 4$ and $R(x)$ as described in text.

In the mixed surfactant case the model works as expected for an ideal mixture of surfactants. The model also provides a reasonable description of the more complicated case of $C_{10}OBS$ with $C_{10}E_9$ where mixing is nonideal and more dramatic competition and exclusion/activation effects might be expected.

Our analysis suggests that the observed rate loss upon adding $C_{10}E_9$ to $C_{10}OBS$ is a result of both competitive inhibition and exclusion effects. Although the nature of the physical phenomena responsible for differences in exclusion remain to be elucidated, the ability to separate loss in activity due to exclusion from that due to classical inhibition will allow progress to be made in that direction. Hopefully the model presented here will provide a framework for such separation to be made.

The products of the hydrolysis reaction of these simple esters, namely fatty acid and alcohol, can themselves be surface active and co-micellize with substrate and surfactant once formed. In the study presented here the decanoate ion would be expected to partition between the micelle and solution while the alcohol, from either substrate, probably resides in the aqueous phase. By measuring and analyzing only initial rates, where little product is present, it is hoped that this added complexity is successfully avoided; however, the limitation of the current approach to initial velocities of enzyme catalyzed reactions in mixed micelles should be explicitly recognized.

Acknowledgments

We would like to thank Jim Thompson and Larry Sickman for the synthesis, purification and characterization of both the surfactant substrates and the nonhydrolyzable surfactants and Cathy Oppenheimer and Debbie Thaman for the characterization of NOVO's lipolase.

Literature Cited

1. Carter, P. and Wells, J.A.; Nature, 322, 564 (1988).
2. Walsh, C. *Enzymatic Reaction Mechanisms,* W. H. Freeman and Co, San Fransico, (1979).
3. Segal, I.H. *Enzyme Kinetics,* John Wiley and Sons, NY (1975).
4. Cornish-Bowden, A., *Principles of Enzyme Kinetics,* Butterworths, London (1975).
5. Brockman, H.L. in *Lipases,* Borgstrom B. and Brockman, H. L. Eds., Elsevier, Amsterdam, p4 (1984)
6. Sarda, L., and Desnuelle, P., Biochim. and Biophys. Acta., 30, 513 (1958).
7. Brady, L., et. al., Nature, 343, 767 (1990)
8. Brzozowski, A. M., Nature, 351, 491 (1991).
9. Dennis, E. A., in *The Enzymes,*Boyer, P., Ed. 3rd edition, vol. 16, *Lipid Enzymology,* Academic Press, NY, p307 (1983).
10. Scott, D. L., Science, 250, 1541 (1990).
11. Davies, J. T. and Rideal, E. K. in *Interfacial Phonomena,* Academic Press, NY, p42 (1963).
12. Jacobsen, C. F., Leonis, J., Linderstrom-Lang, K., Ottensen, M. in *Methods in Biochemical Analysis,* Vol. IV, Interscience Publishers Inc., NY, p171 (1957).

13. Deems, R. A.,Eaton, B. R., Dennis, E. A., The Journal of Biological Chem., 250, 9013 (1985).
14. Burns, R. A. Jr., El-Sayed, M. Y., Roberts, M. F., Proc.Nat. Acad. Sci., 79, 4902 (1982).
15. Ransac, S., Riviere, C., Soulie, J. M., Gancet, C., Veger, R., de Haas, G. H., Biochemica et Biophysica Acta, 1043, 57 (1990).
16. Fersht, A., *Enzyme Structure and Mechanism,* 2nd ed., W. H. Freeman and Co., NY, p103 (1985).
17. Rubingh, D. N. in *Solution Chemistry of Surfactants,* Mittal, K.L. ed., Plenum Press, NY, vol. 1, p337 (1979).
18. Motomura, K., Yamanda, M., Aratono, M., Colloid and Polym. Sci., 262, 948 (1984).
19. Clint, J., J. Chem. Soc., 71, 1327 (1975).
20. Paul Holland, Adv. in Colloid and Interface Science, 26, 111 (1986).

RECEIVED January 8, 1992

Chapter 13

Binding of Bromide Ion to Mixed Cationic—Nonionic Micelles

Effects on Chemical Reactivity

Sally Wright[1], Clifford A. Bunton[1], and Paul M. Holland[2]

[1]Department of Chemistry, University of California—
Santa Barbara, CA 93106
[2]General Research Corporation, Santa Barbara, CA 93111

Micellar enhancement of chemical reactivity has been used to probe counterion binding to mixed cationic/nonionic micelles composed of CTABr and $C_{10}E_4$ by examining micellar rate effects on the reaction of bromide ion with methyl napthalene-2-sulfonate. The CTABr/$C_{10}E_4$ mixed micellar system was also characterized by using conductivity measurements. Results show the addition of the nonionic surfactant $C_{10}E_4$ leads to a marked decrease in the overall rate of demethylation of methyl napthalene-2-sulfonate by bromide ion, and that a simple pseudophase model can account for this effect.

Micelles and other surfactant aggregates can affect chemical reaction rates via interactions with ionic reactants and hydrophobic substrates. The reaction rates can be enhanced (so called micellar catalysis), or inhibited depending on the charge type of the reaction. Because of the importance of interactions of ions at colloidal surfaces, micellar rate effects can be useful in probing ion binding (*1-3*).

In the present work we consider the S_N2 reaction of Br⁻ and methyl naphthalene-2-sulfonate (MeONs) in water (*4*).

The rate of this reaction is increased by cationic micelles which concentrate the two reactants at their surface. Ions and polar molecules do not enter the hydrocarbon-like micellar core but reside at the micelle-water interface in a region that is often identified with the Stern layer. This concentration of reactants at colloidal surfaces

0097–6156/92/0501–0227$06.00/0

is the major source of enhancements of rates of bimolecular reactions by micelles, microemulsion droplets, vesicles and similar colloidal self assemblies. The data can be described quantitatively by assuming that water and the micelles behave as discrete reaction media, and that reactants partition between them as shown in Scheme 1, below.

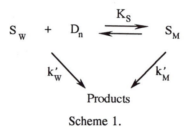

Scheme 1.

The overall rate of reaction is the sum of the rates in each pseudophase (1-3). In this scheme, S is the substrate and D_n is micellized surfactant (detergent), and subscripts W and M denote aqueous and micellar pseudophases respectively. The rate depends on the distribution of the nonionic substrate, expressed by its binding constant

$$K_s = \frac{[S_M]}{[S_W][D_n]} \tag{1}$$

The first order rate constants (with respect to substrate) are k'_W and k'_M, and K_s is the binding constant of substrate with micellized surfactant D_n. The corrected overall first order rate constant, k_ψ^c, is given by

$$k_\psi^c = \frac{k'_W + k'_M K_s[D_n]}{1 + K_s[D_n]} \tag{2}$$

where the superscript c on the overall rate constant indicates that a small correction has been made for reaction of MeONs with water molecules. The first order rate constants k'_W and k'_M depend upon the concentrations of Br$^-$ in the two pseudophases. For reaction in water:

$$k'_W = k_W [Br_W^-] \tag{3}$$

where quantities in square brackets are molarities written in terms of total solution volume. It is convenient to write the concentration of Br$^-$ in the micellar pseudophase as a mole ratio, so that the second order rate constant, k_M, has the dimensions of sec^{-1},

$$k'_M = \frac{k_M[Br_M^-]}{[D_n]} \tag{4}$$

and substitution into equation 1 then gives

$$k_\psi^c = \frac{k_W[Br_W^-] + k_M K_s[Br_M^-]}{1 + K_s[D_n]} \qquad (5)$$

This treatment predicts that reaction rates in aqueous quaternary ammonium bromide surfactants should increase with increasing surfactant concentration as substrate becomes bound to the micellar surface where the concentration of Br⁻ is much higher than in water. The problem is then that of estimating the concentration of Br⁻ at the micellar surface, and it can be calculated in terms of various theoretical models (4,5). An alternative approach is to determine the fractional ionic coverage of the surface, β

$$\beta = \frac{[Br_M^-]}{[CTABr_M]} \qquad (6)$$

and $\beta = 1 - \alpha$, where α is the fractional micellar ionization which can be determined experimentally, for example from the variation of conductance with surfactant concentration (4-8).

The reaction in aqueous solutions of CTABr fits the pseudophase model (equations 1-5) and values of k_ψ^c become independent of CTABr when all the substrate is micellar bound. Addition of a moderately hydrophobic alcohol, 1-butanol, slows the reaction by decreasing β and by increasing the volume of the micellar pseudophase. Both these effects decrease the concentration of Br⁻ at the micellar surface but the value of k_M is almost unaffected by addition of 1-butanol (7). This result was unexpected because the properties of a micellar surface should be affected by incorporation of a hydrophobic alcohol and reaction rates in solution are sensitive to changes in solvent properties. We therefore examined the reaction of Br⁻ with MeONs in a solution of CTABr plus the nonionic surfactant, $C_{10}E_4$ $(C_{10}H_{21}O(CH_2CH_2O)_3CH_2CH_2OH)$ and measured α conductimetrically so that we could test the pseudophase model (equations 1-5) and we compare values of k_M in solutions of CTABr, CTABr + 1-butanol, and CTABr + $C_{10}E_4$. This allows us to assess the effects of mixed CTABr/$C_{10}E_4$ micelles on the reaction rate.

Experimental Section

We carried out all our experiments in solutions of cetyltrimethylammonium bromide (CTABr, n-$C_{16}H_{33}NMe_3Br$) at 25.0 °C. Preparation and purification of 2-MeONs, CTABr, and CTA(SO₄)$_{1/2}$ have been previously described. The $C_{10}E_4$ used in this study was the single-species surfactant with a purity of 99.69% as detemined by gas chromatography.

The reaction of Br⁻ with MeONs in micellar solution was followed spectrophotometrically at 326 nm. These experiments were designed such that surfactant concentrations were always much higher than the CMC. The fractional micellar ionization, α, was determined conductimetrically (6-8).

Results

Kinetics. Reactions were followed under conditions such that MeONs was essentially fully micellar bound ($K_s(D_n) >> 1$) and reactions in the aqueous pseudophase could be neglected (4). We made a (small) correction for reaction of micellar-bound substrate with water by following the reaction in $CTA(SO_4)_{1/2}$.

Reaction rates in mixed $CTABr/C_{10}E_4$ micelles at constant CTABr concentrations of 0.05 M and 0.025 M are shown in Figure 1. These show a decrease in the reaction rate with the addition of $C_{10}E_4$. In Figure 2 the reaction rate at a constant total surfactant concentration of 0.05 M shows a similar decrease with increasing $C_{10}E_4$ in the surfactant ratio. The solid lines in Figures 1 and 2 show the theoretical fits for the variation of k_ψ^c with concentrations of CTABr and $C_{10}E_4$ (see discussion).

Estimation of α. Addition of nonionic solutes increases the fractional ionization, α, of ionic micelles by decreasing the charge density of the micellar surface. The conductivity increases as counterions are released from the micelles and α can be calculated from the slopes of plots of conductance against [CTABr] above and below the critical micelle concentration (CMC). This method involves assumptions regarding the contribution of micelles to conductivity and an alternative method of calculation includes a correction based on micellar size (9). We do not know the aggregation number or hydrodynamic radius of the mixed $CTABr-C_{10}E_4$ micelles and therefore we used the simple "ratio of slopes" calculation which is a good indicator of changes in α (6). The conductivity method becomes unsatisfactory as $\alpha \rightarrow 1$ and we limited our measurements to $[C_{10}E_4]/[CTABr] \leq 2.0$. Results from the conductivity data using the ratio of slopes method are given in Table 1. The observed increase in fractional charge α is consistent with recent results on the change of overall micellar charge when β-dodecylmaltoside is added to sodium dodecyl sulfate (10).

Table 1. Fractional Micellar Ionization, α, from Conductivity Data

$[C_{10}E_4]/[CTABr]$	α
0^a	0.25
0.5	0.47
1.0	0.54
2.0	0.59

[a] Taken from *ref 7*.

Discussion

The pseudophase model treats the aqueous and micellar pseudophases as distinct reaction regions (Scheme 1), but provided that substrate is fully micellar bound and

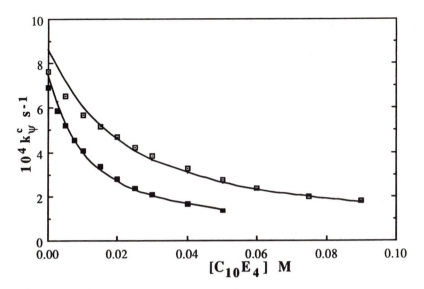

Figure 1. Dependence of corrected first-order rate constants of reaction of Br with 2-MeONs upon $[C_{10}E_4]$ in constant [CTABr]. The solid lines are theoretical fits to Equation 7. (■) 0.025 M CTABr; (⊡) 0.05 M CTABr.

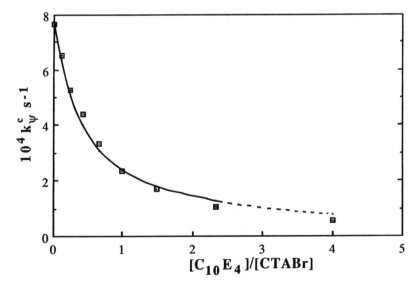

Figure 2. Dependence of corrected first-order rate constants of reaction of Br with 2-MeONs upon ratio of $[C_{10}E_4]/[CTABr]$ in constant total surfactant concentration of 0.05 M. The solid line is a theoretical fit to Equation 7 (the dashed portion is based on an extrapolated value for β).

surfactant concentrations are much larger than the CMC, as in our experiments, equation 4 simplifies to (2-4)

$$k_\psi^c = k_M \beta R \tag{7}$$

where

$$R = \frac{[CTABr]}{[CTABr] + [C_{10}E_4]} \tag{8}$$

The solid lines in Figures 1 and 2 are calculated by using Equation 7 and interpolated values of β and $k_M = 1.07 \times 10^{-3} \text{ s}^{-1}$. The results for reactions with a constant [Br$^-$] of 0.05 M and variable [surfactant], data not shown, are fitted reasonably with $k_M = 1.30 \times 10^{-3} \text{ s}^{-1}$. We think that the difference between the two sets of values of k_M is due to our using values of β, based on our conductivity experiments, that are low in the latter case because of the difficulty in estimating α at high [Br$^-$]. In water, addition of Br$^-$ to CTABr slightly increases β so that our values of β that are determined conductimetrically in the absence of added Br$^-$ are probably too low in experiments with added NaBr. The data point at [C$_{10}$E$_4$]/[CTABr] = 4 (Figure 2) was not fitted because we have no value of α at this concentration ratio.

The simple pseudophase model (Scheme 1 and equations 1-5) fits rate effects of mixed micelles of CTABr and C$_{10}$E$_4$ over a range of surfactant concentration. Values of k_M are little affected by marked changes in micellar composition. For aqueous CTABr (4) $k_M = 9.6 \times 10^{-4} \text{ s}^{-1}$, in CTABr + 1-butanol (7) $k_M = 9.1 \times 10^{-4} \text{ s}^{-1}$, and in CTABr + C$_{10}E_4$ $k_M = 10.7 \times 10^{-4} \text{ s}^{-1}$ and $13.0 \times 10^{-4} \text{ s}^{-1}$ in the absence and presence of added NaBr, respectively.

These comparisons are based on our writing concentrations of Br$^-$ at the micellar surface as mole ratios. It should be pointed out that S$_N$2 reactions of nucleophilic anions are not very sensitive to small changes in the properties of the medium such as polarity or dielectric constants, so that changes in the polarity of the micellar surface by addition of C$_{10}$E$_4$ are probably not very important kinetically. None the less, these similarities in values of k_M were unexpected because addition of C$_{10}$E$_4$ should change the properties of the micellar surface, and we need more evidence before we can decide whether they apply only to S$_N$2 reactions or are more general.

Conclusions

The addition of the nonionic surfactant C$_{10}$E$_4$ to micellar solutions of CTABr leads to a marked decrease in the overall rate of demethylation of 2-MeONs by bromide ion. At constant CTABr concentration there is a monotonic decrease in the rate on addition of C$_{10}$E$_4$. The good fit of a simple psuedophase model to the results suggests that the observed effect of C$_{10}$E$_4$ on the rate is due to a combination of Br$^-$ being expelled from the micellar region as C$_{10}$E$_4$ is added and dilution of the phase volume of the reaction region at the micellar surface. The observation that the second order rate constants at the micellar surface are nearly the same for CTABr, CTABr + 1-butanol, and

CTABr + $C_{10}E_4$ was unexpected, in view of differences in the structures of surfactants and the alcohol.

Legend of Symbols

D_n	micellized surfactant
K_s	binding constant of substrate based on concentration of micellized surfactant
k'_W	first order rate constant with respect to substrate in aqueous phase
k'_M	first order rate constant with respect to substrate in micellar pseudophase
k_ψ^c	corrected overall first order rate constant
k_W	second order rate constant with respect to substrate in aqueous phase
k_M	second order rate constant with respect to substrate in micellar pseudophase
R	ratio of CTABr in total surfactant
S_W	substrate in aqueous phase
S_M	substrate in micellar pseudophase
α	degree of fractional micellar ionization
β	fractional counterion binding (neutralization) of micelle, $\beta = 1 - \alpha$

Acknowledgements

Support by the National Science Foundation (Organic Chemical Dynamics Program) is gratefully acknowledged. The authors also wish to thank Dr. Robert G. Laughlin of The Procter & Gamble Company for supplying us with the $C_{10}E_4$ used in this study.

Literature Cited

1. Romsted, L. S. In *Surfactants in Solution;* Mittal, K. L.; Lindman, B.; Eds.; Plenum Press: New York, NY, 1984; Vol.2., p. 1015.
2. Bunton, C. A.; Savelli, G. *Adv. Phys. Org. Chem.* **1986,** *22,* 213.
3. Bunton, C. A. In *Cationic Surfactants: Physical Chemistry;* Rubingh, D. N.; Holland, P. M.; Eds.; Surfactant Science Series 37; Marcel Dekker, Inc.: New York, NY, 1990; pp 323-405.
4. Bacaloglu, R.; Bunton, C. A.; Ortega, F. *J. Phys. Chem.* **1990,** *94,* 5068.
5. Bunton, C. A.; Moffatt, J. R. *J. Phys. Chem.* **1988,** *92,* 2896.
6. Zana, R. *J. Colloid Interface Sci.* **1980,** *78,* 330.
7. Bertoncini, C. R. A.; Nome, F.; Cerichelli, G.; Bunton, C. A. *J. Phys. Chem.* **1990,** *94,* 5875.
8. Neves, F. de F.S.; Zanette, D.; Quina, F.; Moretti, M. T.; Nome, F. *J. Phys. Chem.* **1989,** *93,* 4166.
9. Evans, H. C. *J. Chem. Soc.* **1956,** 579.
10. Bucci, S.; Fagotti, C.; Degiorgio, V; Piazza, R. *Langmuir* **1991,** *7,* 824.

RECEIVED January 6, 1992

Chapter 14

Antimixing Micelles
of Dimethyldodecylamineoxide and Nonionic
Surfactants

Jiulin Xia, Paul L. Dubin[1], and Huiwen Zhang[2]

**Department of Chemistry, Indiana University, Purdue University
at Indianapolis, Indianapolis, IN 46205-2810**

Quasi-elastic light scattering studies of mixed surfactants of dimethyldodecylamineoxide (DMDAO) and octaethylene glycol monododecyl ether ($C_{12}E_8$) have been carried out. Bimodal distributions of relaxation spectra resolved by inverse Laplace transformation of autocorrelation functions at pH<7 were found to be related to the diffusion processes of pure $C_{12}E_8$ and DMDAO-rich micelles. Above pH 7, mixed micelles are formed.

An understanding of mixed micelles is important. Surfactants which are used in practical applications are highly heterogeneous, containing impurities such as the chemical precursors of the desired detergents. In many cases, co-surfactants are added intentionally because mixed micelles can have superior properties to those of the respective pure surfactant components (1,2). These synergistic properties of mixed micelles have stimulated research on interactions among surfactants. Many studies have been reported, including measurements of critical micelle concentrations (CMC) (3,4), dye solubilization (5,6), and other solution properties (7-12) by means of NMR, ESR, fluorescence probe method, etc. However, most of these studies focused on the measurement of the CMC of mixed micelles and thermodynamic explanations of these results (13-14).

It is commonly assumed that mixed surfactant systems form mixed micelles (15,16), i.e. that the composition of the micelles resembles that of the total system, especially as the surfactant concentration becomes large compared to the mixed CMC. Thus, the possibility of very large compositional polydispersity, including the presence of more than one distinct micellar population, is generally neglected. Cases in which such an unfavorable mixing energy has been confirmed are few. Dennis et al. (17) have reported the existence of more than one population of thermodynamically stable mixed micelles in systems of nonionic detergents and

[1]Corresponding author
[2]Current address: R. I. D. C. I., Taiyuan, Shanxi, People's Republic of China

0097–6156/92/0501–0234$06.00/0

phospholipids. Abe et al. (18) observed micelle de-mixing for sodium 3,6,9-trioxaicosanoate and alkyl polyoxyethylene ethers, as was also observed in other systems (19, 20). In this paper, we report on the incompatibility of N,N-dimethyldodecylamine-N-oxide (DMDAO) and octaethylene glycol monododecyl ether $C_{12}E_8$ based on Quasi-Elastic Light Scattering (QELS) measurements.

QELS has been extensively applied to micellar solutions. Kato et al (21) has obtained conclusions about intermicellar interactions in $C_{12}E_6$ solutions from their QELS diffusion data. Nilsson et al (22, 23) and Brown et al (24) have studied not only the mutual diffusion coefficient but also self-diffusion coefficient, hence the hydrodynamic size, as a function of temperature of $C_{12}E_8$ and $C_{12}E_6$ solutions. The application of QELS to micellar systems have been reviewed by Corti (25) and by Candau (26).

In contrast to the nonionic $C_{12}E_8$, DMDAO is completely cationic (protonated) at low pH (<3) and completely nonionic at high pH (>9). At 3<pH<9, DMDAO forms partially protonated micelles containing both the cationic and nonionic forms of surfactant (27). When DMDAO and $C_{12}E_8$ are combined in aqueous solution at any pH, they form an optically clear phase, which suggests that the DMDAO and $C_{12}E_8$ micelles are compatible. However, our QELS results show that DMDAO and $C_{12}E_8$ molecules are not miscible in their micellar form in neutral or acidic solution.

Experimental

Both DMDAO and $C_{12}E_8$ (Puriss grade) were purchased from Fluka (Happague, NY). The DMDAO was found to contain 1.3% H_2O by Karl Fisher titration. pH titration in acetic acid with $HClO_4$ to the acid end-point only consumed 97.2% of the theoretical amount of acid, from which it was concluded that the surfactant was partially in the salt form, perhaps because of adsorption of CO_2. Therefore, solutions of DMDAO were adjusted by addition of NaOH to pH *ca* 11 prior to further preparation. HCl and NaOH standard solutions and NaCl (analytical grade) were from Fisher (Pittsburgh PA).

To prepare solutions, desired amounts of DMDAO and $C_{12}E_8$ were weighted, and then dissolved by deionized water that was made from glass distilled water subsequently passed through one carbon and two ion-exchange filters. Solutions were sonicated for 30 minutes and stirred for couple of hours to ensure equilibrium of mixing. The solution pH was adjusted by 0.5 M HCl or NaOH. Scattering solutions were made dust-free by filtration through 0.2 μm Acrodisc filters from Gelman Sciences (Ann Arbor, MI). The concentrations of the DMDAO/$C_{12}E_8$ (1:1 or 7:3) in I=0.4 M NaCL solution was 20 mM.

QELS measurements were made at scattering angles from 30° to 150° with a Brookhaven system equipped with a 72 channel digital correlator (BI-2030 AT) and using a Jodon 15 mW He-Ne laser (Ann Arbor, MI). We obtain the homodyne intensity-intensity correlation function G(q,t), with q, the amplitude of the scattering vector, given by $q=(4\pi n/\lambda)\sin(\theta/2)$, where n is the refractive index of the medium, λ is the wavelength of the excitation light in a vacuum, and θ is the scattering angle. G(q,t) is related to the time correlation function of concentration fluctuations g(q,t) by:

$$G(q,t) = A (1 + b\, g(q,t)^2) \qquad (1)$$

where A is the experimental baseline and b is the fraction of the scattered intensity arising from concentration fluctuations. The quality of the measurements were verified by determining that the difference between the measured value of A and the calculated one was less than 1%. General discussions of QELS data analysis may be found in refs. (28) and (29).

Results and Discussion

In general, the normalized intensity autocorrelation function for homodyne detection arising from a single diffusive process with a decay constant Γ, can be written as:

$$g(t)^2 = e^{-2\Gamma t} \tag{2}$$

with $\Gamma=Dq^2$, where D is the mutual diffusion coefficient (reduce to translation diffusion coefficient with very low concentration. Shown in Figure 1 are the logarithm of the autocorrelation functions obtained for pure DMDAO (10 mM) and $C_{12}E_8$ (10 mM) at pH=3.96 in 0.40 M NaCl. The linear plots show that pure diffusion processes for DMDAO and $C_{12}E_8$ micelles are detected. However, the single exponential function can not fit the autocorrelation function obtained for mixed DMDAO and $C_{12}E_8$ solution (1:1) at pH=3.68, as shown in Figure 2a. The logarithm of g(t) vs. t is not linear, which suggests that more than one process is involved.

Several techniques for analysis of QELS data have been developed to analyze different correlation functions from different sample systems. These include the methods of cumulants, histogram, constrained regularization, Laplace transform inversion, nonnegatively constrained least squares, etc. The advantages and disadvantages of different methods have been discussed by Stock and Ray (29). In general, the correlation function can be expressed as an integral sum of exponential decays weighted over the distribution of relaxation times $\rho(\tau)$:

$$\left[\frac{G(t)-A}{A}\right]^{1/2} = b^{1/2}g(t) = \int_0^\infty e^{-t/\tau} \rho(\tau) \ d\tau \tag{3}$$

In principle, it is possible to obtain the distribution $\rho(\tau)$ by integral transformation of the experimental $\{G(t)/A-1\}^{1/2}$, but in practice this presents a formidable problem for numerical analysis as taking the inverse Laplace transform is numerically an ill-posed problem. Several numerical methods developed so far are devoted to calculating $\rho(\tau)$. In present work, we analyze the autocorrelation function of DMDAO-$C_{12}E_8$ mixed micelles by using the CONTIN program, which employs the constrained regularization method (30).

From equation (3), the mean relaxation time, $<\tau>$, defined as the area of g(t), is given by

$$<\tau> = \int_0^\infty g(t) \ dt$$

$$= \int_0^\infty \tau \rho(\tau) \ d\tau / \int_0^\infty \rho(\tau) \ d\tau \tag{4}$$

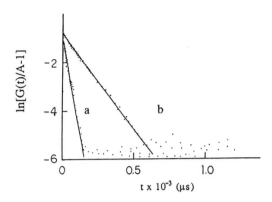

Figure 1: The logarithm of correlation function vs time for (10 mM) DMDAO (a) and (10 mM) $C_{12}E_8$ (b) micelles at pH=3.96.

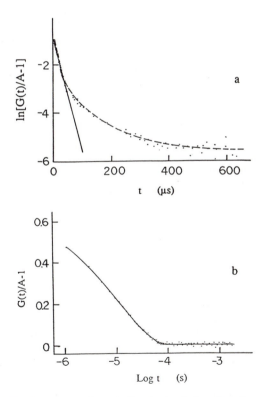

Figure 2: (a) The logarithm of normalized correlation function plotted vs time for (20 mM 1:1) DMDAO-$C_{12}E_8$ micelle at pH 3.68. One note that it is not a linear function. (b) Normalized correlation function at pH 3.68 for DMDAO-$C_{12}E_8$ (20 mM 1:1). Dots are experimental data. Curve is the CONTIN fit.

Shown in Figure 2b is the autocorrelation function of DMDAO-$C_{12}E_8$ (1:1) mixed micelles in pH=3.68 aqueous solution. Dots are the experimental data, while the curve is the CONTIN fit. The mean relaxation time constants are deduced from the fitting. One notes the good quality of the fit.

The calculated relaxation time spectra obtained using CONTIN for DMDAO, $C_{12}E_8$ and DMDAO-$C_{12}E_8$ (1:1) mixtures at varying pH are displayed in Figure 3. One notes that the shape of the relaxation time spectrum for pure and mixed micelles appears different at pH<7.0. Instead of a single relaxation as for pure DMDAO or $C_{12}E_8$, DMDAO-$C_{12}E_8$ mixtures show bimodal distributions at pH<7 (see Figure 3c).

As given by equation (4), the mean relaxation time, $<\tau>$, is the first moment of the normalized relaxation spectrum. For multi-modal relaxation, $<\tau>$ may resolved from each mode using CONTIN. Therefore, two $<\tau>$ values are obtained simultaneously for DMDAO-$C_{12}E_8$ micelles: the mode with small $<\tau>$ refers to a fast relaxation; the other is a slower process.

In order to verify that the relaxation times are related to the diffusion processes, the angle dependence measurement of $<\tau>$ was performed, as shown in Figure 4. All the angle dependence curves are linear with zero intercept, which confirms that the measured decay constants in both single mode and bimodal distribution arise from diffusion processes.

The diffusion coefficients of the micelles can be calculated using

$$D = \frac{\Gamma \lambda^2}{16 \pi^2 \sin^2(\theta/2)} \tag{5}$$

where Γ is the reciprocal of the diffusion time constant, which is obtained either from the slope of the linear plots in Figure 1 or by equation (4). The diffusion process for multi-component system has been theoretically described by the Onsager theory (31), which takes into account coupled diffusion. For the present study, this coupling vanishes due to high dilution. Therefore, the diffusion coefficient, D, is directly related to the hydrodynamic radius, R_s, by Stokes' equation

$$D = \frac{kT}{6\pi\eta R_s} \tag{6}$$

where k is Boltzmann's constant, T is the absolute temperature and η is the viscosity of the solvent. This equation is only true for systems of non-interacting particles. For interacting particles, Pusey et al (28) and Ackerson (32) have shown that the measured value of D is also a function of the structure factor S(q) and hydrodynamic interaction H(q). Here we assume non-interacting particle and obtain R_s values for DMDAO, $C_{12}E_8$ and DMDAO-$C_{12}E_8$ micelles via equation (6).

Figure 5 shows the pH dependence of R_s for DMDAO, $C_{12}E_8$ and DMDAO-$C_{12}E_8$ micelles. While R_s is independent of pH for $C_{12}E_8$ as expected, the size of DMDAO micelles strongly depends on pH. The maximum in R_s vs pH for DMDAO has been interpreted in terms of the strong hydrogen bonding between protonated and unprotonated head groups (33).

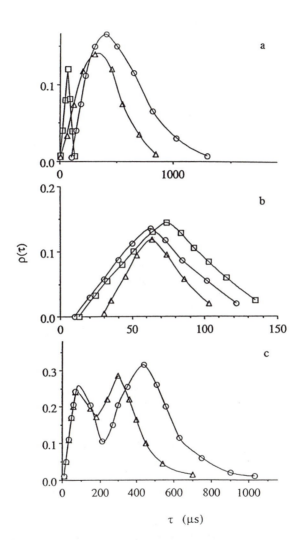

Figure 3: Relaxation spectra $\rho(\tau)$ calculated by using the CONTIN program plotted vs time for (a) 10 mM DMDAO at pH 7.95 (□); 5.51 (Δ) and 4.46 (O). (b) 10 mM $C_{12}E_8$ micelle at pH 4.46 (Δ) and 20 mM DMDAO-$C_{12}E_8$ (1:1) mixed micelles at pH 10.54 (O) and pH 7.00 (□). (c) DMDAO-$C_{12}E_8$ (1:1) mixed micelles at pH 3.68 (Δ) and 5.03 (O). Note the bimodal distributions with one mode at about the same position as pure $C_{12}E_8$.

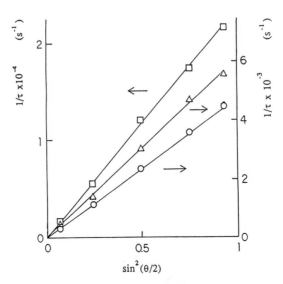

Figure 4: The reciprocal of mean relaxation times plotted as a function of $\sin^2(\theta/2)$ for DMDAO at pH 4.67 (O); and DMDAO-$C_{12}E_8$ (1:1) at pH 5.45 (□ fast mode, Δ slow mode). These linear functions with zero intercepts suggest that the relaxation times are only related to diffusion processes.

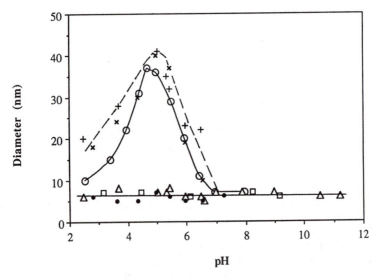

Figure 5: The apparent hydrodynamic size plotted as a function of pH for DMDAO (O), $C_{12}E_8$ (□) and DMDAO-$C_{12}E_8$ fast and slow modes (Δ, +; 1:1), (●, x; 7:3).

For DMDAO-$C_{12}E_8$ mixed system with composition of 1:1 or 7:3, a single mode with R_s values identical to those of pure $C_{12}E_8$ are observed for pH>7. This suggests that the two surfactants are compatible. The molecular miscibility of DMDAO and $C_{12}E_8$ surfactants cannot be resolved from QELS alone since both surfactants form micelles with about the same hydrodynamic size at this pH region. At lower pH we observe two modes, one with the same R_s as pure $C_{12}E_8$, and the other which displays a maximum at pH about 5 (corresponding to 50% protonation, $\beta=0.5$) as does pure DMDAO.

To interpret this unusual behavior, we first consider the model of DMDAO micelles. At intermediate pH DMDAO is a binary mixture of cationic (DMDAOH$^+$) and nonionic (DMDAO) surfactants. Thus, the protonated and neutral species can be treated as separate surfactants, the composition of which is varied by adjusting the solution pH. The maximum in R_s close to $\beta=0.5$ suggests that the H-bonded dimer (DMDAO-DMDAOH$^+$) also be considered as a component. This "double tailed species" plays an important role in stabilizing low surface curvature, i.e. rods or bilayer (33).

The mixing of DMDAO and $C_{12}E_8$ can be considered as the substitution for DMDAO molecules by $C_{12}E_8$ in DMDAO micelles. This process is much easier at high pH since the H-bond formed between DMDAO and DMDAOH$^+$ is not disrupted. Thus, at high pH only a single diffusant with r=7 nm is detected. At pH around 5, almost all the DMDAO molecules may be dimerized (33). The substitution of DMDAO molecules by $C_{12}E_8$ is not energetically favorable. Therefore, two diffusing species are observed, one with R_s equal to pure $C_{12}E_8$. Furthermore, we note that the difference in R_s between DMDAO and the slow species in DMDAO-$C_{12}E_8$ reaches a minimum around $\beta=0.5$ of DMDAO at 2.5<pH<7. We suggest that the molecular miscibility of DMDAO and $C_{12}E_8$ is nearly prohibited in the micellar form at pH<7, where the $C_{12}E_8$ may only be placed between dimers to stabilize the mixed micelles. The resistance to mixing at the microscopic level is further demonstrated by the results shown in Figure 5, in which fast and slow mode diffusivities are compared for 0, 50 and 70 mol.% DMDAO (all 20 mM total surfactant). The fact that R_s values in the mixed systems are insensitive to the bulk composition, suggests that changes in the latter variable lead not to changes in the microscopic composition of the mixed micelles, but rather to a redistribution of surfactant between two populations of two very stable micellar species, one rich in DMDAO, the other close to pure $C_{12}E_8$.

Summary

QELS studies of DMDAO-$C_{12}E_8$ system were carried out. Completely compatible micelles are formed only at pH>7. At lower pH, the two surfactants are partially mixed and two micellar forms, with one identical to pure $C_{12}E_8$ micelle, are observed.

Acknowledgment

This research was supported by NSF grant DMR-9014945.

Literature Cited

1. Rosen, M. J.; Zhu, B. Y. *J. Colloid Interface Sci.* **1984**, *99*, 427.
2. Miyazawa, K.; Ogawa, M.; Mitsui, T. *J. Soc. Cosmet. Chem. Japan* **1984**, *18*, 96.
3. Holland, P. M. *Adv. Colloid Interface Sci.* **1986**, *26*, 111.
4. Funasaki, N.; Hada, S. *J. Phys. Chem.* **1979**, *83*, 2471.
5. Smith, G. A.; Christian, S. D.; Tucker, E. E.; Scamehorn, J. F. *J. Colloid Interface Sci.* **1989**, *130*, 254.
6. Abe, M.; Kubata, T.; Uchiyama, H.; Ogino, K. *Colloid Polym. Sci.* **1988**, *30*, 335.
7. Uchiyama, H.; Abe, M.; Ogino, K. *J. Colloid Interface Sci.* **1990**, *138*, 69.
8. Uchiyama, H.; Christian, S. D.; Scamehorn, J. F.; Abe, M.; Ogino, K. *Langmuir* **1991**, *7*, 95.
9. Kameyama, K.; Takagi, T. *J. Colloid Interface Sci.* **1990**, *140*, 517.
10. Zhu, B. Y.; Rosen, M. J. *J. Colloid Interface Sci.* **1984**, *99*, 435.
11. Stellner, K. L.; Scamehorn, J. F. *Langmuir* **1989**, *5*, 77.
12. Scamehorn, J. F.; Schechter, R. S.; Wade, W. H. *J. Colloid Interface Sci.* **1982**, *85*, 494.
13. Rubingh, D. N. *Solution Chemistry of Surfactants;* Mittal, K. C., Ed.; Plenum Press: New York, 1979, Vol. I, pp 337-354.
14. Clint, J. *J. Chem. Soc. Faraday* **1975**, *71*, 1327.
15. Sundler, R.; Alberts, A. W.; Vagelos, P. R. *J. Biol. Chem.* **1978**, *253*, 4175.
16. Yedgar, S.; Barenholz, Y.; Cooper, V. G. *Biochim. Biophys. Acta* **1974**, *363*, 98.
17. Robson, R. J.; Dennis, E. A. *Accounts of Chemical Research* **1983**, *16*, 251.
18. Abe, M.; Tsubaki, N.; Ogino, K. *J. Colloid Interface Sci.* **1985**, *107*, 503.
19. Lake, M. *J. Colloid Interface Sci.* **1983**, *91*, 496.
20. (a) Puig, J. E.; Franses, E. I.; Miller, W. G. *J. Colloid Interface Sci.* **1982**, *89*, 441; (b) Tsujii, K. *Yukagaku* **1982**, *31*, 981.
21. Kato, T.; Seimiya, K. *J. Phys. Chem.* **1986**, *90*, 3159.
22. Nilsson, P. G.; Wennerstrom, H; Lindman, B. *J. Phys. Chem.* **1983**, *87*, 137.
23. Nilsson, P. G.; Wennerstrom, H; Lindman, B. *Chemica Scripta* **1985**, *25*, 67.
24. Brown, W.; Johnson, R.; Stilbs, P.; Lindman, B. *J. Phys. Chem.* **1983**, *87*, 4548.
25. Corti, M. *Physics of Amphiphiles: Micelles, Vesicles and Microemulsions;* Degiorgio, V.; Corti, M., Eds.; North-Holland: Amsterdam, 1985, pp 122-151.
26. Candau, S. J. *Surfactant Solutions;* Zana, R., Ed., Marcel Dekker: New York, 1987, pp 147-207.
27. Ikeda, S.; Tsunoda, M.; Maeda, H. *J. Colloid Interface Sci.* **1979**, *70*, 448.
28. Pecora, R. *Dynamic Light Scattering: Applications of Photon Correlation Spectroscopy;* Plenum Press: New York, 1976; Schmitz, K. S. *An Introduction to Dynamic Light Scattering by Macromolecules;* Academic Press: New York, 1990.
29. Stock, R. S.; Ray, W. H. *J. Polym. Sci. Polym. Phys. Ed.* **1985**, *23*, 1393.
30. Provencher, S. W. *Comput. Phys. Commun.* **1982**, *27*, 229.
31. Fitts, D. D. *Nonequilibrium Thermodynamics;* McGraw Hill: New York, 1962.
32. Ackerson, B. J. *J. Chem. Phys.* **1976**, *32*, 1626.
33. Zhang, H.; Dubin, P. L.; Kaplan, J. I. *Langmuir*, in press.

RECEIVED January 6, 1992

SURFACTANT MIXTURES WITH UNUSUAL SURFACTANT TYPES

Chapter 15

Micellization in Mixed Fluorocarbon—Hydrocarbon Surfactant Systems

Wen Guo, Bing M. Fung[1], Sherril D. Christian, and Erika K. Guzman

Department of Chemistry and Institute for Applied Surfactant Research, University of Oklahoma, Norman, OK 73019–0370

Mixed surfactant systems containing an anionic fluorocarbon surfactant (sodium perfluorooctanoate) and various types of hydrocarbon surfactants have been studied comprehensively by means of surface tension, H-1 NMR, and F-19 NMR. When both fluorocarbon and hydrocarbon surfactants are anionic, the systems show strong positive deviation from ideal behavior. If the chains are long so that their mutual phobicity is large enough, two types of micelles can coexist in the same solution for intermediate mole fractions of the fluorocarbon component. For an anionic fluorocarbon surfactant mixed with nonionic, zwitterionic, or cationic hydrocarbon surfactants, the systems deviate negatively from ideal behavior. These observations are interpreted in terms of different interactions between the head groups and the hydrophobic chains. The experimental cmc data are compared with calculated results.

Fluorocarbon surfactants are more efficient than hydrocarbon surfactants in lowering the surface tension of aqueous solutions (1), and they can spread on hard-to-wet surfaces such as fluorocarbon and silicon polymers. The study of the micellization of fluorocarbon surfactants mixed with hydrocarbon surfactants (2) is important for both basic research and industrial applications.

The earliest study of a mixed fluorocarbon-hydrocarbon surfactant system was reported by Klevens and Raison in 1954 (3). Since then, a number of authors have investigated various fluorocarbon-hydrocarbon mixed surfactant systems. Recently, Funasaki has made a comprehensive review on this subject (2). In this article we will focus our attention on the mixtures of a single anionic fluorocarbon (FC) surfactant with various hydrocarbon (HC) surfactants:

Anionic fluorocarbon surfactant: sodium perfluorooctanoate (SPFO),
$$CF_3(CF_2)_6COONa$$

[1]Corresponding author

0097–6156/92/0501–0244$06.00/0

Anionic hydrocarbon surfactants:	sodium decyl sulfate (SDeS),
	$CH_3(CH_2)_9OSO_3Na$
	sodium dodecyl sulfate (SDS),
	$CH_3(CH_2)_{11}OSO_3Na$
Nonionic hydrocarbon surfactant:	N-triethoxylated nonanamide (HEA8-3),
	$CH_3(CH_2)_7C(O)NH(CH_2CH_2O)_3H$
Zwitterionic hydrocarbon surfactant:	N-decyl-N,N-dimethyl-3-ammonio-1-propane-sulfonate (DEDIAP),
	$CH_3(CH_2)_9N(CH_3)_2(CH_2)_3SO_3$
Cationic hydrocarbon surfactant:	N-octyl,N,N,N-trimethylammonium bromide (OCTAB),
	$CH_3(CH_2)_7N(CH_3)_3Br$

We have made comprehensive investigations on these mixed surfactant systems by means of surface tension, H-1 NMR, and F-19 NMR measurements. The results are summarized in the following, and the effects of head group charge and inter-chain interactions are discussed.

Anionic Fluorocarbon - Anionic Hydrocarbon Surfactant Mixtures

Fluorocarbons are extremely insoluble in water. They also have low affinities towards hydrocarbons; the solubility of fluorocarbons in hydrocarbons is limited, and their solutions are highly non-ideal. Therefore, although two hydrocarbon surfactants with the same type of head group usually obey the thermodynamics of ideal mixing (4), the situation for fluorocarbon and hydrocarbon surfactant mixtures is different. A mixture of an anionic fluorocarbon surfactant and an anionic hydrocarbon surfactant is expected to show positive deviation from ideal behavior due to the lack of affinity between the two types of hydrophobic chains. However, there is no consensus on the extent of the deviation (2). Some investigators suggest that the positive deviation can be explained by regular solution theory (5); others believe that the deviation is usually quite large, and two types of micelles, one rich in hydrocarbons and another rich in fluorocarbons, are formed (6); some investigators even suggest that hydrocarbon surfactants are practically not incorporated into the fluorocarbon micelles (7). We have examined the mixed systems of SPFO/SDeS and SPFO/SDS carefully, and the results are discussed in the following. It should be pointed out that perfluoroalkyl carboxylic acids are strong acids, and solutions of SPFO are neutral (8).

For single surfactant systems, the plots of surface tension (γ) versus log(c) generally contains two linear segments, the interception point of which corresponds to log(cmc) (Figure 1). In contrast, the γ-log(c) plot for the SPFO/SDS system has a very different characteristic: beyond the first cmc, the plot shows a curved part that is concave downward, followed by a slow linear decrease similar to those of the pure surfactants (Figure 1). Thus, a second cmc value can be obtained for each mole fraction of SPFO (X_F). We will call the smaller cmc value cmc(1), and the larger one cmc(2). They could be determined for $0.3 \leq X_F \leq 0.85$ because in this range two breaks are clearly distinguishable in the surface tension plots. For $X_F > 0.85$, there is no obvious curved part in the plot; for $X_F < 0.3$, the high-concentration segment is still concave upward, but does not show a clear-cut second break point.

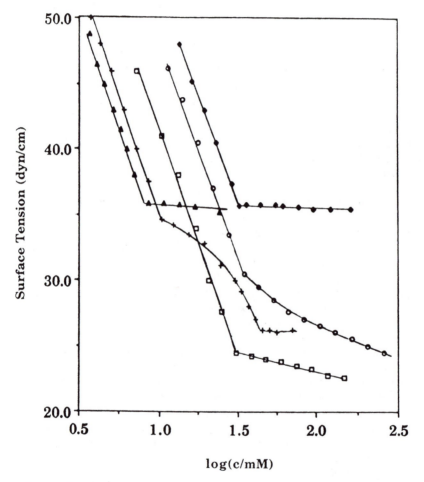

Fig. 1   Plots of surface tension versus log(c) for various surfactants systems. Pure SDS, (▲); pure SPFO, (□); pure SDeS, (♦); mixture of SPFO/SDS with $X_F = 0.33$, (+); and mixture of SPFO/SDeS with $X_F = 0.25$, (o).

To determine the cmc from NMR data, the chemical shift is plotted as a function of reciprocal concentration; the interception of two linear segments corresponds to the concentration at the cmc. For the SDS/SPFO system, the values of the cmc's obtained from H-1 NMR data (9) agree well with those of cmc(1) determined from surface tension measurements (Figure 2). However, the F-19 NMR data yield accurate cmc values only for mixtures with $0.5 \leq X_F \leq 1$, because for $X_F < 0.5$ the plots of the F-19 chemical shifts show a curved region around the cmc rather than a sharp break (9). The cmc values thus obtained agree well with those of cmc(2) determined from surface tension measurements (Figure 2). Based upon these observations, we suggest that cmc(1) corresponds to the formation of HC-rich micelles, and cmc(2) corresponds to the formation of FC-rich micelles (9).

For the SPFO/SDeS system, the segment in the surface tension plot (Figure 1) is not linear beyond the cmc, and its slope is considerably more negative than those of the pure surfactant segments. The plots of H-1 chemical shifts versus 1/c yield definite cmc data only for small values of X_F, and the plots of F-19 chemical shifts versus 1/c yield definite cmc data only for large values of X_F (9). The corresponding cmc values agree well with those obtained from surface tension measurements (Figure 2). Upon a critical examination of all the cmc data, we conclude that there is no compelling evidence for the formation of two types of mixed micelles in the SPFO/SDeS mixed system. On the other hand, the surface tension, H-1 NMR and F-19 NMR data show smooth changes near and above the cmc. This is probably an indication that at each value of X_F, only one type of mixed HC/FC micelle is formed, and the composition of the mixed micelle changes continuously as the total surfactant concentration increases.

To compare the experimental data with theoretical calculations, the "pseudo-phase separation" model is considered. In this model, the aggregation of monomers to form micelles is treated as a phase separation, similar to the condensation of vapor into liquid. For a two-component system, Shinoda developed the following equations (5):

$$C_m = C_{1m} + C_{2m} \tag{1}$$
$$C_{1m} \cdot (C_{1m} + C_{2m})^k = C_1^{1+k} \cdot X_{FM} \cdot \gamma_1 \tag{2}$$
$$C_{2m} \cdot (C_{1m} + C_{2m})^k = C_2^{1+k} \cdot (1 - X_{FM}) \cdot \gamma_2 \tag{3}$$

where C_m is the cmc, C_{1m} and C_{2m} are the monomer concentrations of components 1 (HC) and 2 (FC), respectively, at the cmc; C_1 and C_2 are the cmc's of the pure components; X_{FM} is the mole fraction of the FC component in the micellar phase; γ_i (i = 1, 2) is the activity coefficient of the surfactant in the mixed micelle; k is the micelle counterion binding parameter. The parameter k is slightly dependent on the structure of the surfactant, and has a value between 0.58 and 0.65 for most single chain FC or HC anionic surfactants (10). A fixed value of k = 0.60 was used in our calculation in order to avoid the introduction of variable parameters. The activity coefficients γ_1 and γ_2 can be evaluated based on the concept of group contribution. One part of the contribution is due to molecular interactions and the other part is due to molecular size. Details of the treatment have be given by Asakawa *et al.* (10) and will not be repeated here.

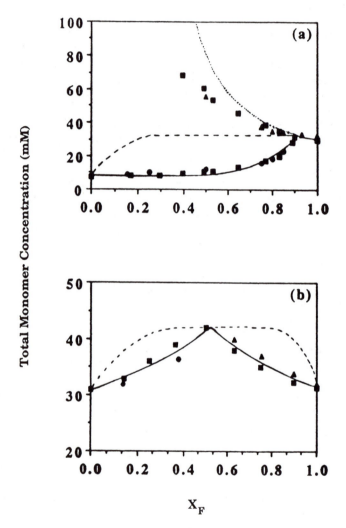

Fig. 2   Values of the cmc's for (a) the SDS/SPFO system and (b) the SDeS/SPFO system obtained from different measurements: surface tension, (■); H-1 NMR, (●); F-19 NMR, (▲). The curves are calculated values; see text for details of the calculation.

The results of the calculations are presented as smooth curves in Figure 2. In these plots, X_F (the mole fraction of FC in the whole system) is the abscissa for the solid lines, which represents the calculated cmc; X_{FM} (the mole fraction of FC in the micellar phase) is the abscissa for the dashed lines, which represent the composition of the micellar phase.

For the SPFO/SDS system with $X_F > 0.90$, the dashed line and the solid line are almost parallel to the abscissa and coincide with each other. This implies that the composition of the micellar phase is close to that of the monomer phase, and only one kind of micelle is formed. For intermediate mole fractions of the FC component, the dashed line has a horizontal portion, which implies that there are two types of mixed micelles existing simultaneously in the solution (2). The composition of the FC-rich micelles is $X_{FM} \approx 0.90$, and that of the HC-rich micelle is $X_{FM} \approx 0.25$-0.30. The dependence of the second cmc on X_F calculated from this model is shown as a dotted line in Figure 2. It agrees well with the experimental values of cmc(2) data for $X_F > 0.7$, but becomes substantially higher than the experimental result for $X_F < 0.6$. This might be due to the fact that the model calculation does not take into account the significant changes in the ionic strength at such high concentrations.

The calculated solid curve for the SPFO/SDeS system also agrees well with the experimental cmc data (Figure 2). The dashed curve represents the composition of the micellar phase. The curve has a rather flat portion, but no unambiguous horizontal part. This is consistent with the interpretation that only one type of mixed micelle exists, and its composition changes with X_F as well as the total surfactant concentration.

It should be pointed out that the results of the calculations are only semi-quantitatively correct because some simplified assumptions have been made (5,10). For example, the existence of two types of micelles in the SPFO/SDS system should lead to two sharp breaks in the dashed curve, but the calculated curve in Figure 2 shows a rather smooth change. One possible reason for this is that the value of k actually varies with composition and ionic strength, rather than being a constant as treated in the calculation. Furthermore, the calculation would be more complicated if the size of the mixed micelles is dependent on X_F. Nevertheless, our work shows that the experimental results can be reasonably accounted for by simple modelling of the mixed surfactants.

In short, the lack of affinity between the fluorocarbon chain and the hydrocarbon chain leads to a strong deviation from ideal behavior for the anionic fluorocarbon-anionic hydrocarbon surfactant mixtures. If the chains are long enough so that the mutual phobicity is large, two types of micelles can coexist in the same solution for intermediate values of X_F.

To investigate the effect of the head-group charge on the formation of mixed micelles, we have replaced the anionic hydrocarbon surfactant by nonionic, zwitterionic, and cationic hydrocarbon surfactants.

Anionic Fluorocarbon - Nonionic Hydrocarbon Surfactant Mixtures

The nonionic hydrocarbon surfactant we used was N-triethoxylated nonanamide (HEA8-3), which has a smaller head group than most other nonionic surfactants. Its cmc is 14.8 mM at 298 K. The cmc's of mixtures of HEA8-3 with SPFO were also determined as a function of X_F by means of surface tension and NMR (9,11). In this system, a single cmc was detected for each series of measurements, indicating that only one type of mixed micelles is formed. Unlike the cases of SPFO/SDeS and the SPFO/SDS systems, the SPFO/HEA8-3 system deviates negatively from ideal mixing (Figure 3). This behavior can be understood by the following consideration. When the nonionic surfactant molecules are incorporated into the micelles formed by SPFO, there is a reduction in the repulsive force between the negatively charged head groups of the latter. The decrease in coulombic repulsion overcomes the mutual phobicity between the fluorocarbon and hydrocarbon chains. Therefore, the cmc of a mixed system in the fluorocarbon-rich region decreases rapidly upon addition of HEA8-3 to the mixtures. When the composition of the mixed micelles is hydrocarbon-rich, the repulsive force between the anionic head groups becomes smaller, and the deviation from ideal behavior is less obvious.

Quantitatively, the formation of mixed micelles with differently charged head groups can be described by an alternate form of Equations (1) - (3):

$$C_m = (1 - X_{FM}) \cdot \gamma_1 \cdot C_1 \cdot I_1 + X_{FM} \cdot \gamma_2 \cdot C_2 \cdot I_2 \qquad (4)$$

where X_{FM} is the mole fraction of the fluorocarbon component in the mixed micelle, and I_1 and I_2 are factors accounting for varying ionic strengths. They are related to the mole fraction of the fluorocarbon component in the bulk solution, X_F, by (7)

$$I_1 = \{C_1/[C_m \cdot (1 - X_F)]\}^k \qquad (5)$$
$$I_2 = [C_2/(C_m \cdot X_F)]^k \qquad (6)$$

where k is the coefficient related to the counterion binding on the micelles, which is taken to be 0.59 in the present mixed systems (12). According to the regular solution theory, the activity coefficients can be expressed by the equations (13)

$$\gamma_1 = \exp[\beta \cdot (1 - X_{FM})^2] \qquad (7)$$
and $$\gamma_2 = \exp(\beta \cdot X_{FM}^2) \qquad (8)$$

where the parameter β is a parameter characterizing the extent of deviation from ideal behavior. However, the experimental data of the SPFO/HEA8-3 system and the system to be discussed below could not be fitted very well by the use of Equations (4)-(8). The most likely reason for this is that the activity coefficients expressed by Equations (7) and (8) are based upon the assumption of equal partial molar volumes for the two components. This is not the case for hydrocarbon and fluorocarbon surfactants. Therefore, another adjustable parameter δ is added to these expressions to account for the inequivalence and mutual phobicity of the two components (9):

$$\gamma_1 = \exp[X_{FM}^2 \cdot (\beta - \delta/2 + \delta \cdot X_{FM})] \qquad (9)$$
and $$\gamma_2 = \exp[(1 - X_{FM})^2 \cdot (\beta + \delta \cdot X_{FM})] \qquad (10)$$

The form of these equations is consistent with the Gibbs-Duham equation.

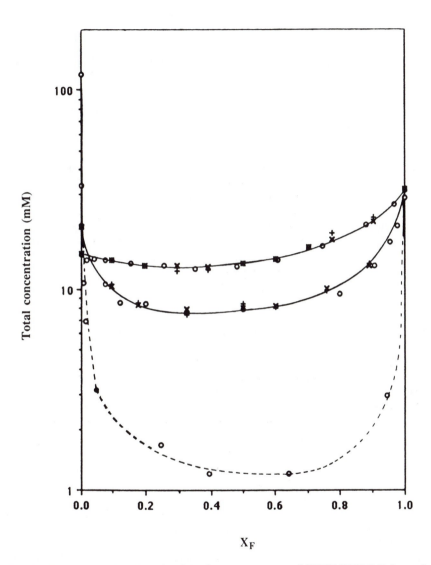

Fig. 3 Cmc values for the mixed surfactant systems of SPFO/HEA8-3 (upper), SPFO/DEDIAP (middle) and SPFO/OCTAB (lower) obtained from surface tension measurements (o), H-1 NMR (+), and F-19 NMR (x). The solid curves are calculated values.

Assuming that the cmc of the nonionic surfactant is not affected by the ionic strength of the solution, γ_1 in Equation (9) can be taken as unity. Using these conditions, Equation (4) gives a very good fit to the experimental data (Figure 3); the parameters obtained from the calculation are $\beta = -4.0$ and $\delta = 6.0$.

Anionic Fluorocarbon - Zwitterionic Hydrocarbon Surfactant Mixtures

For the zwitterionic surfactant DEDIAP, the positive charge carried by the tertiary ammonium group is located next to the hydrocarbon chain, while the negative charge carried by the sulfonate group is located at the end of the zwitterionic head group. When DEDIAP forms a mixed micelle with the anionic fluorocarbon surfactant SPFO, the tertiary ammonium group might be close to the carboxylic group of SPFO, and the sulfonate group is likely to extend farther into the aqueous medium. Therefore, there may be a small coulombic attraction between the head groups of the two types of surfactants in spite of the overall neutral nature of the zwitterionic surfactant. Consequently, the negative deviation from ideal behavior is more pronounced than the SPFO/HEA8-3 system (Figure 3). Again, only one set of cmc's was observed, and the cmc's data could be fitted to Equations (4) very well with the activity coefficients expressed by (9) and (10). The parameters obtained are $\beta = -9.5$ and $\delta = 14.7$. Compared to the previous system, the more negative value of β corresponds to a larger negative deviation. The more positive value of δ is rather unexpected, and is probably due to the fact that the hydrocarbon chain in DEDIAP is longer than that in HEA8-3.

Anionic Fluorocarbon - Cationic Hydrocarbon Surfactant Mixtures

Strong coulombic attractions between the head groups of cationic and anionic hydrocarbon surfactants always lead to very large deviations of these mixed surfactant systems (14). This is also true for anionic fluorocarbon surfactant and cationic hydrocarbon surfactant mixtures, even though there is a lack of affinity between the two types of hydrophobic chains. For the SPFO/OCTAB system, the deviation is so large that the cmc drops precipitously from 120 mM at $X_F = 0$ to 33 mM at $X_F = 0.001$ (11). The surface tension of the mixtures is also reduced considerably. While the values of surface tension are 22 and 46 nN/m^2, respectively, for the pure components, the surface tension at the cmc is 14-15 nN/m^2 for all mixtures with $0.01 \leq X_F \leq 0.94$ (9). Unfortunately, it is difficult to evaluate the activity coefficients analytically in this mixed system. Therefore, the dashed curve in Figure 3 is drawn to represent the experimental data, but it is not a calculated curve.

A characteristic of the mixed anionic fluorocarbon-cationic hydrocarbon surfactant mixtures not found in the other mixed systems discussed above is the formation of large aggregates at high surfactant concentrations. These aggregates may actually be tiny droplets dispersed in the aqueous phase instead of being large micelles, which by definition exist in a homogeneous solution. The evidence for this came from the H-1 and F-19 NMR study (11). For surfactant concentrations below 4-5 times the cmc, the spectra resemble those of normal surfactant systems. However, the peaks broaden abruptly at higher surfactant concentrations, and a second peak for the CF$_3$ group is observed in the F-19 spectrum. By the use of line-shape analysis (15), the average mean life time of the surfactant molecule in such a mixed system is found to be 5-7 x

10^{-4} s (*11*). This is to be compared with the mean life time of 10^{-6} to 10^{-7} s (*16*) for the exchange rate between monomers and normal-size micelles. The slow exchange is caused by the need for the surfactant molecules to cross the phase boundary of the aggregates during exchange. When these solutions were centrifuged at 14000 G for 15 minutes, the peaks in the NMR spectra became much sharper, and the extra CF_3 peak disappeared (*11*). This is due to the sedimentation of the aggregates, which have a larger density due to the presence of the fluorocarbon surfactant. Actually, the formation of coacervates and precipitates in mixed anionic hydrocarbon-cationic hydrocarbon surfactants is quite common (*17*), but the similar phenomenon for fluorocarbon hydrocarbon mixtures has not received much attention in the literature. We believe that our work is the first to address this aspect in some detail.

Summary

In mixed surfactant systems in which both the fluorocarbon component and the hydrocarbon component are anionic, the mutual phobicity between the chains is the dominant interaction, and the systems show strong deviation from ideal behavior. For the SFPO/SDS system in the intermediate mole-fraction range, two types of mixed micelles are formed, one rich in the fluorocarbon component, and the other rich in the hydrocarbon component. For the SPFO/SDeS system, only one type of micelles is formed. For mixtures of the anionic fluorocarbon surfactant SPFO with a nonionic, zwitterionic, or cationic hydrocarbon surfactant, the reduction in coulombic repulsion between the head groups dominates the interactions, and the systems deviate negatively from ideal behavior. The deviation is such that cationic > zwitterionic > nonionic. Again, only one type of micelles is formed. To account for the phobicity between the fluorocarbon and the hydrocarbon chains in calculating the cmc, a second parameter is added to the expression of activity coefficient developed from the regular solution theory. At high surfactant concentrations, the anionic fluorocarbon-cationic hydrocarbon mixtures form large aggregates which can be precipitated by centrifugation.

Acknowledgment. The authors gratefully acknowledge the assistance of industrial sponsors of the Institute for Applied Surfactant Research, including E. I. Du Pont de Nemours & Co., Kerr-McGee Corporation, Sandoz Chemicals Corp., and Union Carbide Corporation.

References

1. Tiddy, G. L. T., in *Mordern Trends of Colloid Science in Chemistry and Biology*, ed. Eicke, H. S., Birkhauser: Basel, **1985**, 148.
2. Funasaki, N., in *Mixed Surfactants*; ed. Ogino, K. and Abe., M., Marcel Dekker: N. Y., in press.
3. Klevens, H. B. and Raison, M., *J. Chimie*, **1954**, 51, 1.
4. Scamehorn, J. F., in *Phenomena in Mixed Surfactant System*, ed. Scamehorn, J. F., The ACS Symposium Series **1986**, 311,1.
5. Shinoda, K., Nomura, R., *J. Phys. Chem.*, **1980**, 84, 365.
6. Mukerjee, P., Yang, A. Y. S., *J. Phys. Chem.*, **1976,** 80, 1388; Sugihara, G., Nakamura, D., Okawaucm, M., Sakai, S., Kuriyama, K., Tanaka, M.,

Fukuoka, Univ. Sci. Reports, **1987**, 17, 31; Asakawa, T., Johten, K., Miyagishi, S., Nishia, M., *Langmuir*, **1988**, 4, 136.
7. Zhu, B. Y., Zhao, G. X., *Acta Chimica Sinica*, **1981**, 39, 493.
8. Guo, W., Brown, T. A., and Fung, B. M., *J. Phys. Chem.*, **1991**, 95, 1829.
9. Guo, W., *Ph. D. Dissertation*, University of Oklahoma, 1991.
10. Asakawa, T., Johten, K., Miyagishi, S., Nishida, M, *Langmuir*, **1985**, 1, 347; Asakawa, T., Mouri, M., Miyagishi, S., Nishida, M., *Langmuir*, **1989**, 5, 343; Asakawa, T., Miyagishi, S., Nishida, M., *J. Colloid Interface Sci.* **1985**, 104, 279; Asakawa, T., Imae, T., Ikeda, S., Miyagishi, S., Nishida, M., *Langmuir*, **1991**, 7, 262.
11. Guzman, E. K., *M. S. Thesis*, University of Oklahoma, 1989.
12. Lin, I. J., Somasundaran, P., *J. Colloid Interface Sci.*, **1971**, 37, 731.
13. Holland, P. M. and Rubingh, D. N., *J. Phys. Chem.*, **1983**, 87, 1984.
14. Holland, P. M., *Advances in Colloid and Interface Science*, **1986**, 26, 111.
15. Williams, K. C., Brown, T. L., *J. Am. Chem. Soc.*, **1966**, 88, 4134
16. Aniansson, E. A. G., Wall, S.N., Almgren, M., Hoffmann, H., Kielmann, I., Ulbricht, W., Zana, R., Lang, J., Tondre, C., *J. Phys. Chem.*, **1976**, 80, 905.
17. Stellner, K. L., Amante, J. C., Scamehorn, J. F., Harwell, J. H., *J. Colloid Interface Sci.*, **1988**, 123, 186; Stellner, K.L., Scamehorn, J. F., *Langmuir*, **1989**, 5, 70; Stellner, K. L., Scamehorn, J. F., *Langmuir*, **1989**, 5, 77.

RECEIVED January 6, 1992

Chapter 16

Coexisting Mixed Micelles
Pulse Radiolysis Studies

Mohammed Aoudia[1], Stephen M. Hubig[2], William H. Wade[3], and Robert S. Schechter[4]

[1]B[d] Abdellah El Aichi BP 270, University of Blida, Algeria
[2]Center for Fast Kinetics Research, University of Texas, Austin, TX 78712
[3]Department of Chemistry, University of Texas, Austin, TX 78712
[4]Department of Chemical Engineering, University of Texas, Austin, TX 78712

Mixed fluorocarbon-hydrocarbon surfactant micelles may exhibit two distinct compositions which, in the phase separation limit, can be considered as coexisting mixed micelles in equilibrium with the same monomer. The concentrations in excess of the mixture CMC for which two distinct types of mixed micelles coexist have been determined using pulse radiolysis. The principle upon which the determination relies is that the decay of hydrated electrons is a function of the fluorocarbon monomer concentration even in the presence of other surfactant monomers and mixed micelles. The mixture CMC determined by pulse radiolysis is confirmed by surface tension measurements.

Since Mukerjee and Yang (1) suggested that mixed hydrocarbon and fluorocarbon surfactant micelles may exhibit a distinct biomodal distribution over a range of solution compositions and concentrations, a number of papers have appeared [Zhoa and Zhu (2), Funasaki and Hada (3), Carlfers and Stieb (4), Funasaki and Huida (5), Zhao and Zhu (6), Ueno et al. (7), and Smith and Ottewill (8)] confirming the existence of two distinct types of micelles, one rich in fluorocarbon surfactant and the second type composed mainly of hydrocarbon surfactant molecules. The main evidence supporting this hypothesis is the existence of two critical micelle concentrations as the total surfactant concentration is increased keeping the ratio of the two types of surfactant fixed and the primary probe used to detect these two CMC's has been surface tension. In many cases in dealing with this sort of system, it may not be easy to detect both CMC's. Often one is dealing with surface tensions that vary by only a few tenths of a dyne per cm and breakpoints characterizing the CMC are slight. Furthermore, the composition of the monomer in equilibrium with the micelles is often inferred by asserting that the surface tension of a binary surfactant mixture depends on the monomer concentration and composition but is independent of the composition of the micelles. There is evidence to support this argument [Clint (9)]; however, if substantial changes in monomer concentration produce only minor variations insurface tension, which is often the case when

0097–6156/92/0501–0255$06.00/0

mixtures of fluorocarbon and hydrocarbon surfactants are used, then the monomer concentration can only approximately be located.

In this paper we present a new technique based on the decay of hydrated electrons that can be used to measure the monomer concentration of a fluorocarbon surfactant over a wide range of conditions even in the presence of mixed micelles of varying compositions. Other studies of hydrated electron e_{aq}^- reactions in micellar dispersions have been reported but these have generally concentrated on the interactions of the electrons with solubilized substrates (10). Relatively few studies have related to the intrinsic reactivity of the surfactant with e_{aq}^- (11,12) and none to our knowledge have considered the interaction in mixed surfactant systems. Here, we show that this technique can be used for the study of mixed fluorocarbon-hydrocarbon surfactant systems and even in cases where miceller phase separation occurs, the complex phase boundaries can be precisely mapped by observing the decay of hydrated electrons. In fact, as will be seen, this method, when applicable, appears to enjoy significant advantages relative to any other that has been used to date.

The results presented here establish that fluorocarbon-rich and hydrocarbon-rich micelles do coexist within a certain range of compositions and a complete pseudophase diagram is developed. Based on these data, we have tested various models that have been previously used to represent mixed micelle behavior and have found that none of them apply to the system studied here.

Experimental

The fluorocarbon surfactant Neos Ftergent (FC) from Neos Company was recrystallized four times from acetone. Its molecular structure is shown by Figure 1. Sodium dodecyl sulfate (HC) from BHD Chemicals Ltd., Poole, England, was 99% pure and used as supplied. Sodium chloride from Fisher Scientific (certified ACS) was used as supplied. Aqueous surfactant solutions were made with Millipore "Reagent Grade" water.

Surface tension was measured by the spinning drop technique (13).

Electron pulse radiolysis experiments were carried out using a 4 MeV Van-de-Graaff electron accelerator and a time resolved spectrophotometer including a 500 W Xenon Lamp (Oriel), a monochromator (Baush and Lomb), and a photo multiplier tube (Hamamatsu). The samples were renewed after each electron pulse using a remotely controlled syringe driver and a flow cell. With an absorbed dose between 110 and 370 rad, the concentration of hydrated electrons e_{aq}^- produced per electron was ca 0.3 - 1 μM. The decay of e_{aq}^- (at 720 nm) was measured at different surfactant compositions. The experimental data were best fitted by first order kinetics. All samples were deaerated prior to use by blowing a stream of nitrogen over the top of the sample for about one hour while stirring vigorously.

In the absence of sodium chloride, the CMC of the fluorocarbon surfactant is 2 mM. The bimolecular rate constant k_{FC} for the reaction between hydrated electrons e_{aq}^- and the fluorocarbon was determined to be 2.2 x 10^{10} $M^{-1}s^{-1}$. This leads to one observed rate constant k_0 at the CMC of 4.4 x 10^7 s^{-1} which is much faster than the time resolution of our experimental setup (250 ns). To partially compensate for this difficulty, this study has been carried out in 0.08 M NaCl solutions to decrease the CMC's of the anionic surfactants. This concentration of electrolyte also suppresses the influence of the varying surfactant counterion concentration.

Principle of the Method

Pulse radiolysis of deaerated aqueous solutions generates the transient species: hydrated electrons e_{aq}^-, hydrogen atom ($H^{\bullet}$), hydroxyl radicals ($^{\bullet}OH$), and oxyde radicals ($^{\bullet}O^-$). The interaction of these species with an aqueous surfactant solution may be characterized in kinetic terms. The hydrated electron shows a strong absorption maximum (λ_{max} = 720 nm, ε = 19000 $M^{-1}s^{-1}$). Early pulse radiolysis measurements showed that the decay of the hydrated electron is catalyzed by the presence of benzene and that reaction follows first order kinetics (*14*). Our study relies on the strong catalytic action imparted by the monomers of FC surfactant. The decay reaction

$$e_{aq}^- + FC + HC \rightarrow Products \qquad (1)$$

is first order in hydrated electron concentration and the first order rate constant, k_o, can be represented as

$$k_o = K_{intrinsic} + K_{FC}[FC] + K_{HC}[HC] \qquad (2)$$

where $K_{intrinsic}$ is the rate constant in aqueous electrolyte solutions in the absence of added surfactant (6 x $10^4 s^{-1}$) and [FC] and [HC] are the respective concentrations of FC and HC.

Experiments revealed that $K_{HC} \cong 10^6 M^{-1}s^{-1}$ at HC concentration of 3 mM. One should note that $K_{HC} < 2$ x 10^{-5} $M^{-1}s^{-1}$ has been reported for HC concentration of 50 mM (*15*).

The reactivity is greatly enhanced by the presence of aromatic constituents. For example, K for the monosubstituted benzene derivatives $C_6H_5SO_3Na$ and $C_6H_5CF_3$ are 4 x 10^9 $M^{-1}s^{-1}$ and 1.8 x 10^9 $M^{-1}s^{-1}$, respectively (*16*). Furthermore, a rate constant of 5 x 10^8 $M^{-1}s^{-1}$ has been reported for a twin-tailed alkyl benzene sulfonate below its CMC (*12*).

The FC molecules aggregated as micelles do not contribute to the decay constant. The plateau region found at higher FC concentrations in Figure 2 support this contention. The mechanism responsible for shielding aggregated FC molecules appears to be related to their negative charge. Electrons are apparently repelled by the negatively charged Stern layer although we know that the shielding is not complete. We have, for example, previously found that although micelle formation does inhibit the decay of hydrated electrons, the rate continues to increase with increasing concentrations of alkyl benzene sulfonate micelles (*12*). In these systems, the reactive benzene moiety is located near the micellar interface. Thus, it was concluded that diffusion of e_{aq}^- into the Stern layer is not completely hindered by electrostatic forces (*12*). Overall, however, our experiments show that the reactivity of e_{aq}^- is primarily determined by FC monomer.

Based on these values, it is reasonable to expect the decay rate to be most sensitive to the FC monomer concentration. This is confirmed by the results shown by Figure 3. For surfactant concentrations below the mixture CMC, the rate constant is a linear function of FC concentration as represented by equation 2. Furthermore, as expected, k_o is practically independent of the HC concentration. From the slope of the variation of k_o with FC monomer concentration, the bimolecular rate constant K_{FC} for the reaction of the fluorocarbon surfactant with e_{aq}^- is computed and found to be 2.2 x 10^{10} $M^{-1}s^{-1}$, four orders of magnitude larger than K_{HC}.

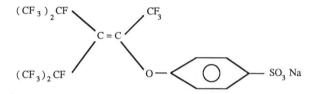

Figure 1. Chemical structure of the fluorocarbon surfactant Ftergent Neos from Neos Company.

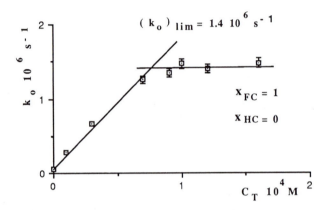

Figure 2. Dependence of the rate constant k_O of the reaction of e_{aq}^- with fluorocarbon surfactant monomer.

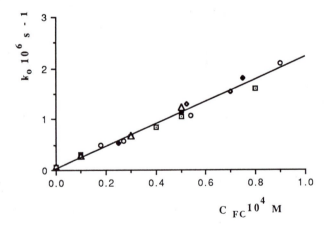

Figure 3. Dependence of the rate constant k_O of the reaction of e_{aq}^- with fluorocarbon surfactant concentration at different system compositions: $x_{FC} = 0.1$ (◊); 0.5 (♦); 0.7 (⊡); 0.9 (o) and 1 (Δ).

Results and Discussion

The decay of hydrated electrons has been measured at different mixture compositions. In some cases, the transition from linearly increasing k_0 with increasing FC surfactant (see Figure 2) to a plateau value is sharp. However, for a wide range of mixture compositions, k_0 deviates from linear behavior at a well-defined concentration but gradually approaches a plateau value. Representative examples of both types of trends are shown by Figures 4 and 5.

To interpret these two different patterns of behavior, it must be understood that in contrast to pure surfactant systems, when mixed micelles form, the monomer concentrations continue to vary as the total surfactant concentration is increased (*17,18*). In all cases, the initial deviation from linear behavior is interpreted to denote the onset of mixed micelle formation. Furthermore, a plateau region (if one exists) corresponds to a region of constant monomer concentration.

Thus, the two curves shown in Figure 4 indicate two transitions with an intermediate zone of varying FC monomer concentration. The plateau values for k_0 in both cases is found to be 2.0×10^6 M^{-1}s^{-1}. The curves in Figure 5 show, in contrast with those in Figure 4, a single sharp transition.

These intriguing results and all of our other observations are consistent with the pseudophase diagram shown by Figure 6. In preparing this diagram, the micelle is treated as a separate phase so that sharp, rather than diffuse, phase boundaries can be drawn and micelles can be assigned distinct composition rather than being considered as a collection of aggregates having a distribution of aggregation numbers and compositions.

The pseudophase diagram is divided into four separate regions. Region I represents compositions below CMC. In this region, an increase of C_T at constant x_{HC} yields a corresponding linear increase in k_0 as defined by equation 2. The broken line separating region I from the other regions is the mixture CMC. Values of the mixture CMC are given in Table I. As noted above, these values are the concentrations at which a sudden change in slope occurs when k_0 is plotted against C_T keeping x_{HC} fixed.

Table I. Critical Micellar Concentration (CMC) of Fluorocarbon (NF) and Hydrocarbon (HC)

Fluorocarbon Molar Fraction x_{FC}	CMC (mM) (Pulse Radiolysis)	CMC (mM) (Surface Tension)
0.0	-----	1.50
0.1	0.70	0.74
0.2	0.45	0.45
0.3	0.25	0.22
0.4	0.22	0.20
0.6	0.15	0.15
0.7	0.15	0.15
*0.8	*0.11	*0.13
0.9	0.10	0.11
1.0	0.074	0.08

*Degenerate point.

In region III, $k_0 = 2.0 \pm 0.1 \times 10^6$ s^{-1} for all compositions in this region. Experimental points are designated in Figure 6. This, therefore, is a region of <u>constant</u> FC monomer concentration and hence constant FC chemical potential.

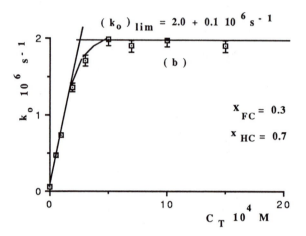

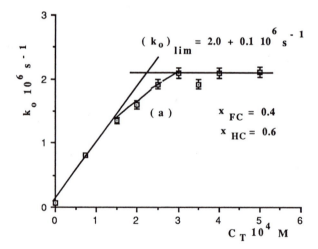

Figure 4.   Dependence of k_O for the reaction of e_{aq}^- with total surfactant concentration at two different fluorocarbon molar fractions: $x_{FC} = 0.4$ (a) and 0.3 (b).

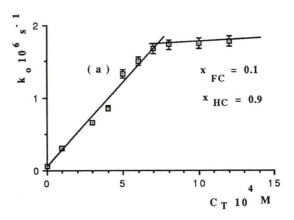

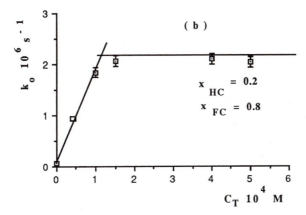

Figure 5. Dependence of the rate constant k_0 for the reaction of e_{aq}^- with surfactant concentration at two different fluorocarbon molar fractions: $x_{FC} = 0.1$ (a) and 0.8 (b).

Such behavior is expected when two different micelles—one HF-rich and the other HC-rich—coexist. In this case,

$$\mu_{monomer} = \mu'_{micelle} = \mu''_{micelle} \tag{3}$$

where "prime" and "double prime" represent HC and FC-rich micelles, respectively. An overall composition represented by a point in region III, therefore, consists of a monomer having a mixed composition, micelles having mole fractions x'_{HC} and x''_{HC}.

Based on these results, the composition of the equilibrium monomer phase can for the first time be accurately obtained. Heretofore, this composition has been inferred based on surface tension measurements. In this approach, the monomer composition can be calculated using equation 2 and the plateau value for k_o found throughout region III. This calculation yields

$$[FC]_{monomer} = 8.8 \times 10^{-5} \text{ M} \tag{4}$$

From Table I, it is seen that this monomer concentration corresponds to a point on the CMC curve which is $CMC^* = 1.1 \times 10^{-5}$ M at $x^*_{HC} = 0.2$. The monomer concentration of HC throughout the region is, therefore, found to be 2.2×10^{-5} M.

The monomer in equilibrium with the two micelles that coexist throughout region III is, therefore, represented as a single point on the mixture CMC curve. This point has been called a "degenerate point" (5) and is labeled here CMC^*. The degenerate point is the only one on the mixture CMC curve that is included in region III. Therefore, increasing C_T at $x_{HC} = 0.2$ will pass directly from region I into III and only one transition should exist when k_o is plotted as a function of C_T. This transition from a linear curve to the plateau should occur with no intervening transition region. Note that this expected behavior is the one realized as shown by Figure 5(b).

The composition of the micelles in region III can also be determined by hydrated electron decay using the following strategy. Select a C_T in region IV where only x_{HC}-rich micelles exist. Keeping C_T constant, decrease x_{HC} until the plateau value of k_o is attained. Two such scans are shown in Figure 7. The compositions on the boundary separating region III from IV are easily distinguished since k_o is a known constant throughout region III. At this boundary monomer and HC-rich micelles of the type that exists throughout region III are in equilibrium. There are no FC-rich micelles. Thus, the total FC inventory is either present as monomer or in HC-rich micelles.

$$x_{FC}C_T = [FC] + (C_T - CMC^*)x''_{FC} \tag{5}$$

Since x_{FC} and C_T at the boundary are measured and [FC] monomer and CMC^* correspond to the conditions at the degenerate point, then x''_{FC} can be calculated. Taking all of our measurements into account, $x''_{FC} = 0.15$. Based on this value and those of the degenerate point, the curve (solid line) separating regions III and IV can be drawn.

The determination of the phase boundary separating regions II and III is slightly less precise because for concentrations much above the CMC, the FC monomer concentration is very nearly that found in the plateau region and a sharp transition from regions II to III is not evident. Thus, we were restricted to measuring this transition at total surfactant concentrations slightly greater than the mixture CMC. The FC-rich micelles were found to have a composition of $x''_{FC} = 0.9$. Using this value, the curve separating regions II and III is calculated.

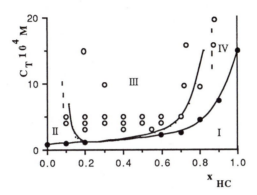

Figure 6. Micellar pseudophase diagram for the mixed FC-HC system, I: < CMC, II: FC-rich micelles only, III: FC-rich and HC-rich micelles determined from material balance (solid curve)and from experimental measurements (o o o), IV: HC-rich micelles only.

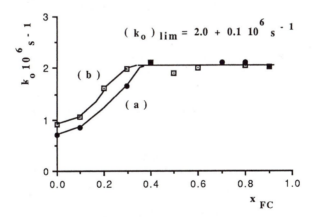

Figure 7. Dependence of the rate constant k_0 for the reaction of e_{aq}^- with fluorocarbon molar fraction of two different overall surfactant concentrations: $C_T = 0.4$ mM (a) and 0.5 mM (b).

In regions II and IV, only one type of micelle exists in equilibrium with a monomer of varying composition.

The phase diagram shown by Figure 6 is entirely consistent with all observations of k_0 as a function of C_T. It also is confirmed by extensive surface tension measurements. Two interesting scans of surface tension as a function of total surfactant concentrations are shown by Figure 8. The upper curve shows the surface tension as C_T is increased keeping the ratio of FC to HC constant. The mixture CMC boundary is crossed from region I into IV. The concentration at crossing (the sudden change in slope of the surface tension) compares well with the value found by pulse radiolysis. Values for the mixture CMC determined by the two different methods are compared in Table I. It is seen that the methods are in good agreement. Furthermore, since in region IV only HC-rich micelles exist, the monomer composition depends on C_T, thus explaining the continued decrease in surface tension seen in Figure 8(a).

The scan shown by Figure 8(b) passes through the degenerate point going, therefore, from region I into III without passing through intermediate regions. Thus, the surface tension is expected to attain a plateau value as it does.

Scans for any other composition terminating in region III must first pass through either region II or IV and thus show two break points. Examples of this type of behavior are shown by Figure 9.

All of the surface tension results confirm the phase diagram shown by Figure 6, making this the most complete micellar phase diagram including micellar demixing yet published.

Final Remarks

Since we have a reasonable picture of the behavior of mixed FC and HC micelles, it is of interest to determine whether any of the models previously applied to describe mixed micelle behavior are applicable to this system. The conditions to be satisfied by a model are unusually stringent. The mixture CMC curve has in the past been successfully modeled using a regular solution approximation for the activity coefficient of surfactant molecules in mixed micelles (19,20,21). Furthermore, regular solutions may exhibit phase separation for certain values of the interaction parameter and, therefore, apply when micelle demixing occurs. However, this model is not applicable to FC and HC micelles because, at phase separation, regular solution models predict a symmetry not observed in this system. The compositions of the coexisting micelles are not symmetric since $x_{HC} = 0.85$ and $x''_{HC} = 0.9$. Thus, the simple regular solution model which has been used with some success in the past will not be useful here. Funasaki and Hada (3) have applied a nonsymmetric model which is an extension of regular solution theory. By an appropriate choice of the two parameters, this model can predict both of the region III micellar compositions but then using these two parameters, the theoretical degenerate point is incorrectly positioned along the mixture CMC curve. The deviation from the measured degenerate is unacceptably large.

Thus, we conclude that the phase behavior of the system studied here will be difficult to model. Certainly, some of the models which are useful for systems that do not exhibit micelle demixing will not apply here.

Conclusions

A micellar pseudophase diagram for a binary surfactant mixture has been constructed using a new technique, pulse radiolysis, that measures the monomer concentration of one of the surfactant even in the presence of mixed micelles of varying composition. This is the first diagram which includes micelle demixing that

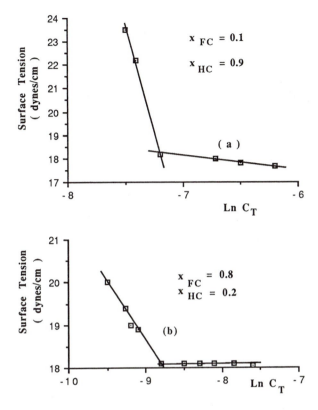

Figure 8. Variation of the surface tension with Ln C_T at different system compositions: $x_{FC} = 0.1$ (a) and 0.8 (b).

has been constructed without relying on the interpretation of the surface tension changes with composition. However, extensive surface tension measurements have also been carried out and these confirm position of the pseudophase boundaries found by pulse radiolysis.

The observed first order decay rate of hydrated electrons has been found to be constant within a certain domain of mixture compositions thereby providing the direct evidence for micellar demixing. In this domain, the monomer concentration is constant—a condition necessary for micelle demixing.

The degenerate point which is often difficult to determine based on surface tension is accurately determined by this technique.

The existence of a well-defined pseudophase diagram will challenge those modeling mixed micelle behavior. Neither a regular solution model nor an assymmetric regular solution model can correctly map the observed phase boundaries. It will be difficult to find a comprehensive model representative of this system studied here.

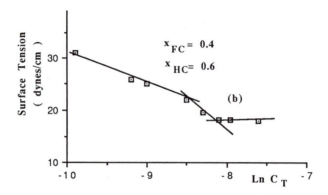

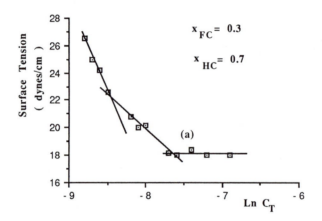

Figure 9. Variation of the surface tension with Ln C_T at different system compositions: $x_{FC} = 0.3$ (a) and 0.4 (b).

Acknowledgments

Pulse radiolysis experiments were performed at the Center for Fast Kinetics Research, which is supported jointly by the Biomedical Research Technology Program of the Division of Research Resources of the National Institutes of Health (RR00886) and by The University of Texas at Austin. This research was also sponsored by the Robert A. Welch Foundation.

Literature Cited

1. Mukerjee, P.; Yang, A. *J. Phys. Chem.* 1976, *80*, 1388.
2. Zhoa, G. X.; Zhu, B. Y. *Colloid Polymer Sci.* 1983, *261*, 89.
3. Funasaki, N.; Hada, S. *J. Phys. Chem.* 1983, *87*, 342.
4. Carlfors, J.; Stilbs, P. *J. Phys. Chem.* 1984, *88*, 4410.
5. Funasaki, N.; Hada, S. *J. Phys. Chem.* 1980, *84*, 736.
6. Zhoa, G. X.; Zhu, B. Y. In *Phenomena in Mixed Surfactant Systems*; Scamehorn, J. F., Ed.; ACS Symp. Series 311; American Chemical Society: Washington, DC, 1986, Chapt. 30.

7. Ueno, M.; Shioya, K.; Nakamura, T.; Meguro, K. In *Colloid and Interface Science*; Kerker, M., Ed.; Academic Press: New York, NY, 1977, 2; 411.
8. Smith, I.; O'Hewill, R. *In Surface Active Agents*; SCI Symp. Proc.; London, 1979; 77.
9. Clint, J. H. *J. Chem. Soc. Faraday Trans. I* 1973, *17*, 1327.
10. Farhataziz; Rodgers, M.A.J. In *Radiation Chemistry: Principles and Applications*; VCH Publishers, Inc.: 1987.
11. Gratzel, M.; Thomas, J. K. Patterson, L. K. *Chem. Phys. Lett.* 1974, *29*, 393.
12. Aoudia, M.; Rodgers, M.A.J.; Wade, W. H. In *Surfactant Solutions*; Mittal, K. L.; Bothorel, P., Eds.; Plenum Press, New York, NY, 1986; 103.
13. Cayias, J. L.; Wade, W. H.; Schechter, R. S. In *Adsorption at Interfaces*; Mittal, L. K., Ed.; ACS Symposium Series 234; American Chemical Society: Washington, DC, 1975, Chapt. 8.
14. Fendler, J. H.; Patterson, L. K. *J. Phys. Chem.* 1970, *74*, 4608.
15. Bansal, K. M.; Patterson, L. K.; Fendler, E. J.; Fendler, J. H. *J. Radiol. Phys. Chem.* 1971, *3*, 321.
16. Anbon, M.; Hont, E. J. *J. Amer. Chem. Soc.* 1964, *86*, 5633.
17. Lange, H.; Beck, K. H. *Kolloid Z u. Z. Polymere* 1973, *251*, 424.
18. Franses, E. I.; Bidner, M. S.; Scriven, L. E. In *Micellization, Soubilization, and Microemulsions*; Mittal, K. L., Ed.; Plenum Press, New York, NY, 1977; 855.
19. Rubingh, D. N. In *Solution Chemistry of Surfactants*; Mittal, K. L., Ed.; Plenum Press, New York, NY, 1979; 337.
20. Scamehorn, J. F.; Schechter, R. S.; Wade, W. H. *J. Disp. Sci.*, 1982, *3*, 2611.
21. Holland, P. M.; Rubingh, D. N. *J. Phys. Chem.*, 1983, *87*, 1984.
22. Holland, P. M.; *Advances in Colloid Interface Sci.*, 1986, *26*, 111.

RECEIVED January 6, 1992

Chapter 17

NMR Spectroscopic and Neutron Scattering Studies on Ammonium Decanoate–Ammonium Perfluorooctanoate Mixtures

R. M. Clapperton[1], B. T. Ingram[2], R. H. Ottewill[1], and A. R. Rennie[1]

[1]School of Chemistry, University of Bristol, Bristol BS8 1TS, United Kingdom
[2]Procter & Gamble Limited, Newcastle Technical Centre, Newcastle upon Tyne, NE12 9TS, United Kingdom

The composition of mixed micelles of ammonium perfluoro-octanoate and ammonium decanoate was previously investigated using small angle neutron scattering. In this paper nmr was used to examine micellar composition using ^{1}H and ^{19}F resonances respectively from the hydrocarbon and fluorocarbon chains. The results from nmr experiments were compared with those from neutron scattering. Good correspondence between the two techniques was obtained.

In the case of mixed micelle formation an important parameter is the mole fraction of the two species of surface active agent in the micelle. A particularly interesting case occurs with mixtures of hydrocarbon and fluorocarbon surfactants where a point of interest is how the stiffer fluorocarbon chains can arrange or admix in the presence of the more flexible hydrocarbon chains. In previous work (1,2,3) we examined by small angle neutron scattering mixed surfactant systems composed of ammonium perfluoro-octanoate (APFO) and ammonium decanoate (AmDec) in NH_4OH : NH_4Cl buffer at pH 8.8 and an ionic strength of 0.1. These experiments using a variable contrast approach indicated that mixed micelles were formed over the range of molar concentration ratio's examined. Analysis of the scattering data gave the composition of the micelles at various mol ratio's.

An alternative approach to determine the micellar composition is to use a combination of proton ^{1}H nmr and ^{19}F nmr and in this contribution we present information on results obtained by this technique (4). The results obtained by nmr appear to be in good agreement with those obtained from neutron scattering.

This work has been carried out in conjunction with other studies designed to determine the composition of the surface film at the air-water interface existing in equilibrium with surfactant species in solution both below and above the critical micelle concentration. This has involved a detailed study of the surface tension of mixed systems (5) and use of the neutron reflectivity method to determine the surface composition.   Again the difference in coherent scattering length density between hydrocarbon chains and fluorocarbon chains makes them ideal systems for this technique (6).

Experimental

Materials.   Water was obtained by double distillation from an all-Pyrex apparatus.

Decanoic acid was Fluka puriss grade material with an estimated g.l.c. purity of greater than 99%.   This was converted to the ammonium salt, ammonium decanoate (AmDec), by slow neutralisation with ammonium carbonate.   Ammonium perfluorooctanoate was Rimar material which was recrystallised from diethyl ether before use.

All experiments were carried out in ammonium chloride - ammonium hydroxide solutions with an ionic strength of 0.1 at a pH of 8.8.

Small Angle Neutron Scattering.   All the scattering expriments were carried out using the neutron diffractometer D17 at the Institut Laue Langevin, Grenoble, using neutrons of wavelength 1 nm.   Details of the experimental procedures and data processing have been given elsewhere (1,2,7).   All experiments were carried out at 25°.

Nuclear Magnetic Resonance.   [19]F nmr spectra were obtained using an FX90 Q Fourier Transform spectrometer and ^{1}H spectra were obtained using a JNM GX 270 FT spectrometer.   All the samples contained 10% D_2O as a reference material and were examined at a temperature of 20° (4).

Results

Single Component Systems.   The ^{19}F spectrum of APFO and the ^{1}H spectrum of AmDec are shown in Figure 1.   The assignments of the peaks are marked on the figures together with the chemical formulae of the compounds.

The APFO ^{19}F spectra show three sharply defined peaks for the units with carbon atoms C_2, C_7 and C_8.   All of these show a down-field shift above the c.m.c. indicating a reduction in the shielding effect.   The magnitude of the shifts is in the order, $C_8 > C_7 \gg C_2$ which can be explained by the greater reduction in water contact in the micelle for those groups in the fluorocarbon chain far removed from the head group.

The observed chemical shift, δ_{obs}, can be related to the chemical shift of the monomer molecules, δ_{mo}, and the micelles, δ_{mi}, as well as to the concentrations of monomer, micelles and total concentration, C_{mo}, C_{mi} and C_T respectively by the equation:

$$\delta_{obs} = \frac{C_{mo}}{C_T}\delta_{mo} + \frac{C_{mi}}{C_T}\delta_{mi} \qquad \ldots \quad (1)$$

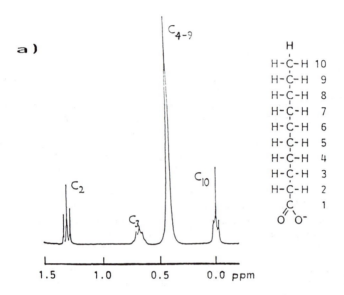

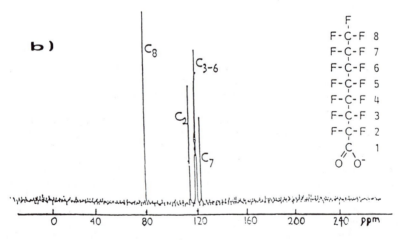

Figure 1.   a) ^{1}H spectrum of AmDec.
             b) ^{19}F spectrum of APFO.

Moreover,

$$C_{mi} = C_T - C_o \qquad \ldots (2)$$

where C_o = the c.m.c. and all concentrations are expressed as concentrations of monomeric surfactant in mol dm^{-3}. Above the c.m.c. we obtain

$$\delta_{obs} = \frac{C_o}{C_T} (\delta_{mo} - \delta_{mi}) + \delta_{mi} \qquad \ldots (3)$$

and below the c.m.c.

$$\delta_{obs} = \delta_{mo}$$

Some results for solutions of APFO are shown in Figure 2. Extrapolation of the two linear portions of the curve, using the C_8 unit data to give maximum accuracy, gave a value for the c.m.c. of 8.3×10^{-3} mol dm^{-3}. This can be compared with the value of 9.6×10^{-3} mol dm^{-3} obtained from surface tension measurements of solutions using the same buffer (5).

The results obtained for AmDec using the ^{1}H chemical shifts and expressing the ordinate in terms of a frequency shift are shown in Figure 3. These data, using the C_{10} unit results, gave a value of 5.7×10^{-2} mol dm^{-3}, for the c.m.c. which can be compared with the value of 4.07×10^{-2} mol dm^{-3} obtained from surface tension measurements (5).

Mixed Systems

The stoichiometric mol fraction, α, of the mixed surfactants was defined by the expression,

$$\alpha = \frac{[AmDec]}{[APFO] + [AmDec]} \qquad \ldots (4)$$

with [AmDec] and [APFO] the concentrations of the two components expressed in mol dm^{-3}.

For the mixed systems, δ_{obs} is again the sum of the micellar (δ_{mi}) and monomer (δ_{mo}) contributions (see equations 1 and 3). It can be assumed that the monomer chemical shift above the mixed c.m.c. is constant and equal to the value below the c.m.c. The micelle chemical shift δ_{mi} is dependent on the composition of the micelle. In a mixed micelle containing both hydrocarbon and fluorocarbon molecules, each fluorine and hydrogen atom, to a lesser or greater extent, is in the proximity of hydrogen atoms from hydrocarbon chains and fluorine atoms from fluorocarbon chains. The chemical shift of each atom will depend upon the average shielding effect of the surrounding chains, and hence upon micelle composition.

For a mixed system of a given composition and concentration, the micelle composition, x, can be determined from the observed ^{19}F and ^{1}H chemical shift using the equation,

$$\delta_{obs}^{HC} = \left[1 + \left(\frac{\alpha-1}{\alpha}\right)\left(\frac{x}{1-x}\right)\left(1 - \frac{\delta_{obs}^{FC} - \delta_{mic}^{FC}}{\delta_{mo}^{FC} - \delta_{mic}^{FC}} \right) \right]\left(\delta_{mo}^{HC} - \delta_{mic}^{HC}\right) + \delta_{mic}^{HC} \qquad \ldots (5)$$

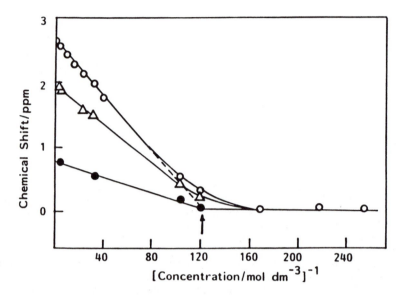

Figure 2.   Chemical shift (ppm) against reciprocal concentration
of APFO, ● , C_2; △ , C_7; ○ , C_8; ↑ , intercept
value.

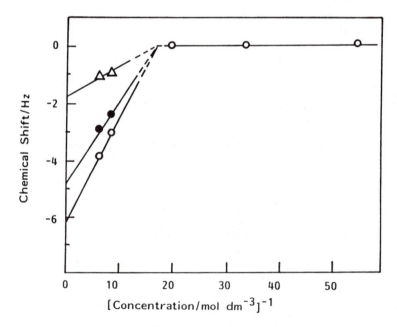

Figure 3.   Chemical shift (H_z) against reciprocal concentration
of AmDec, △ , C_3; ● , C_{4-9}; ○ , C_{10}.

This equation can be solved numerically to give x if δ_{mic}^{FC} and δ_{mic}^{HC} are known as a function of micelle composition. This relationship can be established from δ_{obs} at high mixture concentrations, at which point the micelle composition must tend to the solution composition α.

The ^{19}F chemical shifts of the C_8 units are shown in Figure 4 for mixtures with α = 0, 0.56, 0.72 and 0.89. The curve of δ_{obs} against C_T^{-1} for the mixtures is non-linear, suggesting a change of micelle composition with total surfactant concentration. The monomer chemical shift is identical to that observed for the single component system and indicates negligible contact between molecules in the monomeric state. α = 0.56 and α = 0.72 show association of the APFO at concentrations of 1.34 x 10^{-2} amd 2.00 x 10^{-2} mol dm^{-3} respectively. These values can be compared with c.m.c. values for the mixed systems of 1.48 x 10^{-2} mol dm^{-3} and 2.27 x 10^{-2} mol dm^{-3} estimated from surface tension measurements (5). In the case of α = 0.89, however, the curve was linear (Figure 4) at high concentrations but then changed gradient and decreased more rapidly to give an intercept value of 182.0 ppm corresponding to an association concentration of 5.5 x 10^{-3} mol dm^{-3}.

The ^1H chemical shift data for α values of 0.56, 0.89 and 1.0 (AmDec) are shown in Figure 5. In this case, the proton spectrum at α = 0.89 has two linear portions suggesting a sharp association of hydrocarbon chains and indicating that the micelle composition attains a constant value. Extrapolation indicates association at a mean concentration value of ca. 5.65 x 10^{-2} mol dm^{-3}; this can be compared with the value of 4.3 x 10^{-2} mol dm^{-3} obtained from surface tension data.

The α = 0.56 curve is non-linear on both of the plots of ^1H and ^{19}F chemical shifts (Figures 4 and 5) suggesting that the composition of the micelle changes steadily with change in surfactant concentration. At concentrations close to the c.m.c. the micelles appear to be fluorocarbon rich. However, at high concentration the hydrocarbon content increases and eventually tends towards x = 0.56.

Small Angle Neutron Scattering. An alternative approach to obtaining the composition of the micelles is to use small angle neutron scattering. This, as shown earlier (2), is particularly appropriate for the hydrocarbon-fluorocarbon systems since the two types of chains have different coherent scattering length densities. By examining the scattering of the micellar systems in $H_2O:D_2O$ mixtures, it is possible to find the medium composition at which the scattered intensity goes to zero (2); location of this point enables the composition to be obtained. Also from the scattering data it was possible to obtain the micellar mass as a function of α. This was found to be 18,000 at α = 0 [APFO] reaching a value of 40,000 at α = 0.40 and then decreasing again to 13,000 for α = 1.0 [AmDec].

Figure 6 shows a plot of the micelle composition, x, as a function of α. The values obtained independently from nmr experiments and from small angle neutron scattering are shown on the same plot. The agreement between the two sets of results is excellent.

For ideal mixing the mol fraction of AmDec in a mixed micelle is given by (9),

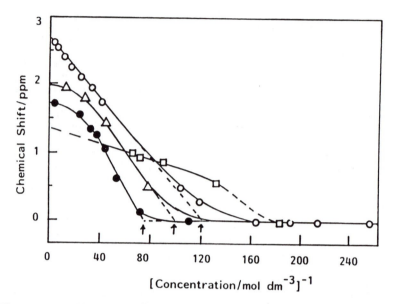

Figure 4.   Chemical shift (ppm) against reciprocal concentration
             of APFO; $\bigcirc$, $\alpha = 0$; $\triangle$, $\alpha = 0.56$; $\bullet$, $\alpha = 0.72$;
             $\square$, $\alpha = 0.89$; $\uparrow$, intercept values.

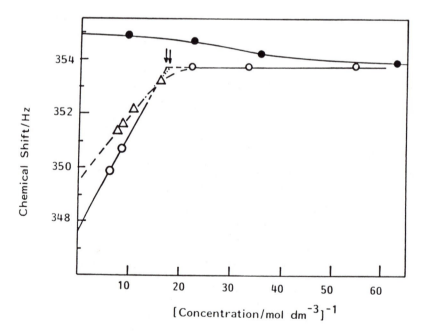

Figure 5.   Chemical shift (H_z) against reciprocal concentration
             of AmDec: $\bullet$, $\alpha = 0.56$; $\triangle$, $\alpha = 0.89$; $\bigcirc$, $\alpha = 1.0$;
             $\uparrow$, intercept values.

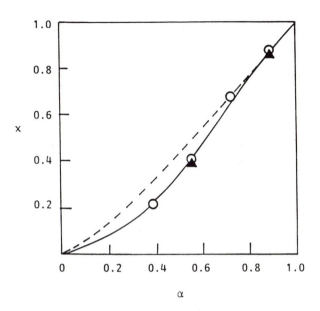

Figure 6.   x against α, - - - - , x_{IDEAL}; $\bigcirc$, x_{SANS}; $\blacktriangle$, x_{NMR}.

$$x_{ideal} = \frac{\alpha\, C_T - C_1^m}{C_T - C_2^m - C_1^m} \qquad \ldots \, (6)$$

where C_1^m and C_2^m are the concentrations of unassociated monomeric surface active agents 1(AmDec) and 2(APFO) with C_1^m given by

$$C_1^m = \left[-(C_T - \Delta) \pm \sqrt{(C_T - \Delta)^2 + 4\alpha\, C_T\, \Delta} \right] \Bigg/ 2 \left[\frac{C_{M_2}}{C_{M_1}} - 1 \right] \ldots (7)$$

and

$$C_2^m = \left[1 - \frac{C_1^m}{C_{M_1}} \right] C_{M_2} \qquad \ldots \, (8)$$

with $\Delta = C_{M_2} - C_{M_1}$

C_{M_1} and C_{M_2} are respectively the c.m.c. values of AmDec and APFO.

Table I. Composition of Mixed Micelles

α	c_T /mol dm^{-3}	c_1^m /mol dm^{-3}	c_2^m /mol dm^{-3}	x_{SANS}	x_{NMR}	x_{IDEAL}
0	0.120	0	0.0096	0	0	0
0.20	0.125	0.0067	0.0080	-	-	0.12
0.39	0.130	0.0136	0.0069	0.21	-	0.34
0.56	0.135	0.0204	0.0048	0.40	0.39	0.50
0.72	0.141	0.0274	0.0031	0.58	-	0.67
0.89	0.147	0.0353	0.0013	0.86	0.86	0.87
1.00	0.150	0.0407	0	1.00	1.00	1.00

Discussion

Small angle neutron scattering provides an excellent means of determining the composition of mixed micelles of fluorocarbon and hydrocarbon surfactants (2). Considerable sensitivity is obtained as a
consequence of the substantial difference in coherent scattering
lengths of the fluorocarbon and hydrocarbon surfactants. However,
neutron facilities are located only in a few key centres, whereas
multinuclear nmr facilities are laboratory based. It was hence of
interest to explore the use of nmr methods as a quantitative means
of probing the composition of mixed micelles and to compare the
results obtained by this technique with those obtained by neutron
scattering. The agreement between the results from small angle
neutron scattering and nmr, although directly comparable only
at α = 0.56 and α = 0.89, is excellent. Since the two techniques
used have a very different basis, one measuring spatial correlations
and the other local temporal correlations, the agreement obtained
gives considerable confidence that both methods can be used and the
present nmr results extended, to examine the composition of mixed
micelles over a wider range of compositions. Studies by small angle
neutron scattering can also provide additional information on micellar
size and shape.

A comparison of the measured values of x and x_{IDEAL} suggests
that at α = 0.89, corresponding to a high ratio of AmDec the mixing
process is close to ideal and that fluorocarbon chains, in a small
proportion can be accommodated in the hydrocarbon environment of
the AmDec micelle. At α values between 0.39 and 0.72 the mixing
process is not ideal (10,11) and the measured value of x is less
than x_{IDEAL}. This suggests that in this region the micelles are
fluorocarbon rich and that the fluorocarbon chains tend to exclude
hydrocarbon chains.

There appears to be no indication from these results that
separate hydrocarbon and fluorocarbon micelles are formed as has
been suggested from some systems in the literature (12,13). All the
evidence obtained in this and previous work (1,2,3) indicates that
for the APFO-AmDec system at the concentrations examined mixed
micelles are formed. However, a point which needs to be studied in
more detail is the variation of composition with the concentration of
the surfactants above the c.m.c. Further neutron scattering studies
will be directed towards examining this point.

Acknowledgments

We wish to thank Procter and Gamble Limited for support of this work. We also wish to thank SERC for support of RMC and for support of the neutron studies. We thank the Institut Laue Langevin for the use of neutron facilities.

Literature Cited

1. Burkitt, S.J.; Ottewill, R.H.; Hayter, J.B. and Ingram, B.T.; Colloid and Polymer Sci., 1987, 265, 619.
2. Burkitt, S.J.; Ottewill, R.H.; Hayter, J.B. and Ingram, B.T.; Colloid and Polymer Sci., 1987, 265, 628.
3. Burkitt, S.J.; Cebula, D.J. and Ottewill, R.H.; In Surfactants in Solution; K.L. Mittal, Ed.; Plenum Press, New York, 1989, Vol.7, 211.
4. Clapperton, R.M.; Ph.D. Thesis, University of Bristol, 1990.
5. Burkitt, S.J.; Ingram, B.T. and Ottewill, R.H.; Prog.Colloid and Polymer Sci., 1988, 76, 247.
6. Simister, E.A.; Lee, E.M.; Thomas, R.K.; Ottewill, R.H.; Rennie, A.R. and Penfold, J.; to be published.
7. Ghosh, R.E.; A Computing Guide for Small Angle Scattering Experiments, Institut Laue Langevin, 1989, 89GH02T.
8. Inoue, H. and Nakagawa, T.; J.Phys.Chem., 1966, 70, 1108.
9. Clint, J.H.; J.Chem.Soc.Faraday Trans.I, 1975, 71, 1327.
10. Rubingh, D.N.; In Solution Chemistry of Surfactants, K.L. Mittal, Ed.; Plenum Press, New York, 1979, Vol.1, 337.
11. Holland, P.M. and Rubingh, D.N.; In Cationic Surfactants, D.N. Rubingh and P.M. Holland, Eds.; Marcel Dekker, Inc., New York 1991.
12. Mukerjee, P. and Mysels, K.J.; A.C.S. Symposium Series, 1975, 9, 239.
13. Mukerjee, P. and Young, A.Y.S.; J.Phys.Chem., 1976, 80, 1388.

RECEIVED January 22, 1992

Chapter 18

Interactions between Siloxane Surfactants and Hydrocarbon Surfactants

Randal M. Hill

Dow Corning Corporation, Midland, MI 48640–0994

The surface active properties of mixtures of siloxane surfactants and hydrocarbon surfactants are described. Our results include combinations of surfactants which differ in <u>both</u> their hydrophobic and hydrophilic groups. Previous studies have only looked at combinations which differ in one or the other. We found mixing behavior varying from antagonistic (positive-nonideal) to synergistic (negative-nonideal). We attempted to model the mixed CMC's in the usual way, using the formalism of the regular solution approximation, but found that the regular solution approach fails to account for our results in three significant ways: (i) the magnitude of the interaction depends on the proportions of the two constituents, (ii) the symmetry of the predicted CMC curves is incorrect, and (iii) the measured CMC values for the nonionic/nonionic mixtures are much higher than the model can account for. Antagonistic mixing of fluorocarbon/hydrocarbon surfactant mixtures has been attributed to phobicity of these two moieties. However, since low molecular weight silicones and hydrocarbon solvents are generally miscible, we propose another explanation for our systems based on molecular size and shape arguments.

Siloxane surfactants have received increasing interest recently because of their unique surface active properties *(1, 2)*. They find applications in such diverse areas as polyurethane foam additives, textile manufacture, cosmetic formulations, agricultural adjuvants, and paint additives *(2)*. Most <u>aqueous</u> applications of siloxane surfactants actually involve their use in combination with organic surfactants and polymers. For instance, they are incorporated into cosmetic formulations containing a variety of other surface active ingredients. They are used to enhance spreading and plant penetration in agro-chemical formulations which also contain organic surfactants. Since it is well known that different classes of surfactants can interact strongly *(3, 4)*, it becomes vital to understand the behavior of mixtures of siloxane surfactants and organic surfactants. How do such mixtures influence

0097–6156/92/0501–0278$06.00/0

the surface activity and performance of the formulation? Can the performance of products incorporating siloxane surfactants be improved by judicious choice of the other surfactants in the formulation? Are the unique surface active properties of siloxane surfactants simply additive to complex formulations, or do they interact, as other surfactants do, to determine the end-result? These are the questions we set out to answer in this study.

Classification of Surfactants. Surfactants can be usefully classified in terms of the types of hydrophobic and hydrophilic groups they contain. The most common hydrophobic groups are hydrocarbon (linear and branched alkyl and alkylphenyl groups) and fluorocarbon (mostly branched alkyl groups containing various amounts of C-F functionality). Surfactants in which the hydrophobic group consists of dimethyl siloxane moieties are called siloxane surfactants. *Siloxane* is preferred over *silicone* because of the extremely wide use of the latter term to describe polydimethylsiloxane oils. Many siloxane surfactants are oligimers or polymers containing a broad distribution of molecular species. One very important exception to this is the trisiloxane-based surfactants which are the primary focus of this paper.

Surfactants can also be classified in terms of their hydrophilic groups as nonionic, anionic, or cationic surfactants. Zwitterionic, or catanionic surfactants contain both an anionic and a cationic group. Common nonionic groups include polyethylene oxide (PEO), glucose and sucrose, amine oxide and phosphine oxide. Anionic groups include sulfate, sulfonate, carboxylate and phosphate. Cationic groups are usually quaternary ammonium salts of various structures. Siloxane surfactants have been prepared containing most of these hydrophilic groups *(5)*. We will use the term *siloxane polyethylene oxide* (SPEO) *surfactants* to refer to the class of nonionic siloxane surfactants which contain polyethylene oxide hydrophilic groups. Nonionic surfactants may also contain polypropylene oxide (PPO) groups as well as PEO groups. PPO is actually a hydrophobic group, and siloxane surfactants which contain both PEO and PPO are referred to as *siloxane polyalkylene oxide* (SPAO) *surfactants* in order to distinguish them from SPEO surfactants. SPAO surfactants are widely used as polyurethane foam additives.

Purpose of Surfactant Classification. These surfactant classifications are useful because they represent the key to understanding the properties of surfactant mixtures - the behavior of surfactant mixtures depends on these class differences. For instance, the behavior of mixtures of nonionic hydrocarbon surfactants and anionic hydrocarbon surfactants is determined by interactions between the nonionic and anionic hydrophilic groups *(3)*. The behavior of mixtures of anionic fluorocarbon surfactants and anionic hydrocarbon surfactants is controlled by the phobic interactions between the hydrocarbon and fluorocarbon tail groups *(6-9)*. Previous studies of surfactant mixtures have investigated combinations which differ only in either their hydrophobe class or their hydrophile class, but not both. Mixtures of siloxane nonionic surfactants with anionic and cationic hydrocarbon surfactants differ in both their hydrophobic and hydrophilic groups. Therefore, previous work gives us no sure guidance to predict the behavior or our systems.

Surfactant Mixing Behavior. Mixtures of surfactants belonging to the same hydrophile class and hydrophobe class, such as a pair of homologous nonionic hydrocarbon surfactants, mix ideally *(10, 11)*; the properties of the mixture can be predicted from the properties of the individual components by treating the micelle as a pseudo-phase and using the ideal solution equations *(3)*:

$$C_{mixed} = C_M = \frac{CMC_A \; CMC_B}{Y_A \, CMC_B + Y_B \, CMC_A}$$

$$X_A = \frac{Y_A \; C_M}{CMC_A}$$

where C_M is the total monomer concentration, Y_A and Y_B are the monomer mole fractions, and X_A and X_B are the micellar mole fractions.

These equations contain no empirical parameters. Lange and Beck (10) showed mixed CMC's for mixtures of three alkyl ethoxylate nonionic surfactants. The measured values were almost exactly given by the ideal solution equations. Their data also showed that as the mixed CMC decreased, the value of the surface tension at the CMC also decreased. Meguro, Ueno and Esumi (11) also show data which displays this trend. They state that ideal mixing behavior is observed for mixtures of alkyl ethoxylates of varying alkyl chain length and ethoxylate chain length, and for mixtures of alkyl sulfoxides.

In all these examples the hydrophobic portions of the molecules are similar, and consist of hydrocarbon moieties, whether linear or branched alkyl groups, or alkyl phenyl groups. In contrast to this, mixtures of siloxane and hydrocarbon surfactants involves mixing siloxane and hydrocarbon hydrophobic groups together in the micelle. The size, shape, and chain flexibility of these two types of hydrophobic group are markedly different, and this will certainly affect how they mix in the micelle.

A well-studied example of a mixed surfactant system containing different types of hydrophobic groups is the combination of fluorocarbon and hydrocarbon surfactants. Such mixtures show strong positive deviations from ideal behavior (antagonistic mixing), and sometimes form immiscible hydrocarbon and fluorocarbon micelles rather than mixed micelles (6-9). This behavior is generally attributed to the mutual phobicity between fluorocarbon and hydrocarbon moieties.

Non-ideal surfactant mixing effects have been successfully modelled by treating the micelle as a pseudo-phase and using the formalism of the regular solution approximation (RSA) (3, 12-15):

$$C_1 = X_1 \; C_1^0 \exp (\beta_{12} \, X_2^2) = Y_1 \, C_M$$

$$C_2 = X_2 \; C_2^0 \exp (\beta_{12} \, X_1^2) = Y_2 \, C_M$$

$$C_{mixed} = C_M = C_1 + C_2$$

$$Y_2 = \frac{C_2}{C_1 + C_2}$$

Where β_{12} is an emprical interaction parameter. This approach has been shown to work for mixtures of nonionic and ionic hydrocarbon surfactants (3, 14) and for mixtures of fluorocarbon anionic and hydrocarbon anionic surfactants (6, 9). The RSA is based on pairwise, short-range interactions (12, 13, 16) and, as applied to modelling classical liquid-vapor equilibria, works best when there are no volume of mixing effects, in other words, for pairs of similar-sized molecules. Its use to model surfactant mixtures has been criticized because (i) micelle formation and hydrophobic interactions in general are highly cooperative, rather than pairwise, (ii) the interactions responsible for the greatest non-ideality are electrostatic shielding effects for nonionic/anionic mixtures and electrostatic interactions are long-range rather than short-range, and (iii) calorimetric measurements are completely at odds with calculated interaction parameters (12-16).

In contrast to mixtures of fluorocarbons and hydrocarbons, low molecular weight polydimethylsiloxanes are miscible with many hydrocarbon solvents such as toluene and hexane. The immiscibility of fluorocarbon/hydrocarbon mixtures is due in part to the relative stiffness of the fluorocarbon chain compared with hydrocarbons, and mutual phobicity between CF and CH. In contrast, the siloxane backbone is *more* flexible than the hydrocarbon, and contains CH_3 groups which, of course, interact favorably with other hydrocarbons. A closer analogy to siloxane/hydrocarbon surfactant mixing might therefore be mixtures of polypropylene oxide/polyethylene oxide block copolymer surfactants with conventional alkyl hydrocarbon surfactants. However, no work on mixed micelle formation by mixtures of such materials has been reported. Thus, previous work with surfactant mixtures does not enable us to predict the behavior of mixtures of siloxane surfactants and hydrocarbon surfactants.

EXPERIMENTAL

Materials. Three nonionic siloxane surfactants were included in this study: two low molecular weight trisiloxane SPEO surfactants and one polymeric SPEO surfactant. All three were prepared at Dow Corning Corp. Their structures are:

Name	Structure
HMTSE7	MD'(R1)M
	$R1 = (CH_2)_3(OCH_2CH_2)_7OH$
HMTSE12	MD'(R2)M
	$R2 = (CH2)_3(OCH_2CH_2)_{12}OH$
RP120	$MD_{100}D'_{18}(R4)M$
	$R4 = (CH_2)_3(OCH_2CH_2)_{12.5}OH$

M stands for $(CH_3)_3SiO-$, and D stands for $-(CH_3)_2SiO-$. D' represents a di-organofunctional siloxane unit in which one of the methyl groups has been replaced with the hydrophilic group specified. The first two materials are based on the heptamethyltrisiloxy (HMTS), or trisiloxane, hydrophobe which is very close in its hydrophobicity to a C_{11} hydrocarbon group *(5)*. RP120 is a rake-type polymeric surfactant with an average degree of polymerization of 120; it contains a broad distribution of siloxane chain lengths. The PEO portion of all three surfactants contain the usual broad distribution of chain lengths.

Three hydrocarbon surfactants were chosen as representative to be included in this study: heptaethyleneoxide dodecyl ether (C12E7), sodium docecyl sulfate (SDS), and dodecyltrimethyl ammonium bromide (DTAB). C12E7 is a monodisperse alkyl ethoxylate obtained from Nikko Chemical Co. of Japan. SDS was obtained from Aldrich Chemical Co. and recrystallized from ethanol before use. DTAB was obtained from Aldrich Chemical Co. Except for SDS, all surfactants were used without further purification. As used, none of the individual surfactants exhibited a significant minimum in its surface tension near the CMC. Sodium chloride and sodium bromide were obtained from Aldrich and used as received. De-ionized water was taken from a four-stage Milli-Q water purification system fed by a reverse osmosis system. This water typically had a conductance of 18 MegOhms, and organic impurities less than 1 ppb. Its surface tension was regularly checked and was typically 72.3 ± 0.2 dynes/cm.

Experimental Procedures. Surface tensions were measured using a Cahn DCA 322 balance and a platinum Wilhelmy blade. Measurements were made on solutions placed in a thermostatted jacketed beaker held at 25 ± 0.5 C. Temperature control was provided by a Lauda RM4 circulating water bath. All surfactant solutions, even for the nonionic surfactants, were prepared in 0.1 M NaCl or 0.1 M NaBr to provide a constant ionic strength background. In order to avoid hydrolysis problems, solutions were prepared and run on the same day, preferably within 1-2 hours. Hydrolysis in a known problem for SDS, and may also be a problem with trisiloxane surfactants *(2)*.

A minimum of five points below the CMC and five points above the CMC were obtained. The CMC and area/molecule were determined from straight line fits of the pre-CMC and post-CMC portions of the Gibbs plot. The surface tension at the CMC was also calculated from the intersection of these two lines and taken as the minimum surface tension. In some cases, most notably HMTSE7 and SDS, a significantly lower actual minimum was present for certain ratios. Systems containing the polymeric siloxane surfactant have a relatively large uncertainty in the CMC values because the Gibbs plots did not always show linear regions, before and after a clear break. For such a Gibbs plot, calculation of the CMC becomes uncertain, and somewhat arbitrary. This will be noted below in the discussion of the results.

RESULTS

Figure 1 shows a Gibbs plot for a mixture of HMTSE7 and DTAB (0.25 mole fraction DTAB). This plot is representative of the results we obtained in this study, except for mixtures involving RP120 which were less well-behaved. Although none of the individual surfactants had a minimum in its Gibbs plot, some of the combinations did exhibit a minimum at low proportions of the siloxane surfactant. We believe this is due to a substantial deviation of the monolayer composition from that of the bulk solution-preferential adsorption of the siloxane component.

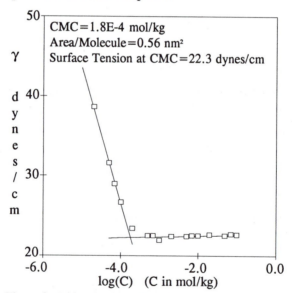

Figure 1. Gibbs plot for 0.25 mole fraction DTAB/ HMTSE7 mixture. 0.1 M NaBr, 25 C.

Nonionic Siloxane Surfactant / Cationic Hydrocarbon Surfactant.

HMTSE7 and DTAB. Figure 2 summarizes the mixed CMC results for mixtures of HMTSE7 and DTAB. The CMC of HMTSE7 is much lower than that of DTAB, and the CMC values for the mixtures fall within experimental uncertainty of the solid curve predicted by ideal mixing. Thus, unlike mixtures of hydrocarbon cationics and nonionics which show negative nonideal (or synergistic) mixing *(3)*, this mixture of a siloxane nonionic with a hydrocarbon cationic surfactant behaves as an ideal mixture. The value of

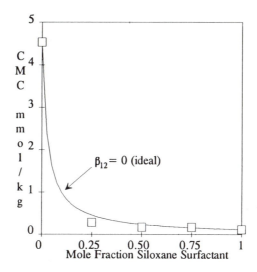

Figure 2. Mixed CMC's for DTAB/HMTSE7 system.
0.1 M NaBr, 25 C.

the surface tension at the CMC for this surfactant pair decreased monotonically from 36 dynes/cm for DTAB to 21 dynes/cm for HMTSE7.

RP120 and DTAB. Mixed CMC results for mixtures of DTAB with the polymeric siloxane nonionic surfactant RP120 are shown in Figure 3. Within the limits of experimental uncertainty (which is substantially higher for this system), this system also shows ideal mixing. The minimum surface tension for RP120 is about 31 dynes/cm, and mixtures with DTAB vary monotonically from this value to 36 dynes/cm (for DTAB).

Nonionic Siloxane Surfactant / Anionic Hydrocarbon Surfactant
HMTSE7, HMTSE12 and SDS. Figures 4 and 5 show the mixed CMC results for HMTSE7/SDS and HMTSE12/SDS. Combining these HMTS based siloxane surfactants with SDS leads to a drastic lowering of the total surfactant concentration necessary for micelle formation. The CMC of the mixture decreases as the proportion of siloxane surfactant increases. For example, 0.025 mole fraction of HMTSE7 decreases the CMC of SDS from 1.3 mmol/kg to 0.36 mmol/kg. This is a large negative deviation from what ideal mixing of the surfactants predicts. When we attempted to fit the mixed CMC data using the RSA, we found that lower proportions of siloxane surfactant required a substantially larger interaction parameter (a more negative value of β_{12}) than higher proportions - a single value of β_{12} could not account for all of the results. Thus, the interactions responsible for the nonideality appear to be stronger at smaller proportions of siloxane surfactant than at higher proportions. This is a very unexpected result which has not been observed before.

By comparing the value of the interaction parameter, β_{12}, that best fits the low mole fraction data for HMTSE12/SDS with the value for HMTSE7/SDS in the same region, it is evident that the degree of synergy is somewhat greater for HMTSE12. This indicates that increasing the number of ethylene oxide (EO) groups in the siloxane surfactant increases the strength of the interaction with SDS, or the degree of synergy. This dependence on EO chain length is also observed for mixtures of anionic and nonionic hydrocarbon surfactants.

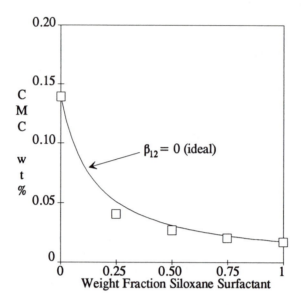

Figure 3. Mixed CMC's for DTAB/RP120 system.
0.1 M NaBr, 25 C.

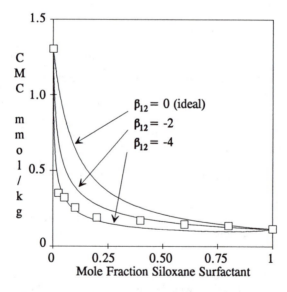

Figure 4. Mixed CMC's for SDS/HMTSE7 system.
0.1 M NaCl, 25 C.

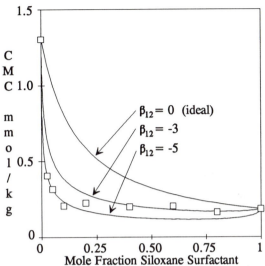

Figure 5. Mixed CMC's for SDS/HMTSE12 system. 0.1 M NaCl, 25 C.

Figure 6 shows the variation of the surface tension at the CMC for these two systems. It can readily be seen that a small proportion of HMTSE7 in combination with SDS produces a large lowering of the surface tension at the CMC compared with SDS alone. For example, 0.025 mole fraction of HMTSE7 reduces the surface tension at the CMC by about 4 dynes/cm. The effect is even larger if the actual minimum in the surface tension curve is considered; in this case 0.025 mole fraction of HMTSE7 lowers the minimum surface tension by about 8 dynes/cm.

Thus, mixtures of these two HMTS siloxane nonionic surfactants and SDS show the same general type of behavior as is observed for hydrocarbon nonionic/anionic systems: negative nonideal, or synergistic, mixing; and the magnitude of the interaction coefficient increases with increasing EO chain length *(3)*. However, these systems also show a strong dependence of the interaction coefficient on composition which has not been previously observed.

RP120 and SDS. Mixed CMC results for mixtures of RP120 and SDS are shown in Figure 7. On average, the experimental data points seem to lie below the ideal mixing line, indicating that this system also mixes non-ideally, but we cannot precisely determine the magnitude of the interaction coefficient, β_{12}. The surface tension at the CMC shows the same general trends as the two HMTS surfactants with a sharp decrease in surface tension at low mole fractions of RP120. However, the surface tension of RP120 itself is much closer to that of SDS, so the magnitude of the decrease is not as great.

Nonionic Siloxane Surfactant / Nonionic Hydrocarbon Surfactant
HMTSE7, HMTSE12 and C12E7. Figures 8 and 9 show the mixed CMC results for HMTSE7/C12E7 and HMTSE12/C12E7. The measured CMC's are higher than the values predicted by ideal mixing, which is in marked contrast with mixtures of different nonionic hydrocarbon surfactants which mix ideally. We are unable to model these data using the RSA. There are three problems: (i) as above, the magnitude of the interaction coefficient

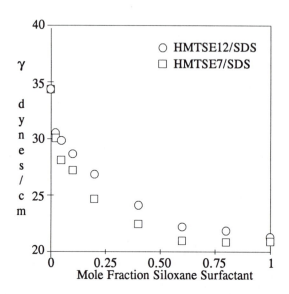

Figure 6. Surface tension at the CMC for SDS/HMTSE7 system and SDS/HMTSE12 system. 0.1 M NaCl, 25 C.

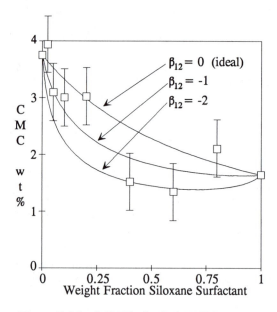

Figure 7. Mixed CMC's for SDS/RP120 system. 0.1 M NaCl, 25 C.

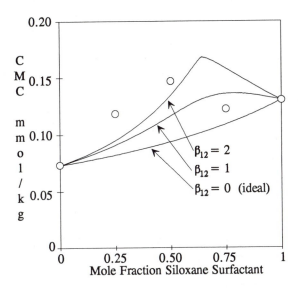

Figure 8. Mixed CMC's for C12E7/HMTSE7 system.
0.1 M NaCl, 25 C.

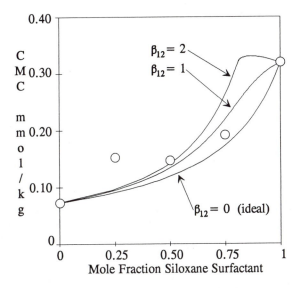

Figure 9. Mixed CMC's for C12E7/HMTSE12 system.
0.1 M NaCl, 25 C.

varies with composition, increasing with decreasing mole fraction siloxane surfactant, (ii) no value of β_{12} can account for the 0.25 and 0.5 mole fraction data points, and (iii) the symmetry of the predicted curves is obviously wrong - the calculated curves go through a maximum shifted toward the component with the higher CMC, whereas the data shows the opposite trend. If β_{12} is increased above $+2$, a singularity (corresponding to phase separation) appears at the apex of the curve, the remainder of the curve does not change much. Thus, these data indicate that mixtures of these siloxane nonionic surfactants and C12E7 may be forming separate immiscible micelles, as do some mixtures of fluorocarbon and hydrocarbon surfactants. The observation that the symmetry of the predicted curves does not match that of the data indicates a fundamental failure of the RSA as applied to this system. We have recently measured mixed CMC's for combinations of these two trisiloxane surfactants with Triton X-100 and found the same results.

The surface tension at the CMC for these two systems decreases monotonically from 33 dynes/cm for C12E7 to 21 dynes/cm for HMTSE7 and to 22 dynes/cm for HMTSE12. The surface tension at the CMC of the HMTS based surfactants is 12-13 dynes/cm lower than that of C12E7, and the mixtures give substantially lower values than C12E7, which is surprising in view of the disruption of micellar mixing exhibited by this system. This is another indication that the phenomena responsible for the atypical effects in these systems influences micellar mixing more than adsorption at the air/water interface.

DISCUSSION

We have shown that interactions between nonionic siloxane surfactants and hydrocarbon surfactants range from negative nonideal to positive nonideal depending on the ionic character of the hydrocarbon surfactant. The affects appear to influence micellar mixing more than adsorption at the air/water interface.

siloxane surfactant	hydrocarbon surfactant	β_{12} (CMC)
nonionic	nonionic	+
nonionic	cationic	0
nonionic	anionic	-

A similar trend is, in fact observed for mixtures of hydrocarbon nonionic surfactants with other hydrocarbon nonionic, cationic and anionic surfactants:

hydrocarbon surfactant	hydrocarbon surfactant	β_{12} (CMC)
nonionic	nonionic	0
nonionic	cationic	-
nonionic	anionic	- -

where "- -" reflects the observation that β_{12} values for nonionic/anionic pairs are usually more negative than corresponding values for nonionic/cationic pairs (3). The similar direction of these two trends indicates that part of the interaction is due to the same physical interaction, while some other effect is contributing which has not been recognized before. The common interaction between these two systems is most likely the hydrophilic group interaction: namely the decrease in electrostatic repulsion which occurs when a nonionic headgroup is incorporated into an ionic micelle. This is supported by the observation that the interaction coefficient increases with increasing EO chain length.

Although our systems display important similarities with other surfactant systems, they also exhibit two phenomena not previously observed: (i) the interaction depends on the composition, or ratio of the surfactant mixture, and (ii) the apparent formation of separate immiscible micelles by a pair of surfactants containing hydrophobic groups which are not generally immiscible. The regular solution model is based on short-range, pairwise interactions. Such interactions cannot depend on the proportions of the two surfactants. As mentioned above, unlike fluorocarbon/hydrocarbon mixtures, low MW silicones are quite miscible with organic solvents such as normal alkanes. We therefore propose to explain the atypical effects by consideration of the size and shape of the molecules, based on the type of micellar and liquid crystalline aggregates that they normally form.

Many single-tail hydrocarbon surfactants, including the three in this study, form spherical micelles at low concentrations *(17, 18)*. Gradzielski, et al. *(5)* have recently reported the existence of globular micelles in solutions of ionic or very soluble nonionic siloxane surfactants. They also found anisometric micelles for one material which also formed a lamellar liquid crystalline phase. We have found that many less soluble nonionic siloxane surfactants form lamellar liquid crystalline phases over a very wide concentration range, in some cases, from just above the CMC to 80-85 wt% surfactant in water. Now the type of aggregate or liquid crystalline phase formed by a surfactant in solution is determined by the effective size and shape of the surfactant molecule. This is illustrated in Figure 10. The size and shape of a surfactant molecule can be quantified using the so-called surfactant parameter *(19-21)*:

where V=volume of hydrophobic chain, A=headgroup area, L=optimal chainlength. When this ratio is less than 1/3 spherical micelles are formed, between 1/3 and 1/2 rod-shaped aggregates are formed (hexagonal phase liquid crystal), and between 1/2 and 1 vesicles or bilayers are formed (lamellar phase liquid crystal).

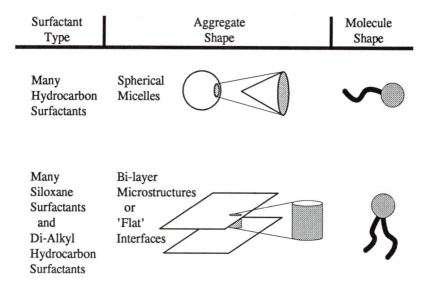

Surfactant Type	Aggregate Shape	Molecule Shape
Many Hydrocarbon Surfactants	Spherical Micelles	
Many Siloxane Surfactants and Di-Alkyl Hydrocarbon Surfactants	Bi-layer Microstructures or 'Flat' Interfaces	

Figure 10. The type of aggregate or liquid crystalline phase formed by a surfactant in solution is determined by the effective size and shape of the surfactant molecule.

Apparently, the HMTS-based surfactants in dilute aqueous solution have a cylindrical shape, while the hydrocarbon surfactants used in this study have more of a conical shape. Ananth., et al. *(22)* have recently published a molecular model of this structure showing it to be rather short and "compact". Thus we would expect HMTS-based surfactants to readily pack into the "flat" air/water interface, but to disrupt the packing of spherical micelles. This explains why the effect is greatest at low proportions of siloxane surfactant and affects micelle formation more than adsorption (surface tension).

An alternative way to think about this packing problem is in terms of the length of the hydrophobic group. It is well known that mixtures of linear alkyl carboxylic acids with chain lengths differing by more than 2 methylene units demix in insoluble monolayers to allow similar chain lengths to be next to each other. If the size of the HMTS hydrophobe is as short and "compact" as shown by Ananth., et al. *(22)*, then this same effect could be operative - siloxane surfactants mixed with linear hydrocarbon surfactants would prefer to pack into their own smaller aggregates (whatever their shape), and would demix at the air/water interface. Gradzielski, et al. *(5)* have reported small micelle sizes for a number of siloxane surfactants. This demixing would not necessarily have a large affect on surface tension, but should disrupt micelle formation, and the effect should vary with the proportions of the two components.

Thus, whatever its exact origin, there appears to be a disfavorable packing between siloxane surfactants and linear hydrocarbon surfactants due to their very different molecular size/shape. The effect is the most noticable in the case of a pair of nonionic surfactants, and is offset in the case of the nonionic/ionic combinations by the highly favorable nonionic/ionic headgroup interaction, although the mixing still depends on composition. The molecular shape is less amenable to calculation in the case of the polymeric silxoane surfactants, but they also form bilayer structures, and rather small micelles, and therefore similar effects must influence their micellar mixing with linear hydrocarbon surfactants.

CONCLUSIONS

Mixtures of several siloxane nonionic surfactants with typical anionic, cationic, and nonionic hydrocarbon surfactants show mixing behavior varying from antagonistic (positive-nonideal) for nonionic/nonionic combinations to neutral (ideal) for nonionic/cationic combinations, to synergistic (negative-nonideal) for nonionic/anionic combinations. Although some aspects of the mixing behavior are shared in common with previously studied systems, other aspects are quite atypical. The regular solution model fails to account for our results in three significant ways: (i) the magnitude of the interaction depends on the proportions of the two constituents, (ii) the symmetry of the predicted CMC curves is incorrect, and (iii) the measured CMC values for the nonionic/nonionic mixtures are much higher than the model can account for. Antagonistic mixing of fluorocarbon/hydrocarbon surfactant mixtures has been attributed to phobicity of these two moieties. However, since low molecular weight silicones and hydrocarbon solvents are generally miscible, we propose another explanation for our systems based on molecular size and shape arguments.

ACKNOWLEDGEMENTS

I would like to thank the Dow Corning Corp. for support to carry out this research.

REFERENCES

1. Schmidt, G. *Tenside Surf. Det.* **1990**, 27, 324.
2. Knoche, M.; Tamura, H.; Bukovac, M.J. *J. Agric. Food Chem.* **1991**, 39, 202.
3. Scamehorn, J.F. *ACS Symposium Ser.* **1986**, 311, 1.
4. Scamehorn, J.F. *ACS Symposium Ser.* **1986**, 311, 324.
5. Gradzielski, M.; Hoffmann, H.; Robisch, P.; Ulbricht, W. *Tenside Surf. Det.* **1990**, 27, 366.
6. Funisaki, N.; Hada, S. *J. Phys. Chem.* **1983**, 87, 342.
7. Mukerjee, P.; Yang, A.Y.S. *J. Phys. Chem.* **1976**, 80, 1388.
8. Muto, Y.; Esumi, K.; Meguro, K.; Zana, R. *J. Coll. Interface Sci.* **1987**, 120, 162.

9. Tadros, Th.F. *J. Coll. Interface Sci.* **1980**, 74, 196.
10. Lange, V.H.; Beck, K.-H. *Kolloid-Z. u. Z. Polymere 2* **1973**, 51, 424.
11. Meguro, K.; Ueno, M.; Esumi, K. In *Nonionic Surfactants*; Schick, M.J., Ed.; *Surfactant Sci. Ser.*, Marcel Dekker: New York, NY, **1987**, Vol. 23; p. 109.
12. Hall, D.G.; Price, T.J. *JCS Faraday Trans. 1* **1984**, 80, 1193.
13. Hey, M.J.; MacTaggart, J.W.; Rochester, C.H. *JCS Faraday Trans. 1* **1985**, 81, 207.
14. Holland, P.M. *ACS Symposium Ser.* **1984**, 253, 141.
15. Holland, P.M. *Adv. Coll. Interface Sci.* **1986**, 26, 111.
16. Hildebrand, J.H.; Scott, R.L. *Regular Solutions*; Prentice-Hall, Inc.: Englewood Cliffs, NJ, 1962, p. 132.
17. Brown, W.; Johnsen, R.; Stilbs, P.; Lindman, B. *J. Phys. Chem.* **1983**, 87, 4548.
18. Nilsson, P.; Lindman, B. *J. Phys. Chem.* **1984**, 88, 5391.
19. Sjoblom, J.; Stenius, P.; Danielsson, I. In *Nonionic Surfactants*; Schick, M.J., Ed.; *Surfactant Sci. Ser.*, Marcel Dekker: New York, NY, **1987**, Vol. 23; p. 369.
20. Mitchell, J.D.; Ninham, B.W. *JCS Faraday Trans. 2* **1981**, 77, 601.
21. Ninham, B.W.; Evans, D.F. *Faraday Discuss. Chem. Soc.* **1986**, 81, 1.
22. Ananthapadmanabhan, K.P.; Goddard, E.D.; Chandar, P. *Colloids Surf.* **1990**, 44, 281.

RECEIVED January 8, 1992

Chapter 19

Mixed Micelles with Bolaform Surfactants

R. Zana

Institut Charles Sadron (CRM-EAHP), Centre National de la Recherche
Scientifique-ULP, 6, rue Boussingault, 67000 Strasbourg, France

This paper reviews studies dealing with the formation of mixed
micelles in solutions of three types of mixtures of bolaform surfactants
and conventional surfactants : (i) of like electrical charges, (ii) of
opposite charges and (iii) where the bolaform surfactant is the counterion of the conventional surfactant. The conditions for mixed micellization and the changes of the CMC and micelle aggregation number or
molecular weight with the relevant parameters of the mixtures are discussed in relation with the results for mixtures of conventional surfactants.

Bolaform surfactants (also named bolaamphiphiles, bolaphiles or α,ω-type surfactants) refer to surfactants where the alkyl chain (or a more complex hydrophobic
moiety) is terminated at both ends by a polar (ionic, zwitterionic or nonionic)
group *(1)*. Bolaform surfactants have a number of properties which distinguish
them from conventional surfactants *(1)*. Complex bolaform surfactants are present
in the lipids constituting the membrane of archaebacteria *(1)* . The micellization of
bolaform surfactants has been much investigated recently as they represent a new
class of surfactants on which can be tested current theories of micellization *(2-8)*.
A controversy exists as to whether, in bolaform surfactant micelles, the alkanediyl
chains are stretched or folded, the folded (wicket-like) conformation being that
adopted at the air-water interface *(2,7)*. Recently the first studies of mixed micellization in solutions of mixtures of bolaform and conventional surfactants have been
reported. Constraints in the packing of the alkyl chains of the bolaform and
conventional surfactants are expected to play an important role in determining
whether mixed micelles can form in these mixtures. This paper provides a review
of some of the results reported thus far and considers three types of mixtures of
bolaform and conventional surfactants (i) having like electrical charges ; (ii) having
opposite electrical charges and (iii) where the bolaform surfactant is the counterion
of a conventional surfactant. These three types of systems are representative of
most of those generated by mixing bolaform and conventional surfactants. Mixed
micellar systems involving nonionic or zwitterionic bolaforms have not been
described yet.

0097–6156/92/0501–0292$06.00/0

Mixed Micellization in Mixtures of Bolaform and Conventional Surfactants of Like Electrical Charge

The occurrence of mixed micelles in mixtures of alkyltrimethylammonium bromides, C_mTAB, and alkanediyl-α,ω-bis(trimethylammonium bromide), $C_n(TAB)_2$, has been systematically investigated as a function of the chain lengths of the conventional and bolaform surfactants, characterized by the carbon numbers m and n of the alkyl and alkanediyl groups, respectively *(9)*. Surfactants with identical head groups were selected in order to eliminate the effect of this parameter on the properties investigated (CMC, micelle aggregation number). Also the micellization of both C_mTAB and $C_n(TAB)_2$ was well characterized *(2,10,11)*. The electrical conductivity method was extensively used. The conductivity κ was measured as a function of the conventional surfactant concentration, C, at constant mole fraction, X, of bolaform surfactant. Comparative measurements where the bolaform was replaced by KBr were also performed to evidence mixed micellization. In mixtures where no mixed micelles formed, the plot of the micelle apparent degree of ionization, α_{ap} (ratio of the slopes of the κ vs C plot above and below CMC) against X found for the conventional -bolaform surfactant mixture was close to that for the conventional surfactant - KBr mixture and generally showed an increase with X. Figure 1 shows typical results for the $C_{14}TAB$-$C_n(TAB)_2$ mixtures (Note that the CMC is expressed in mole of $C_{14}TAB$ per liter. The CMC must be divided by 1-X in order to have it expressed it in mole of total surfactant per liter). In these systems, mixed micelles form at the CMC only in $C_{14}TAB$-$C_{22}(TAB)_2$ mixtures. Table I summarizes the conclusions, valid at the CMC of the mixtures, reached from the conductivity studies. The Table shows that the chain lengths of the two surfactants must fulfill rather strict conditions for the occurrence of mixed micellization at C close to the CMC. In fact the data indicate that when the alkyl chain of the conventional surfactant is increased by one carbon atom, that of the bolaform must be increased by about two carbon atoms in order to retain mixed micellization. The results in Table I suggest that there exist a minimum value and a maximum value of n/m between which mixed micelles form at the CMC of the bolaform - conventional surfactant mixtures. The same conclusion is likely to hold for mixtures of conventional surfactants of like electrical charge.

Table I. Mixed Micellization in C_mTAB - $C_n(TAB)_2$ Mixtures at 25°C [a]

	$C_{12}(TAB)_2$	$C_{16}(TAB)_2$	$C_{22}(TAB)_2$
$C_{10}TAB$	Yes at X > 0.5	Yes	No
$C_{12}TAB$	No	No	Yes at X > 0.25
$C_{14}TAB$	No	No	Yes
$C_{16}TAB$	No	No	Yes at X > 0.5

[a] Adapted from ref.9, valid at the CMC of the mixtures

Other conductivity measurements where the bolaform was added to a solution of conventional surfactant showed that mixed micelles formed at higher total surfactant concentration, even when mixed micellization did not occur at the CMC *(9)*. In this respect mixtures of bolaform and conventional surfactants behave similarly to mixtures of two conventional surfactants.

The micelle aggregation numbers N in C_mTAB-C_n(TAB)$_2$ mixtures were measured by time-resolved fluorescence quenching in a restricted range of bolaform mole fraction and compared to the N values measured when KBr substituted the bolaform *(9)*. Figure 2 shows the variations of N upon addition of KBr or C_n(TAB)$_2$ to a 0.1 M C_{14}TAB solution as a function of the concentration C_B, or mole fraction X of the additive. It is seen that additions of C_{12}(TAB)$_2$ increase N just like additions of KBr. This clearly indicates that mixed micelles do not form in C_{14}TAB-C_{12}(TAB)$_2$ mixtures even at concentrations much larger than the CMC of the mixture, at X < 0.4. The same is true for the C_{14}TAB-C_{16}(TAB)$_2$ mixtures investigated. However mixed micelles are present in the C_{14}TAB-C_{22}(TAB)$_2$ mixtures *(9)*. The N values in Figure 2 represent the numbers of C_{14}TAB per micelle. The number of bolaform surfactants per micelle is given by NC_B/C.

More recently mixed micellization was investigated in mixtures of alkyltrimethylammonium bromides, C_mTAB, and chlorhexidine digluconate, CG *(12)*. The latter is a dicationic surfactant with antibacterial properties which forms only small aggregates in water (N 4) at above an operational CMC of 0.0344x10 M *(12)*. The CMC's of CG mixtures with C_{12}TAB, C_{14}TAB and C_{16}TAB were obtained from conductivity measurements. The variation of the CMC with the mixture composition was used to obtain the mixed micelle composition by the method of Mysels and Otter *(13)* and by the thermodynamic treatment of Motomura et al., *(14)*. The two methods yielded results in good agreement, as seen in Figure 3. For the CG-TAB mixtures, the composition of the micellar and monomeric phases was also determined directly by gel filtration chromatography *(12)*. The micelle compositions obtained in this manner were in good agreement with those from the above cited methods at low mole fraction of CG, X_{CG}, but not at high X_{CG}. The authors reported a lack of stability over the time period required for the measurements *12)*. This may reflect the fact that the micelles were then of much smaller size (see below) and, thus, had a much shorter lifetime *(15)*. Static light scattering was used to determine the mixed micelle aggregation numbers N at the CMC, based on the micellar compositions determined as indicated above. The N values at 25°C are listed in Table II, as a function of X_{CG}. The sharp drop of N at $X_{CG} > 0.45$ is noteworthy. Thus for mixtures of C_mTAB with either CG or C_n(TAB)$_2$, two bolaform surfactants of much differing chemical structure, the formation of mixed micelles results in a decrease of aggregation number, from that of C_mTAB micelles to that of the bolaform micelle, as the latter is generally smaller than the former *(2,10,11,15)*. A similar observation has been reported for mixtures of conventional surfactants *(16)*.

Table II. Aggregation Numbers of Mixed CG-C14 TAB Micelles

X_{CG}	0.00	0.21	0.45	0.71	1.00
N	64	51	42	9	4

Mixed Micellization in Mixtures of Conventional and Bolaform Surfactants of Opposite Electrical Charge

Mixed micellar properties of the aqueous mixtures of sodium dodecylsulfate (SDS) and   dodecanedyl-1,12-bis(triethylammonium bromide), C_{12}(TEB)$_2$, have been investigated by Ishikawa et al *(17)*, at an ionic strength of 0.1 M NaBr, using surface tension, viscosity , static and dynamic light scattering, Orange OT solubiliza-

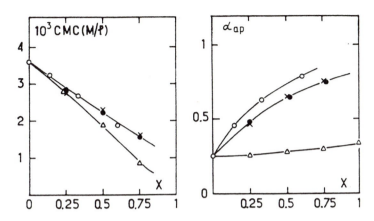

Figure 1.:   Variation of the CMC and α_{ap} in $C_{14}TAB$-$C_n(TAB)_2$ and C_{14} TAB-KBr mixtures with the mole fraction X of (O) KBr ; (x) $C_{12}(TAB)_2$; (●) $C_{16}(TAB)_2$; and (Δ) $C_{22}(TAB)_2$, at 25°C (Reproduced with permission from ref.9. Copyright 1988 Academic Press Inc.).

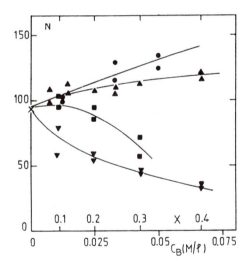

Figure 2.:   Variation of the micelle aggregation number of $C_{14}TAB$ micelles upon   additions of (●) KBr; (▲) $C_{12}(TAB)_2$; (■) $C_{16}(TAB)_2$; and (▼) $C_{22}(TAB)_2$, at 25°C (Reproduced with permission from ref.9. Copyright 1988 Academic Press).

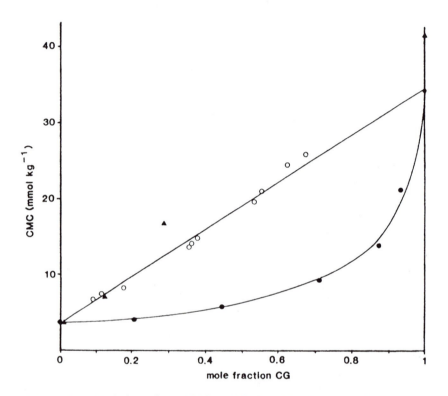

Figure 3.: Variation of the CMC of CG-C$_{14}$TAB mixtures with the mixture composition (●) and with the micelle composition calculated as in ref.13 (O), calculated according to ref.14 (continuous line) and determined by gel filtration (▲), at 25°C (Reproduced with permission from ref.12. Copyright 1989 Academic Press Inc.)

tion and spectrofluorimetry with pyrene as a probe. Surface tension measurements showed that mixed micelles form in the entire composition range above the CMC of the mixture. The change of the CMC with the bolaform mole fraction, X, shown in Figure 4a, indicated that the system is strongly non ideal. The micelle compositions, characterized by the mole fraction X^M of $C_{12}(TEB)_2$, were calculated using Motomura et al treatment *(14)*. Thus it was found that micelles having the same composition as the surfactant mixture formed at the stoichiometric composition, i.e., one bolaform for two conventional surfactants (see Figure 4a). In fact all micellar and solution properties showed a maximum or a minimum at, or close to, this composition. Thus the micellar molecular weight at the CMC, M, the solution viscosity at a total concentration of 50 mM and the Orange OT solubilization, S, at the same concentration all go through a maximum (Figures 4b,4c and 4d) whereas the micelle diffusion coefficient and effective dielectric constant go through a minimum. All these results point to the formation of large neutral micelles at the stoichiometric composition. They are in line with those reported for the currently much investigated mixtures of conventional anionic and cationic surfactants (see This Volume). Note that the change of S with X (bolaform mole fraction in the mixture) is very unusual. However, a similar behavior was recently reported *(18)* in a study of the solubility of benzyl alcohol in mixtures of two cationic surfactants and explained in terms of the effect of one surfactant on the specific interaction probably existing between the aromatic moieties of the solute and of the other surfactant (cetylbenzyldimethylammonium chloride). The same explanation may hold in Ishikawa et al study *(17)*. Indeed, Orange OT probably interacts with the bolaform quaternary ammonium groups, as all aromatic compounds *(19)*.

Mixed Systems where the Bolaform Surfactant is the Counterion of a Conventional Surfactant Ion

The reported studies concern aqueous solution of alkanediyl-α,ω-bis(pyridinium) tetradecane-1-sulfonate, $C_n(Pyr)_2^{2+},2C_{14}SO_3^-$ *(20,21)*. These systems are equivalent to the preceding ones taken at the stoichiometric composition. The solubility, CMC, Krafft range and aggregation behavior of these surfactants were examined as a function of the carbon number n of the alkanediyl group. The Krafft temperature was found to go through a deep minimum for n =8-10. Also, log CMC goes through a flat maximum at around n=6 then decreases linearly at n > 8 (see Figure 5). Finally, the micelle aggregation number is a minimum for n= 8-10 (see Figure 6). These data together with those concerning the molecular area per surfactant at the air-water interface were interpreted on the basis of a change in the location of the alkanediyl chain upon binding to the tetradecanesulfonate micelles. Thus for n < 8 the alkanediyl-α,ω-bis(pyridinium) dications bind to the micelle surface through electrostatic interactions between pyridinium and sulfonate ions, with the alkanediyl chains remaining essentially at the micelle surface, exposed to water. For n $\geq$ 8, the alkanediyl chains would fold and penetrate in the micelle core, thus contributing to the free energy change upon micelle formation and resulting in the observed decrease of CMC *(20)*. Note that from the change of CMC with n the authors calculated a free energy of transfer from water to micelles of 1.2 kT per CH_2 group *(20)*. This value is very close to those found for other homologous surfactant series.

Regarding the presence of a maximum in the log CMC vs n curve of the above surfactants it must be recalled that a similar observation was made in the study of a peculiar class of bolaform surfactants the alkanediyl-α,ω-bis(dodecyl or hexadecyl dimethylammonium bromide), C_n-α,ω-$(C_mN^+(CH_3)_2Br^-)$ *(22)*. The CMC was also found to be a maximum at around n = 6 and the decrease of CMC at n > 6 was attributed to the folding of the alkanediyl chain and its penetration in the micellar core.

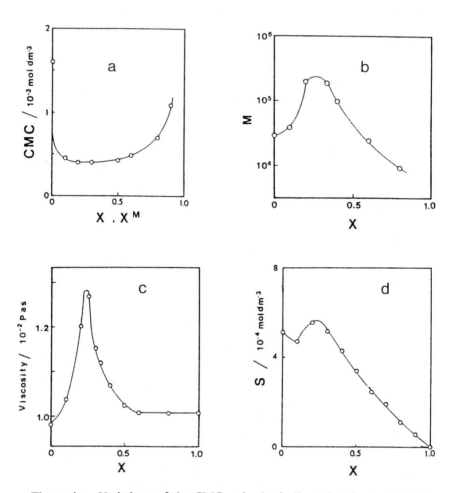

Figure 4.: Variations of the CMC, mixed micelle molecular weight M, solution viscosity, and solubilization S of Orange OT in solution of SDS-$C_{12}(TEB)_2$ mixtures with the mole fraction of the bolaform in the mixture, X, and in the micelle X^M (Reproduced with permission from ref.17. Copyright 1991 Academic Press Inc.).

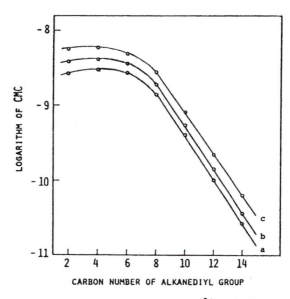

Figure 5.: Variation of the CMC of $C_n(Pyr)_2^{2+}, 2C_{14}SO_3^-$ with the alkanediyl carbon number at 35°C (a), 45°C (b) and 55°C (c). (Reproduced with permission from ref.20. Copyright 1986 Academic Press Inc.).

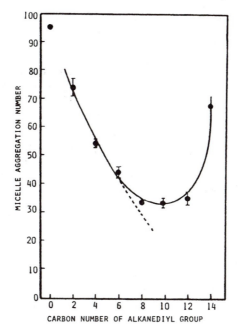

Figure 6.: Variation of the aggregation number of $C_n(Pyr)_2^{2+}, 2C_{14}SO_3^-$ micelles at the CMC with the alkanediyl carbon number n at 35°C (Reproduced from ref. 21. Copyright 1990 American Chemical Society.)

Conclusions

This paper reviewed the most significance studies of mixed micellization between various conventional and bolaform surfactants. As pointed out the reported results are in many respects similar to those found for mixtures of two conventional surfactants and the thermodynamic treatments used for the latter also apply to bolaform-conventional surfactant mixtures. This similarity is also apparent in a recent study *(23)* of the interaction between bolaform surfactants and fairly hydrophobic polyelectrolytes (poly-L-lysine and poly-L-ornithine hydrobromides) which can be considered as an extreme case of mixed micellization. The bolaform surfactant binding induced conformational changes similar to those associated to the binding of conventional surfactants.

Literature Cited

1. Fuhrhop,J.-H.; Fritsch D. *Acc. Chem. Res.* **1986**, *19*, 130 and references therein.
2. Zana, R.; Yiv, S.; Kale, K.M. *J. Colloid Interface Sci.* **1980**, *77*, 456 and references therein.
3. Ikeda, K.; Yasuda, M.; Ishikawa, M.; Esumi, K.; Meguro, K.; Binana-Limbelé, W.; Zana, R. *Colloid Polym. Sci.* **1989**, *267*, 825.
4. Ikeda, K.; Ishikawa, M.; Yasuda, M.; Esumi, K.; Meguro, K.; Binana-Limbelé, W.; Zana, R. *Bull. Chem. Soc. Jap.* **1989**, *62*, 1032.
5. Abid, S.K.; Hamid, S.M.; Sherrington, D.C. *J. Colloid Interface Sci.* **1987**, *120*, 245.
6. Nagarajan, R. *Chem. Eng. Comm.* **1987**, *55*, 251.
7. Wong, T.C.; Ikeda, K.; Meguro, K.; Söderman, O.; Olsson, U.; Lindman, B. *J. Phys. Chem.* **1989**, *93*, 4861.
8. Jayasuriya, N.; Bosak, S.; Regen, S.L. *J. Am. Chem. Soc.* **1990**, *112*, 5844.
9. Zana, R.; Muto, Y.; Esumi, K.; Meguro, K. *J. Colloid Interface Sci.* **1988**,*123*, 502.
10. Zana, R. *J. Colloid Interface Sci.* **1980**, *78*, 330.
11. Lianos, P.; Zana, R. *J. Colloid Interface Sci.* **1981**, *84*, 100.
12. Attwood, D.; Patel, H.K. *J. Colloid Interface Sci.* **1989**, *129*, 222.
13. Mysels, K.; Otter, R. *J. Colloid Interface Sci.* **1961**, *16*, 462.
14. Motomura, K.; Yamanaka, M.; Aratono, M. *Colloid Polym. Sci.* **1984**, *262*, 948.
15. Lang, J.; Zana, R. in *Surfactant Solutions : New Methods of Investigations*, R. Zana, Ed.; M. Dekker, New-York, N.Y. 1987, pp 405-452.
16. Malliaris, A.; Binana-Limbelé, W.; Zana, R. *J. Colloid Interface Sci.* **1986**, *110*, 114.
17. Ishikawa, M.; Matsumura, K.; Esumi, K.; Meguro, K. *J. Colloid Interface Sci.* **1991**, *141*, 10.
18. Bury, R.; Souhalia, E.; Treiner, C. *J. Phys. Chem.* **1991**, *95*, 3824.
19. Lianos, P.; Viriot, M.-L.; Zana, R. *J. Phys. Chem.* **1984**, *88*, 1098 and references therein.
20. Moroi, Y.; Matuura, R.; Kuwamura, T.; Inokuma, S. *J. Colloid Interface Sci.* **1986**, *113*, 225.
21. Moroi, Y.; Matuura, R.; Tanaka, M.; Murata, Y.; Aikawa, Y.; Furutani, E.; Kuwamura, T.; Takahashi, H.; Inokuma, S. *J. Phys. Chem.* **1990**, *94*, 842.
22. Zana, R.; Benrraou, M.; Rueff, R. *Langmuir* **1991**, *7*, in press.
23. Hayakawa, K.; Fujita; M.; Yokoi, S.-I.; Satake, I. *J. Bioactive Compatible Polym.* **1991**, *6*, 36.

RECEIVED January 6, 1992

Chapter 20

Mixed Systems of Bile Salts

Micellization and Monolayer Formation

Shigemi Nagadome, Osamu Shibata[1], Hidenori Miyoshi, Hisao Kagimoto[2], Yoshitomi Ikawa[2], Hirotsune Igimi[3], and Gohsuke Sugihara[4]

Department of Chemistry, Faculty of Science, Fukuoka University, Nanakuma, Jonan-ku, Fukuoka 814–01, Japan

(i) For the mixture of sodium chenodeoxycholate (NaCDC) with its epimer sodium ursodeoxycholate (NaUDC), the critical micelle concentrations (cmc) were determined by means of surface tension measurements (drop volume method) in a Kolthoff buffer solution at pH 9.94 and 30°C., and for the mixture of their taurine conjugates, sodium taurochenodeoxycholate (NaTCDC) and sodium taurourodeoxycholate (NaTUDC), cmc's were determined from the solubilization experiments on monoolein (MO) in physiological saline solution at 37°C.

(ii) Monolayers formed by different bile acids and their mixtures (cholic acid (CA) with β-muricholic acid (βMCA) and CDC with UDC) on the concentrated NaCl solution were investigated by measuring the surface pressure (Π) and the surface potential (ΔV) or the surface dipole moment (μ⊥) as a function of molecular surface area. An additivity rule for A and μ⊥ was found to hold for mixed systems of CA-βMCA and CDC-UDC, implying that these mixed systems seem to form a monolayer with ideal mixing. It was found that the mean molecular surface area and the attitude (the gradient of the plane of steroid skeleton) of the molecules floating on the water surface are dependent mainly on the magnitude of hydrophilicity of α-plane of the steroid skeleton; the surface area increases in the order of βMCA < UDC < CDC < CA. This order corresponds exactly to the order of increasing cholesterol-solubilizing power of the respective bile salts.

[1]Current address: Kyushu University–01, College of Education, Ropponmatsu, Chuo-ku, Fukuoka 810, Japan
[2]Current address: SAN-EI Chemical Industries, Ltd., Department of Research and Development, Sanwa-cho 1–1–11, Toyonaka, Osaka 561, Japan
[3]Current address: Shionogi Research Laboratories, Sagisu 5–12–4, Fukushima-ku, Osaka 553, Japan
[4]Corresponding author

0097–6156/92/0501–0301$06.00/0
© 1992 American Chemical Society

It is noteworthy that functional molecular aggregates found in nature are composed of two or more kinds of surface-active chemical species. From this point of view, therefore, the studies of mixed micelles or mixed films formed from surfactant mixtures are essential to discover some functions, which are unique compared with those of single component aggregates, not only for developing the surfactant abilities in the applied aspect but also for comprehending some functions in biological systems. From the investigation, further, we may obtain certain information with respect to the intermolecular interaction between hydrophobic and/or hydrophilic groups of surface-active molecules. The physicochemical properties of respective bile salts which are surface-active substances in biosystems are rather well known (1-4); however, few fundamental studies have yet been performed for bile salt mixtures.

In this paper, we review at first previous work (5) giving results on the formation of mixed micelles (and mixed adsorbed films) of two bile salt mixtures: (i) sodium chenodeoxycholate (NaCDC) with sodium ursodeoxycholate (NaUDC) and (ii) sodium taurochenodeoxycholate (NaTCDC) with sodium tauroursodeoxycholate (NaTUDC).

The critical micelle concentrations (cmc) and the surface excess of the former combination were investigated as a function of mole fraction of the bile salt mixtures by means of surface tension mesurements (drop volume method, in a Kolthoff buffer solution at pH=9.94 at 30°C) and the cmc's for the latter mixed system were determined from data for monoolein solubilization (in physiological saline solution at 37°C).

There have been only a few studies of monolayers of the bile acids (2,6,7). Recently we have reported mixed monolayer properties of cholic acid (CA) with β-muricholic acid (βMCA) (8); the difference between these trihydroxycholates is attributed to the location as well as the orientation of OH group: 3α, 7α, 12α for CA and 3α, 6β, 7β for βMCA. In addition to this mixed system, chenodeoxycholic acid (CDC) and ursodeoxycholic acid (UDC) were investigated and discussed by comparing results for the respective pure systems of CDC, UDC, CA and βMCA and for the mixed monolayers formed by the mixtures of CDC with UDC and CA with βMCA (9). The hydrophobicity of each bile acid was evaluated in terms of the mean surface area of molecules occupying the air-water interface, and the results were related to the cholesterol-solubilizing power of each bile acid (9). The miscibility of these mixtures in the mixed monolayer was regarded as ideal but, in contrast, the miscibility in the NaCDC-NaUDC mixed micelle showed a slightly positive deviation from ideal mixing. Finally, it should be noted that the only difference between CDC and UDC is the orientation of hydroxy group at the position 7; CDC has 7α-OH while UDC has 7β-OH. The apparently small differences in chemical structure among these bile acids are shown to have a great influence on colloidal and interfacial properties, as will be described in this paper.

Experimental

Materials   Sodium salts of bile acids: taurochenodeoxycholate (TCDC), tauroursodeoxycholate (TUDC), chenodeoxycholate (CDC) and ursodeoxycholate (UDC) were purchased from Calbiochem, CA, USA and used as received. These salts were examined by TLC and found to be about 96% pure.

Highly purified cholic acid (CA) was obtained from Sigma and used without further purification. β-muricholic acid was synthesized as described by Hsia (10), and purified by high performance liquid chromatography (HPLC). The purities of these materials were also checked by TLC and elemental analysis; the observed and calculated values were in satisfactory agreement.

All solvents used were purified by two or three distillations. Inorganic salts were analytical grade and used as received.

Methods

Cmc determinations.   The cmc's for NaCDC-NaUDC mixed systems were determined from surface tension data, obtained with a drop volume apparatus described by Uryu et al. (*11*) at 30°C.

The cmc's for NaTCDC-NaTUDC mixed systems were determined from plots of solubilized monoolein concentration vs. bile salt concentration ; the concentration where the solubilization began to occur was regarded as the cmc. The procedure used to determine monoolein solubilization was described elsewhere (*5*). The temperature of solution was kept constant by circulating water thermostated to ±0.01°C.

Surface pressure and surface potential measurements.   Each bile acid was spread from hexane/ethanol mixture (5:3 in v/v, the former from Merck, Uvasol, and the latter from Nacalai Tesque) at the air/aqueous solution interface.

The substrate solutions of 5 M (1 M = 1 mol dm^{-3}) were prepared using triply distilled water. Sodium chloride (Nacalai Tesque) was roasted at 973 K to remove any surface active impurities. Hydrochloric acid (Nacalai Tesque: ultra-fine grade) was used to bring the substrate to pH 1. The surface pressure was measured by an automated Langmuir film balance, which was that used in the previous study (*8,9,12,13*). During compression of the monolayer, measurements of surface potential were also made by using an ionizing ^{241}Am electrode (Nuclear Radiation Development, New York) at 2-3 mm above the interface with a reference electrode dipped in the subphase. Other details are described elsewhere (*8,9*).

Results and Discussion

Mixed Micelle Formations of NaCDC-NaUDC System and NaTCDC-NaTUDC System.

It is well known that the HLB (hydrophile-lipophile balance) is not a good measure of the hydrophilicity or lipophilicity of bile salts. As a substitute for the HLB, Miyajima et al. have proposed the "Hydrophobic Index (HI)" which can be evaluated from the ratio of computed areas of hydrophobic and hydrophilic surfaces (*14*). For instance, the HI of CDC is 7.27 (α-side, 0.80; β-side, 6.47; total, 7.27), while the HI of UDC is 5.48 (α-side, 0.96; β-side, 4.52; total, 5.48). The larger the HI value, the greater the hydrophobicity, so that CDC is more hydrophobic than UDC. This was confirmed as the following.

At first, we examined in detail the surface activity change with concentration for NaCDC-NaUDC bile salt mixtures and respective pure systems by using the drop volume method. The surface tension of NaCDC-NaUDC bile salt solutions decreases very rapidly until cmc is reached and then decreases slowly. The sharp break signifies the cmc. The results thus obtained clearly show that CDC has a lower cmc and greater surface activity (degree of surface tension depression) as is predicted by the HI values. The cmc's of NaCDC-NaUDC mixtures were found to be between those of the respective pure systems and to change monotonically with change in mole fraction of NaUDC in the surfactant mixture (X_{UDC}); experimental values are denoted by O in Fig. 1.

In this figure, the theoretical cmc curve for ideal mixing as a function of mole fraction in the surfactant mixture, X_2, was calculated from the following equation (*15,16*):

$$X_1 = \frac{1 - (C_m/C_{2a})^{1+\beta_2}}{(C_m/C_{1a})^{1+\beta_1} - (C_m/C_{2a})^{1+\beta_2}}$$ Eq.1

where 1 and 2 correspond to CDC and UDC, and C_m is cmc of the mixed system; C_{ia} is the cmc of pure system i when no salt is added; and β_i is the degree of counterion binding of i. The β value was assumed to be 0.5 for both bile salts by refering to literature (17,18) (The difference in β values does not greatly influence the calculated curve. For $\beta=0$, the calculated curve applies to nonionic surfactant mixtures.).

This figure shows that the cmc-X_{UDC} curve of NaCDC-NaUDC mixed system deviates slightly-positively from ideal mixing.

For analysis of the cmc data to estimate the composition of the mixed micelle and to evaluate the intermolecular interaction in micelle, the following equations have been derived by Shinoda et al. (19).

$$cmc_1 C_g^\beta = C_{1a}^{1+\beta} Y_1 \exp(\omega_{12} Y_2^2/kT)$$
$$cmc_2 C_g^\beta = C_{2a}^{1+\beta} Y_2 \exp(\omega_{12} Y_1^2/kT)$$ Eq.2

where C_g is the concentration of counterion ($C_g = cmc_1 + cmc_2$ when no salt is added), k is the Boltzmann constant and T is the absolute temperature.

$$C_m = cmc_1 + cmc_2 = \left[C_{1a}^{(1+\beta)} Y_1 \exp(\omega_{12} Y_2^2) kT + C_{2a}^{(1+\beta)} Y_2 \exp(\omega_{12} Y_1^2) kT \right]^{1/(1+\beta)}$$ Eq.3

It should be noted that the exponential terms in Eq. 2 have been derived assuming that the molal volumes of surfactants 1 and 2 are the same, and Eq. 3 requires that the degree of counterion binding, β, is common to both surfactants. If these assumptions are true for a given mixed system, one can infer not only the composition of the mixed micelles, Y_2, and that of the singly dispersed phase, $X_2 = cmc_2/(cmc_1 + cmc_2)$, but also the interchange energy per molecule, ω_{12}, which is a measure of the intermolecular interaction.

In the present case, NaCDC and NaUDC are considered to have equal molal volumes and the same degree of counterion binding, so that the assumptions employed for deriving Eqs. 2 and 3 will allow us to apply the Shinoda equations to the NaCDC-NaUDC system. The β value was assumed to be 0.3. By adjusting the ω_{12} value, it was possible to obtain the optimum fit of data, which yielded the value $\omega_{12} = 0.86kT$. In Fig. 1, the computed curves for the micellar phase and the singly dispersed phase are indicated by solid lines. As is shown, the computed cmc vs. X_{UDC} curve coincides well with the observed data. Further, application of the Motomura equations (20) to this mixed system led to a calculated micellar phase curve that was almost completely consistent with that calculated from the Shinoda equations. The derived ω_{12} value is positive, indicating that CDC and UDC interact relatively weakly.

Next, in the NaTCDC-NaTUDC mixed micelle system, the cmc values were clearly determinable from the plot of solubilized amount of monoolein vs. concentration of the bile salt mixture, as shown in Fig. 2, because the solubilization curves based on numerous measured points near the cmc exhibit sizeable increases. From Fig. 2, it is clear that a greater increase in solubilization occurs at the lower concentration of bile salt mixture for the system containing the richer NaTCDC; this fact indi-

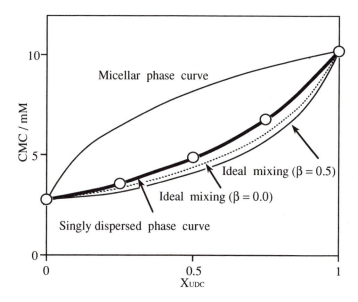

Fig.1   The phase diagram of NaCDC-NaUDC mixed system in a Kolthoff buffer solution at 30°C and pH=9.94. Circles indicate measured points. (Adapted from ref. 5).

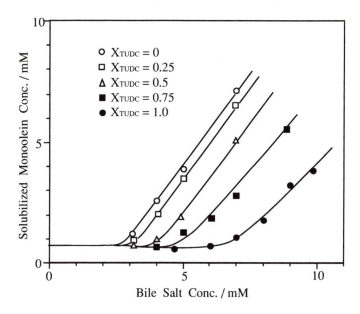

Fig.2   Plots of concentration of solubilized MO against concentration of NaTCDC-NaTUDC mixture in physiological saline solution at 37°C. (Adapted from ref. 5).

cates that NaTCDC is more hydrophobic than NaTUDC. The observed curvature of the plots at the concentrations just above cmc, suggests that the aggregation number changes to a greater extent with increase in concentration as the mixing ratio of NaTUDC increases.

The cmc vs. composition curve was simulated by the Shinoda equation, indicating almost ideal mixing as is shown in Fig. 3, where β is assumed to be 0.3 and ω_{12} is negligibly small within experimental error. But, it should be noted that although the NaTCDC-NaTUDC mixed system apparently gives an ideal cmc curve, the bulk solution contains not only small amounts of singly dispersed monoolein but also glycerol and oleic acid produced from spontaneous hydrolysis during the incubation process (its concentration is 6~7 x 10^{-4} mol dm^{-3}). That is, the NaTCDC-NaTUDC mixed micelle contains not only the two bile salts but also monoolein in hydrolyzed and unhydrolyzed forms. It can only be concluded that the micelles in this case contain three components, and that the mixed micelles of NaTCDC and NaTUDC form ideally with participation of monoolein.

One should examine critically the applicability of Eqs. 2 and 3 to the bile salt mixed ststems. One problem is that the equations are based on the phase separation model, requiring an aggregation number of more than several tens, but the aggregation number of bile salt micelles in general may be too small to apply the phase separation model. However, the NaCDC-NaUDC mixed system is dissolved in a Kolthoff buffer solution at pH=9.94 (0.05M borax : 0.05M disodium carbonate = 246 : 754, in volume), so the aggregation number is greater than that in pure water,while the NaTCDC-NaTUDC mixed micelles incorporate monoolein in 0.15 M NaCl aqueous solution; this mixed system exhibits a large increase in the solubilization with increase in the bile salt concentration. Thus, it is likely that the mixed micelles have a sufficiently large aggregation number enough to be treated with the phase separation model.

An additional problem is that both of the solutions of bile salt mixtures contain Na$^+$ counterions originating from the buffer and saline solution, although Motomura's and Shinoda's equations have been derived for no added salt . If the added salt were to be taken into account, the theoretical equations would become too complicated to be applied. Thus the results shown in Figs. 1 and 3 are only apparent and approximate, but at least the model provides a reasonable fit of the data.

Mixed Monolayer Formations of CDC-UDC and CA-βMCA Mixed Systems.

The Π-A and ΔV-A isotherms for chenodeoxycholic acid and ursodeoxycholic acid at 298.2 K on 5 M NaCl (pH 1) are shown in Fig. 4 (The Π-A and ΔV-A isotherms for CA-βMCA system are given in the previous paper (8).). The isotherms of five mixtures of CDC and UDC, covering the whole range of mole fractions, are also included in Fig. 4. All the isotherms of the mixed monolayers are between those of the pure components, and they successively change with increasing mole fraction in both Π-A and ΔV-A curves.

As expected from the chemical structures, all Π-A curves show a largely expanded molecular area. As seen from Fig. 4, the Π-A curve of UDC shifts toward the smaller area comparing with CDC. The extrapolated limiting areas of CDC and UDC are 1.40 and 1.23 nm^2, respectively. The collapse pressures of CDC and UDC are clearly determinable, and the respective surface areas at the collapse pressures are determined as 0.80 for CDC and 0.73 nm^2 for UDC.

Looking at the Π-A curves, the mixed systems (No.2 ~ 6) show a slight increase with compression after the respective collapse pressures and correspondingly the ΔV-A curves of No.2~ 6 show a slight decrease with compression beyond a maximum.

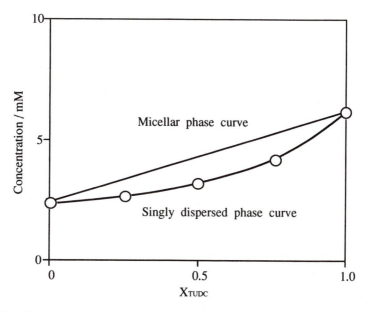

Fig.3   The cmc data determined from MO solubilization and the phase diagram for NaTCDC-NaTUDC mixed system in physiological saline solution at 37°C. (Adapted from ref. 5).

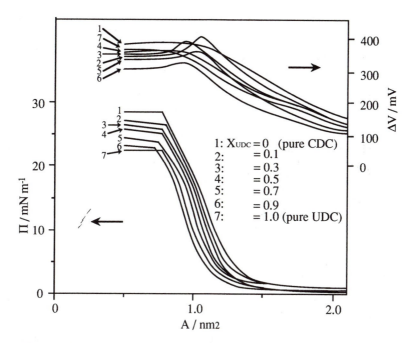

Fig.4   Π-A and ΔV-A isotherms of CDC-UDC mixed system at 25°C on 5M-NaCl (pH=1) subphase.   (Adapted from ref. 9).

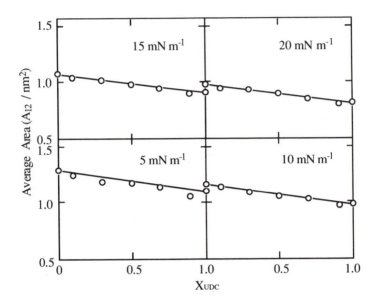

Fig.5   The comparison between the mean molecular area calculated based on
the additivity rule (solid line) and the measured molecular area (open circle) of
CDC-UDC mixed system as a function of X_{UDC} at discrete surface pressures.
(Adapted from ref. 9).

The former characteristic behavior may be attributed to the formation of single component bulk phase (21,22,23), and the latter, though the details are not yet well
known due to the lower reliability of ΔV-A data compared with Π-A data, seems to
be related to the well-mixed monolayer formation.

A better understanding of the interactions between CDC and UDC is provided by
the mean molecular areas calculated by assuming ideality of the mixing, i.e., the additivity rule. The comparison between the mean molecular area of ideal mixing and
the experimental molecular areas is given in Fig. 5, where the experimental mean
molecular areas are plotted as a function of the mole fraction of UDC at four different
surface pressures; 5, 10, 15 and 20 mN m^{-1}. The dashed lines are the mean molecular areas calculated by assuming the additivity rule. The results show that for the four
surface pressures the mixtures are apparently ideal throughout the whole range of
mole fraction. These findings suggest that no contraction of the monolayer takes
place upon mixing of these constituents.

An analysis of the surface dipole moment ($\mu_\perp$) of the monolayers (24) is also made
in terms of the additivity rule; the $\mu_\perp$ results also show that the mixtures behave almost ideally over the whole range of mole fraction at the four surface pressures examined.

In the previous paper (8) on the mixed system of cholic acid and β-muricholic acid
(CA-βMCA), all the data of Π-A and $\mu_\perp$-A indicated that CA forms an ideal mixed
monolayer when mixed with βMCA over the whole range of mole fraction; i.e., the
additivity rule regarding the relations of Π with A and $\mu_\perp$ with A was found to hold
for CA-βMCA. In Table I are listed the basic data of the mean molecular area together with the area at collapse pressure and the dipole moment determined at four
fixed surface pressures for CA, CDC, UDC and βMCA.

The determinations of the area at collapse pressure (A_c) and of the area extrapolated to zero surface pressure, i.e., limiting surface area (A_0), were made as is indicated in Fig. 6 for CA and CDC and in Fig. 7 for βMCA and UDC.

Table I. Monolayer Data for Four Bile Acids

(1) Limiting Areas (A / nm^2) extrapolated to $\Pi=0$ (A_o) and at Collapse Pressure(A_c).

	CA	CDC	UDC	βMCA
A_o	1.50	1.40	1.23	0.95
A_c	ca.1.08	0.80	0.73	ca.0.35

(2) Average Area (A / nm^2) at Each Surface Pressure.

Π / mN m^{-1}	CA	CDC	UDC	βMCA
5	1.39	1.29	1.11	0.85
10	1.24	1.17	1.00	0.76
15	1.13	1.08	0.92	0.66
20	0.87	0.99	0.83	0.50

(3) Surface Dipole Moment ($\mu_\perp$/mD) at Each Surface Pressure.

Π / mN m^{-1}	CA	CDC	UDC	βMCA
5	1349	1114	872	468
10	1334	1104	875	470
15	1285	1060	850	438
20	1031	987	783	330

SOURCE : Adapted from ref. 9.

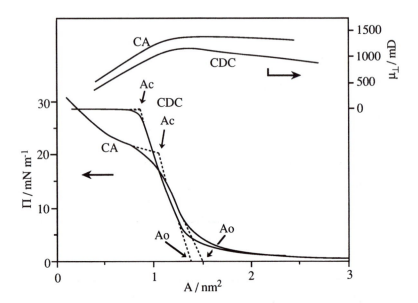

Fig.6   Π-A and μ⊥-A isotherms for CA and CDC.
            A₀ : limiting surface area
            Aᴄ : area at collapse pressure   (Adapted from ref. 9).

In Table I, it may be of interest to look at a comparison between CA and CDC and between βMCA and UDC. In the former combination, the only difference is whether 12α-OH exists or not. Fig. 6 including the respective Π-A curves and μ⊥-A curves for CA and CDC, demonstrates that the Π-A curve of CA, which has one more hydroxyl group at 12α than CDC, shows a slightly larger surface area and a lower but less well defined collapse pressure. This corresponds to a higher solubility due to a stronger hydrophilicity and thus CA dissolves in the subphase during compression (25). Also, the figure suggests that the higher dipole moment together with the slightly larger molecular area causes CA molecules to float flatly at the interface. In contrast, CDC molecules float with a list to one side.

Fig. 7 shows a comparison between βMCA and UDC with respect to Π-A and μ⊥-A relations. The only difference between these molecules is that βMCA has 6β-OH group while UDC does not. As in the case of CA, βMCA has one more hydroxyl group than UDC, so that βMCA exhibits a higher solubility and also an indistinct collapse pressure compared to UDC. Here, let us examine the Π-A and μ⊥-A curves of βMCA. βMCA shows a much smaller surface area as compared to UDC (by ca. 0.25 nm² at Π= 5 ~ 10 mN m⁻¹) and even to CA (by ca. 0.50 nm² at Π= 5 ~ 10 mN m⁻¹) and it has the lowest dipole moment. This behavior suggests that the plane structure of βMCA exists at the water surface with the largest angle of inclination and this as a matter of course leads to the smallest mean molecular surface area.

The gradient of βMCA may be calculated from the ratio of dipole moment values, assumed that CA is oriented horizontally. The angle might be estimated to be cos⁻¹ [μ⊥(βMCA)/μ⊥(CA)], assuming that the dipole moment is the same for these trihydroxy cholates. The calculation shows that βMCA has gradient of about 70° against CA. Based on the same assumption, the gradient of UDC against CDC is about 37°. It is suggested that this inclination results in the smaller mean surface area of UDC as compared to CDC (see Table I).

As mentioned before, CDC is oriented with a small tilt angle when CA is assumed to orient horizontally; therefore for the four bile acids the following gradient order is given: βMCA > UDC > CDC > CA (9). As is seen from Table I, the order of the mean surface area follows the gradient order. Examining the HI values (14), one can see that neither the order of the mean molecular surface area or the gradient is consistent with that of HI, but these quantities agree with the order of the hydrophilic area of the α-plane or of the hydrophobic area of the α-plane: values of the hydrophilic area in nm^2 are UDC (1.24) < CDC (1.35) < CA (1.65) and values of the hydrophobic area are UDC (1.17) > CDC (1.09) > CA (0.89), where data for βMCA are not given. Anyway, it is clear that the behavior concerning the mean molecular area as well as the gradient of molecules at the air-water interface is governed mainly by properties of the α-plane.

It is very important to examine the correlation between solubilizing power of each bile acid and the effective hydropholic surface area inferred from the present study. Concerning cholesterol solubilization by bile salts, the reported solubilizing factors (defined as mole ratio of bile salt to cholesterol, i.e., the number of bile salt molecules required to solubilize one cholesterol molecule at 30 mM bile salt) are 1430 for NaβMCA, 384 for NaUDC and 24 for NaCDC (26,27). Applying a similar definition to the previous work, the solubilizing factor for NaCA is estimated to be 17 (28). This order is the same as the gradient order mentioned above. In other words, the order of the solubilizing power is the same as the order of the mean surface area (see Table I), suggesting that the effective hydrophobic surface area is directly related to the hydrophobic interaction taking place in such a case as lipid solubilization, in which three dimensional structures are formed in water (not yet completely evaluated but it may correspond to the cross section in Fig. 8).

Finally, a phase diagram of surface pressure plotted against mole fraction in the mixed bile acids was obtained by measuring collapse pressures for the CDC-UDC system, as shown in Fig. 9. Measured points (open circles) coincide well with the theoretical curve calculated from the Joos equation (29,30).

$$1 = X_1 \exp\frac{\left(\Pi_{c,m}-\Pi_{c,1}\right)A_{c,1}}{kT} + X_2 \exp\frac{\left(\Pi_{c,m}-\Pi_{c,2}\right)A_{c,2}}{kT} \qquad \text{Eq.4}$$

where, X_1 and X_2 denote the mole fractions of CDC and UDC, respectively; $\Pi_{c,m}$ the collapse pressure of the mixed monolayer; $\Pi_{c,1}$ and $\Pi_{c,2}$, the respective collapse pressures; k, the Bolzmann constant; and T, temperature in K. Not only the average molecular area data and the average surface dipole moment data (both of which show that the additivity rule holds) but also the curve of collapse pressure of the mixture vs. mole fraction, suggest that CDC and UDC mix ideally in the monolayer. As is indicated in the figure, in the region beneath the curve, there coexist the ideally mixed molecules of CDC and UDC in an expanded state, and in the triangular region at higher surface pressure, UDC molecules in bulk (crystalline) state coexist with UDC and CDC molecules in an expanded state. Further increase in surface pressure induces collapse of expanded monolayer of CDC as well as UDC and both are transfered into bulk states.

As mentioned above, it was found that two mixed monolayer bile acid systems, i.e., CA-βMCA and CDC-UDC, exhibit ideal mixing over the whole range of mole fraction. Here, it is worthwhile to compare the present mixed monolayer results with the CDC-UDC mixed adsorbed film which is in equilibrium with bile salt monomers

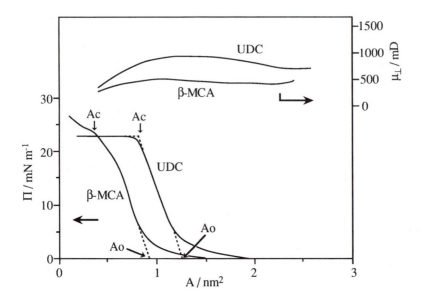

Fig.7   Π-A and $\mu_\perp$-A isotherms for UDC and βMCA.
        A_0 : limiting surface area
        A_c : area at collapse pressure   (Adapted from ref. 9).

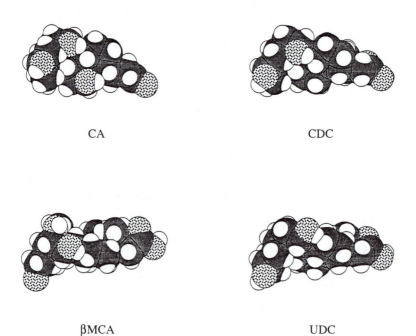

Fig. 8. Projections viewed from the side directly beneath the respective bile
acids floating on the water surface. (Adapted from ref. 9.)

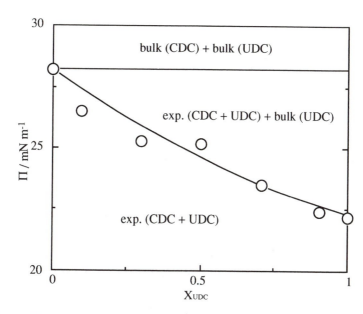

Fig. 9. The phase diagram for the mixed system of CDC-UDC. Open circles indicate measured points and the line, the theoretical curve calculated from the Joos equation. (Adapted from ref. 9)

in bulk solution and with their mixed micelles. In the previous surface tension study of NaCDC-NaUDC mixed system (7), cmc's as well as the surface excess, Γ (mol/m^2), or the mean surface occupying area per molecule, $\overline{A}$, were determined as a function of mole fraction in the NaCDC-NaUDC mixture. In contrast to the mixed monolayer showing ideal mixing, the mixed micelle formation was found to show a slightly positive deviaion from ideal mixing as was shown in Fig. 1.

The small positive deviation comes from the three-dimensional structure formation or micelle formation in bulk phase. This implies that the mode of micelle formation is different from that of the two-dimensional monolayer formation by compressing the floating acid molecules on water surface, because micelle formation accompanies one more complicated factor compared with monolayer formation especially when micelles are formed by such molecules having a complicated structure as bile acid salts. Further, an interesting difference was found between the mixed monolayer and the adsorbed film for the same CDC-UDC mixed system. That is, the mean surface areas per molecule ($\overline{A}$), determined from of the Gibbs isotherm, were estimated to be 0.87 nm^2 for CDC and 0.93 nm^2 for UDC, and did not obey the additivity rule; instead, the curve of $\overline{A}$ vs. X_2 has a minimum at $X_2=0.5$ ($\overline{A}=0.49$ nm^2). Dimerization of bile acids below the cmc has been indicated, not only by our surface tension (5) and electroconductivity studies (unpublished work) but also by the fluorescence study by Kano et al. (31). The existence of dimeric units such as $(UDC)_2$, $(CDC)_2$ and $(UDC-CDC)$ even below the cmc should be further examined by various experiments. For example, the gel filtration chromatography applied by Funasaki et al. (32) will be one of the best ways to solve this problem.

Acknowledgment

This material is based in part on work supported by the Central Research Institute of Fukuoka University.

References

(1) M. C. Carey and D. M. Small, *Arch. Intern. Med.*, **130**, 506 (1972).
(2) D. M. Small, in "The Bile Acids", P. P. Nair, and D. Kritchevsky, Eds.,p.249 Plenum Press, New York (1971).
(3) H. Danielsson and J. Sjovall, Eds., "Sterols and Bile Acids", Elsevier, Amsterdam (1985).
(4) I. M. Aras, Ed., "HEPATOLOGY", Vol.4, No.5, Suppl., American Association for the Study of Liver Diseases (1984).
(5) S. Nagadome, H. Miyoshi, G. Sugihara, Y. Ikawa and H. Igimi, *J. Jpn. Oil Chem. Soc. (YUKAGAKU)*, **39**, 542-547 (1990).
(6) M. C. Carey, J-C. Montet, M. C. Phillips, M. J. Armstrong and N. A. Mazer, *Biochemistry*, **20**, p.3637-3648 (1981).
(7) P. Ekwall, K. Fontell, and A. Sten, "Proceedings of Second International Congress of Surface Activity", London, I. pp. 357-373 (1957).
(8) O. Shibata, H. Miyoshi, S. Nagadome, G. Sugihara and H. Igimi, *J. Colloid Interfase Sci.*, **146**, 594-597 (1991).
(9) H. Miyoshi, S. Nagadome, G. Sugihara, H. Kagimoto, Y. Ikawa, H. Igimi, and O. Shibata, *J.Colloid Interface Sci.*, (in press).
(10) S. L. Hsia, in "The Bile Acids: Chemistry, Physiology and Metabolism., Vol.1, Chemistry." ed. by P. P. Nair and D. Kritchevsky (1971) Plenum Press, New York, pp.95-120.
(11) S. Uryu, M. Aratono, M. Yamanaka, K. Motomura, and R. Matuura, *Bull. Chem. Soc. Jpn.*, **57**, 967-971 (1984).
(12) O. Shibata, S. Kaneshina and R. Matuura, *Bull. Chem. Soc. Jpn.*, **61**, 3077-3082 (1988).
(13) O. Shibata, Y. Moroi, M. Saito and R. Matuura, *J. Colloid Interface Sci.* **142**, 535-543 (1991).
(14) K. Miyajima, K. Machida, T. Taga, H. Komatsu and M. Nakagaki, *J. Chem. Soc. Faraday Trans. 1*, **84**, 2537-2544 (1988).
(15) P. Mukerjee, Private communication to the authors (1981).
(16) G. Sugihara, D. Nakamura, M. Okawauchi, H. Kawamura and Y. Murata, *Fukuoka Univ. Sci. Reports*, **15**, 119-128 (1985).
(17) G. Sugihara and M. Tanaka, *Bull. Chem. Soc. Jpn.*, **49**, 3457-3460 (1976).
(18) Y. Murata, M. Tanaka and G. Sugihara, *Fukuoka Univ. Sci. Reports*, **17**, 23-26 (1987).
(19) K. Shinoda and T. Nomura, *J. Phys. Chem.*, **84**, 365 (1980).
(20) K. Motomura, M. Yamanaka and M. Aratono, *Colloid Polym. Sci.*, **262**, 948-369 (1984).
(21) T. Handa, *Hyomen*, **15**, 162-172 (1977).
(22) T. Handa and M. Nakagaki, *Colloid Pollymer Sci.*, **257**, 374-381 (1979).
(23) M. Nakagaki, K. Tomita and T.Handa, *Biochemistry*, **24**, 4619-4624 (1985).
(24) J. H. Schulman and A. H. Hughes, *Proc. Roy. Soc. (London)*, **A138**, 430-450 (1932).
(25) P. Joos and G. Bleys, *Colloid Pollymer Sci.*, **261**, 1038-1042 (1983).
(26) H. Igimi and M. C. Carey, *J. Lipid Resarch*, **22**, 254-270 (1981).
(27) J-C. Montet, M. Parquet, E. Saquet, A-M. Montet, R. Infante, and J. Amic, *Biochim. Biophys. Acta*, **918**, 1-10 (1987).
(28) G. Sugihara, K. Yamakawa, Y. Murata and M. Tanaka, *J. Phys. Chem.*, **86**, 2784-2788 (1982).
(29) P. Joos, *Bull. Soc. Chem. Belges*, **78**, 207-217 (1969).
(30) P. Joos and R. A. Demel, *Biochem. Biophys. Acta*, **183**, 447-457 (1969).
(31) K. Kano, S. Tatemoto and S. Hashimoto, *J. Phys. Chem.*, **95**, 966-970 (1991).
(32) N. Funasaki, S. Hada and S. Neya., *Chemistry Letters*, pp. 1075-1078 (1990).

RECEIVED January 6, 1992

ADSORPTION OF MIXED SURFACTANTS AT INTERFACES

Chapter 21

Synergism in Binary Mixtures of Surfactants at Various Interfaces

Milton J. Rosen

Surfactant Research Institute, Brooklyn College, City University of New York, Brooklyn, NY 11210

Synergism in mixtures of surfactants depends upon the existence of attractive interactions between the different surfactants. The conditions for synergy, in mathematical terms, can be derived by use of the so-called β-parameters that measure surfactant interactions in mixed monolayers and mixed micelles. Various types of synergy, resulting from mixed monolayer formation at various interfaces and from mixed micelle formation in aqueous solution, can be distinguished. The mathematical conditions for each of these to exist at the aqueous solution/air, aqueous/liquid hydrocarbon, and aqueous/hydrophobic solid interfaces have been derived, together with the surfactant ratios and relevant properties at the points of maximum synergism. When surfactant-surfactant interactions are strong, e.g., in cationic-anionic systems, terms for the change in interfacial areas of the surfactants upon mixing must be included in the relevant equations. Negative synergism can occur when attractive surfactant interactions in the mixed system are weaker than in the individual surfactant systems.

At the 59th Colloid and Surface Science Symposium held in Potsdam, New York, in 1985, I presented the work we had been doing in my laboratory in designing a quantitative framework for the study of synergism in binary mixtures of surfactants. That study was initiated to replace the trial-and-error method used in the past to find synergistic combinations of surfactants and to replace it with an understanding of the principles involved in synergy, an understanding that would permit the selection of surfactant pairs for optimum performance in a rational, scientific fashion. Since then, the search for and study of synergy in mixtures of materials has become even more important because of the difficulties involved in the introduction of new compounds, either for industrial or consumer use. Questions of toxicity, biodegradability, and general environmental impact must now be addressed whenever a new compound is suggested. As a result, in the search for improved performance and properties, a more feasible approach at the present time is often through the use of combinations of existing acceptable materials that show synergy, rather than through the design of new chemical structures. In this chapter, I will discuss what we have been doing in the area of synergism since 1985.

0097–6156/92/0501–0316$06.00/0

Our investigations have moved in three directions: 1) we have extended our study of synergism, which originally was confined to the aqueous solution/air interface, to the aqueous solution/liquid hydrocarbon interface and to the aqueous solution/hydrophobic solid interface, 2) we have modified the basic equations that we have been using to account for the change in interfacial area of the surfactants upon mixing, and 3) we have investigated the phenomenon of negative synergism, which we have discovered in a number of systems.

However, before discussing these developments, let me review some of the principles upon which our work is based. Synergism in binary mixture of surfactants is due to attractive interactions between the two types of surfactants present in the system. We measure the nature and strength of those interactions by the so-called β parameters, first suggested by Donn Rubingh, one of the organizers of this symposium, back in 1978 (1). The equations used by him to evaluate β parameters for mixed micelle formation are:

$$\frac{(X^M)^2 \ln(\alpha C^M_{12}/X^M C^M_1)}{(1-X^M)^2 \ln[(1-\alpha) C^M_{12}/(1-X^M) C^M_2]} = 1 \qquad (1)$$

$$\beta^M = \frac{\ln(\alpha^M C^M_{12}/X^M C^M_1)}{(1-X^M)^2} \qquad (2)$$

Where X^M is the mole fraction of surfactant 1 in the total surfactant in the mixed micelle; α is the mole fraction of surfactant 1 in the total surfactant in the solution phase; C^M_1, C^M_2, and C^M_{12} are the critical micelle concentrations of individual surfactants 1 and 2 and their mixture at a given value of α, respectively; β^M is the interaction parameter for mixed micelle formation.

We extended this treatment, which was developed for mixed micelle formation in aqueous solutions, to mixed monolayer formation at the aqueous solution/air interface (2). The relevant equations are:

$$\frac{X^2 \ln(\alpha C_{12}/X C^o_1)}{(1-X)^2 \ln[(1-\alpha) C_{12}/(1-X) C^o_2)]} = 1 \qquad (3)$$

$$\beta^\sigma = \frac{\ln(\alpha C_{12}/X C^o_1)}{(1-X)^2} \qquad (4)$$

where X is the mole fraction of surfactant 1 in the total surfactant in the mixed monolayer at the interface; C^o_1, C^o_2, and C_{12} are the solution phase molar concentrations of individual surfactants 1 and 2 and their mixture at a given value of α, respectively, required to yield a given surface tension value; β^σ is the interaction parameter for mixed monolayer formation. The experimental evaluation of β^σ is from the surface tension -- concentration curves of the systems (3).

Modification for Change in Interfacial Areas of the Surfactants upon Mixing.

The derivation of the basic equations is based upon the assumption that interfacial areas of the two interacting surfactant molecules do not change significantly when they are mixed with each other. This assumption is tenable when interaction between the two surfactants is weak or moderate, but is not valid when interaction is strong. For example, for mixtures of anionic with cationic surfactants, it has been shown by a number of investigators (4-7) that the molar interfacial areas of the two surfactants decrease markedly when mixed with each other, due to mutual neutralization of their charges. This change makes the value of the β parameter, in these cases, change significantly with change in the ratio of the two surfactants at the interface and in the mixed micelle or with change in the interfacial tension of the system. In such cases, terms must be added to the basic equations to account for the work involved in changing the molecular areas. The relevant equations then become (7) :

$$\frac{X^2}{(1-X)^2} \cdot \frac{\ln(\alpha C12/C^o_1 X) - \gamma(A^o_1 - A_1)/RT}{\ln[(1-\alpha)C_{12}/C^o_2(1-X)] - \gamma(A^o_2 - A_2)/RT} = 1 \qquad (5)$$

$$\beta^\sigma = \frac{\ln(\alpha C_{12}/C^o_1 X) - \gamma(A^o_1 - A_1)/RT}{(1-X)^2} \qquad (6)$$

$\gamma =$ surface or interfacial tension of the system

$A^o_1 =$ surface or interfacial area of surfactant 1 before mixing

$A_1 =$ surface or interfacial area of surfactant 1 after mixing

$A^o_2 =$ surface or interfacial area of surfactant 2 before mixing

$A_2 =$ surface or interfacial area of surfactant 2 after mixing

The values of A_1 and A_2 are obtained by assuming that their ratio after mixing is the same as their ratio before mixing, i.e.,

$$\frac{A1}{A2} = \frac{A^o_1}{A^o_2}$$

We have found that the use of these equations gives β^σ values that do not change significantly with either the ratio of the two surfactants in the system or with the surface or interfacial tension of the system.

Synergism in surface tension reduction efficiency.

The efficiency of surface tension reduction by a surfactant is measured by the solution phase concentration required to produce a given surface tension (reduction). Synergism in this respect is present in a binary mixture of surfactants when a given surface tension (reduction) can be attained at a total mixed surfactant concentration lower than that required of either surfactant by itself.

We have shown *(8)* that the conditions for synergism in this respect are:

1. β^σ must be negative.

2. $|\ln C^o_1/C^o_2| < |\beta^\sigma|$

where C^o_1 and C^o_2 are the solution phase molar concentration of individual surfactants 1 and 2, respectively, required to attain a given surface tension (reduction).

When the equations that include the change in area of the surfactants upon mixing are used, the conditions for synergism in surface tension reduction efficiency become *(9)* :

1. β^σ is negative.

2. $|\beta^\sigma| > |\ln (C^o_1/C^o_2)| + |\gamma [(A^o_1 - A_1) - (A^o_2 - A_2)]/RT|$

The term $|\gamma [(A^o_1 - A_1) - (A^o_2 - A_2)]/RT|$, however is small, generally less than 1, and always less than the increase in the absolute value of β^σ resulting from the use of the extended equation. Consequently, the condition for synergism in this respect will always be fulfilled when use of the simple equations shows that the conditions are met.

At the point of maximum synergism, the mole fraction α^*, of surfactant 1 in the solution phase equals its mole fraction in the mixed monolayer at the aqueous solution/air interface, and is given by the relationship:

$$\alpha^* = X^* = \frac{\ln (C^o_1/C^o_2) + \beta^\sigma}{2\beta^\sigma}$$

where X^* is the mole fraction of surfactant 1 in the mixed monolayer at the point of maximum synergism in this respect. The minimum mixed surfactant concentration in the solution phase, $C_{12, min}$, required to attain a given surface tension (reduction) is given by the expression:

$$C_{12,min} = C^0_1 \exp\left[\beta^\sigma \left(\frac{\beta^\sigma - \ln(C^0_1/C^0_2)}{2\beta^\sigma}\right)^2\right]$$

We have extended *(10)* this treatment to the aqueous solution-hydrocarbon interface and have determined the conditions for synergism in <u>interfacial</u> tension reduction efficiency. The interaction parameter, β^σ_{LL}, for mixed monolayer formation at the liquid-liquid interface is determined from plots of <u>interfacial</u> tension vs. total surfactant concentration in the system at constant phase volume ratio and constant initial ratio of the two surfactants.

When the conditions of either constant initial phase volume ratio (Φ) and constant partition coefficient (K) are met, or $K\Phi << 1$, the equations and the conditions for synergism in this respect, as derived by this treatment, are completely analogous to those obtained for the liquid-air interface:

1. β^σ_{LL} must be negative.

2. $|\beta^\sigma_{LL}| > |\ln\frac{C^0_{1,t}}{C^0_{2,t}}|$

Where $C^0_{1,t}$ and $C^0_{2,t}$ are the total system concentrations of individual surfactants 1 and 2, respectively, required to produce a given interfacial tension in the two-phase systems containing only the individual surfactants.

Under these conditions, the mole fraction, α^*, of surfactant 1 in the total surfactant in the system at the point of maximum synergism equals the mole fraction at the interface and is given by the expression:

$$\alpha^* = X^* = \frac{\ln(C^0_{1,t}/C^0_{2,t}) + \beta^\sigma LL}{2\beta^\sigma LL}$$

The minimum total concentration of mixed surfactant in the system $C_{12,t\ min}$, to produce a given interfacial tension is given by the expression:

$$C_{12,t,min} = C^0_{1,t} \exp\left[\beta^\sigma_{LL}\left(\frac{\beta^\sigma_{LL} - \ln(C^0_{1,t}/C^0_{2,t})}{2\beta^\sigma_{LL}}\right)^2\right]$$

The next extension was to aqueous solution/hydrophobic solid interfaces, e.g., Parafilm, polyethylene, and Teflon.

For a hydrophobic, low energy solid, the solid/vapor interfacial tension γ_{SV}, may be considered to be constant with change in surfactant concentration in the aqueous phase. From Young's equation:

$$\gamma_{SV} - \gamma_{SL} = \gamma_{LV} \cos\Theta,$$

surfactant solutions having the same $\gamma_{LV} \cos\Theta$ (= constant γ_{SL}) value can therefore be used for C^0_1, C^0_2, and C_{12} in the equations given above for X and β^σ. In order to evaluate γ_{SL} at the solid/liquid interface, then, both the surface tension of the aqueous solution and its contact angle on the solid must be known. Alternatively, the adhesion tension, $\gamma_{LV} \cos\Theta$, can be measured directly by the Wilhelmy plate technique, when suitable plates of the solid can be obtained.

Table I. Effect of Interface on β^σ
($\alpha_p = 0.53$; 25°C; $\pi = 32$ mN m^{-1})

Mixture	Interface	β^σ	A_{av}
$C_{12}SO_3Na/C_8P$	0.1M NaCl/air	-3.1	37.9
$C_{12}SO_3Na/C_8P$	0.1M NaCl/Parafilm	-2.9	38.1
$C_{12}SO_3Na/C_8P$	0.1M NaCl/Teflon	-2.5	42.0

C_8P = N-octyl-2-pyrrolidinone

It is interesting to see what happens to the value of β^σ when air is replaced by a hydrophobic condensed phase. The data shown in Table I are typical. The replacement of air by a liquid alkane or a solid nonpolar solid in almost all the mixtures investigated results in a small decrease in the negative value of β^σ. Since air consists mainly of nonpolar N_2 gas, the similarities between the β^σ values against air and against other nonpolar phases should not be surprising. The decrease in the negative value of β^σ is most probably due to the increase in average interfacial area of the surfactant., resulting in weaker interaction between them. In the case of liquid hydrocarbon phases, other work in our laboratory *(10)* has shown that the presence of the hydrocarbon phase increases the area/molecule of the individual surfactants, by themselves, at the interface.

Table II lists some data for mixtures of N-octyl-2-pyrrolidinone (C_8P) and sodium 1-dodecanesulfonate ($C_{12}SO_3Na$) in 0.1 M NaCl, a system that meets the conditions for and shows synergism in surface and interfacial tension reduction efficiency at a number of interfaces *(11)*. As can be seen from the data listed, calculated values for C_{12} agree well with experimental values, even when some of the experimental values of α are not very close to α^* (the calculated mole fraction in the solution phase for optimum efficiency).

Table II. Synergism in γ Reduction Efficiency at Various
Interfaces (11)

$C_8P/C_{12}SO_3Na$: 0.1M NaCl (aqu.); $25^\circ C; \pi = 32$ mN m^{-1}

Interface	$\alpha(\alpha^*)$	C_1^0 x 10^3M	C_2^0 x 10^3M	$\cdot C_{12}$ x 10^3M	$C_{12,calcd.}$ x 10^3M
Air	0.53(0.46)	1.91	1.51	0.78	0.77
Hexadecane	0.23(0.30)	1.32	0.55	0.46	0.47
Teflon	0.53(0.45)	3.31	2.59	1.58	1.55
Parafilm	0.53(0.45)	1.93	1.41	0.81	0.79

C_8P = N-octyl-2-pyrrolidinone

Synergism in mixed micelle formation.

Synergism in this respect is present when the critical micelle concentration of any mixture is lower than that of either individual surfactant by itself. We have found (8) that the conditions for synergism in this respect are:

1. β^M must be negative.
2. $|\ln(C^M_1/C^M_2) < |\beta^M|$

At the point of maximum synergism in mixed micelle formation, the mole fraction, α^{M*}, of the surfactant 1 in the solution phase equals its mole fraction in the mixed micelle, X^{M*} and is given by the relationship:

$$\alpha^{M*} = X^{M*} = \frac{\ln(C^M_1/C^M_2) + \beta^M}{2\beta^M}$$

The cmc at the point of maximum synergism, i.e., the minimum total mixed surfactant concentration in the solution phase required for mixed micelle formation, $C^M_{12,\,min}$, is given by the relationship:

$$C^M_{12,\,min} = C^M_1 \exp\left[\beta^M\left(\frac{\beta^M - \ln(C^M_1/C^M_2)}{2\beta^M}\right)^2\right]$$

Here, also we have extended our treatment of synergism to 2-phase liquid systems *(10)* and have derived equations that are completely analogous to those obtained for solutions in contact with air, when the nonaqueous phase is a hydrocarbon. The interaction parameter, β^M_{LL}, is obtained from the breaks in the plots of interfacial tension vs. total surfactant concentration in the system, indicating the onset of micellization in the aqueous phase. Some data for mixtures of N-dodecyl-N-benzyl-N-methylglycine ($C_{12}BMG$) and $C_{12}SO_3Na$ in distilled water are shown in Table III. Here, again, there is good agreement between experimental and calculated C_{12}^M values.

Table III. Synergism in Mixed Micelle Formation at Various Interfaces
$C_{12}BMG/C_{12}SO_3Na$: H_2O, 25°C

Interface	$\alpha_1(\alpha_1^*)$	C_1^M x10^4 M	C_2^M x10^4 M	C_{12}^M x10^4 M	C_{12}^M,calcd x10^4 M
Air	0.69 (0.8)	5.5	124	4.1	4.5
n-Heptane	0.95(0.93)	4.4	98	2.7	2.9
n-Hexadecane	0.95(0.87)	5.3	106	4.3	4.6

$C_{12}BMG$ = N-dodecyl-N-benzyl-N-methylglycine

Synergism in surface tension reduction effectiveness.

This exists when the mixture of surfactants of its cmc reaches a lower surface tension than that obtained at the cmc of either component of the mixture by itself.

The conditions for this to occur are *(12)* :

1. $\beta^\sigma - \beta^M$ must be negative.

2. $|\beta^\sigma - \beta^M| > |\ln \frac{C^o_1,cmc.C^M_2}{C^o_2,cmc.C^M_1}|$

where $C^o_1,{}^{cmc}$, $C^o_2,{}^{cmc}$ are the molar concentrations of individual surfactants 1 and 2, respectively, required to yield a surface tension value equal to that of any mixture of the two surfactants at its CMC. Extension of this treatment to the aqueous solution/liquid hydrocarbon and aqueous solution/hydrophilic solid interfaces yields analogous relationships. Some data are listed in Table IV.

Table IV. Synergism in Interfacial Tension Reduction Effectiveness at Various Interfaces (13)

$C_{12}BMG/C_{12}SO_3Na$: H_2O, 25°C

Interface	$\alpha_1(\alpha_1^*)$	$\gamma_1{}^{cmc}$ (mN m^{-1})	$\gamma_2{}^{cmc}$ (mN m^{-1})	$\gamma_{12}{}^{cmc}$ (mN m^{-1})	calc $\gamma_{1\,min}$ (mN m^{-1})
Air	.028(.047)	32.8	39.0	28	29.9
n-Heptane	.051(.042)	1.8	7.7	1.1	0.6
n-Hexadecane	.051(.047)	3.5	9.9	1.4	1.4
Isooctane	.051(.042)	2.0	8.3	1.0	1.3
Heptamethyl-nonane	.051(.048)	2.8	9.6	1.3	1.1

The next extension of our work was due to an experience I had during a lecture tour of the People's Republic of China several years ago. I was asked for an explanation of the well-known fact that soap decreases the foam of alkylbenzenesulfonate-based laundry detergents. I offered the commonly-accepted explanation that the $Ca^{\#}$ in the water formed insoluble Ca soap that acted as a foam-breaker, only to learn that this foam reduction also occurs in distilled water! In an attempt to find the correct reason for this foam-reducing action, upon my return home we measured the β parameters for a number of long-chain carboxylate (i.e., soap)-alkylbenzenesulfonate mixed systems and obtained a most interesting result. The β^σ and β^M values for long-chain carboxylate - alkylbenzenesulfonate mixtures are positive, not negative, in contrast to what we had found in all previous systems studied! This means that molecular interaction between the two surfactants when mixed is weaker than surfactant-surfactant interactions in the individual compounds by themselves. Going back to our basic equations, we found that this could produce negative synergism in the properties that we had previously investigated: surface or interfacial tension reduction, mixed micelle formation. For example, the CMC of the mixture of two surfactants might be greater than the CMC of either surfactant by itself. To date, only a few systems have shown positive β values, notably mixtures of hydrocarbon-chain and fluorocarbon-chain surfactants (14) . The only systems that we have discovered to date containing hydrocarbon-chain surfactants only are mixtures of long-chain carboxylates (soaps) and long-chain sulfonates. Under the proper conditions, these systems can show negative synergism.

The conditions for negative synergism in the phenomena previously discussed are (12) :

In surface of interfacial tension reduction efficiency:

   1. β^σ must be positive.

(For interfacial tension reduction, β^σ_{LL} or β^σ_{LS} must be positive).

2. β^σ (or β^σ_{LL} or β^σ_{LS}) must be greater than $|\ln (C^0_1/C^0_2)|$.

In mixed micelle formation:

1. β^M (or β^M_{LL} or β^M_{LS}) must be positive.

2. β^M (or β^M_{LL} or β^M_{LS}) must be greater than $|\ln (C^M_1/C^M_2)|$.

In surface interfacial tension reduction effectiveness:

1. β^σ (or β^σ_{LL} or β^σ_{LS}) - β^M (or β^M_{LL} or β^M_{LS}) must be positive.

2. $\beta^\sigma - \beta M$ (or $\beta^\sigma_{LL} - \beta^M_{LL}$, or $\beta^\sigma_{LS} - \beta^M_{LS}$) must be greater than $|\ln (C^0_1,^{cmc}. C^M_2/C^0_2,^{cmc}.C^M_1|$.

The condition that $\beta^\sigma - \beta^M$ must be positive means that attractive interaction between the two different surfactants in the mixed micelle is greater than interaction in the mixed monolayer at the interface. As a result, upon micellization, surfactant is removed from the interface into the micelle, with a consequent reduction in surfactant concentration at the interface and increase in interfacial tension. This same phenomenon is the cause of the well-know "dip" in surface tension - concentration curves in the vicinity of the CMC of surfactants containing an impurity that is more surface-active. Upon micelle formation, the impurity is removed from the interface and solubilized into the micelles because of stronger interaction with the micellized surfactant than with the unmicellized surfactant at the interface, with a resulting increase in the surface tension of the system.

We have found *(15)* that long-chain carboxylate- alkylbenzenesulfonate mixtures meet the conditions for and show negative synergism in surface tension reduction effectiveness, i.e., the surface tension of their mixtures at the CMC can be higher than that either the soap or the alkylbenzenesulfonate. We have also found *(16)* a good correlation between synergism in surface tension reduction effectiveness and initial foam height as measured by the Ross-Miles foaming technique. Systems that show synergism in surface tension reduction effectiveness also show synergism in foaming effectiveness (that is, the initial foam height of the mixture at a fixed surfactant concentration in the aqueous phase can be greater than that of either component by itself at the same surfactant concentration). On the other hand, when a system shows negative synergism in surface tension reduction effectiveness, then it also shows negative synergism in foaming effectiveness. The reduction in the foam of alkylbenzenesulfonate solutions by soap in the absence of $Ca^{\#}$, consequently, appears to be is due to the increase in the surface tension of the system, due to negative synergism in surface tension reduction effectiveness.

Literature Cited

1. Rubingh, D. In *Solution Chemistry of Surfactants*, Mittal, K.L., Ed.; Plenum: New York, NY, 1979, Vol 1; pp 337-354.
2. Rosen, M.J.; Hua, X.Y. *J. Colloid Interface Sci.*. **1982**, 86, pp. 164.
3. Rosen, M.J. *Surfactants and Interfacial Phenomena;* 2nd ed. John Wiley: New York, NY, 1989, pp 394-397.

4.   Zhao, G.-X.; Chen, Y.-Z.; Ou, J.-G.; Tien, B.-S.; Huang, Z.-M. *Acta Chimica Sinica* **1980**, *38*, 409.
5.   Ding, H.-J.; Wu, X.-L.; Zhao, G.-X. *Acta Chimica Sinica* . **1985**, *43*, 603.
6.   Yang, W.-S.; Zhao, X.-G. *Acta Chimica Sinica.* **1985**, *43*, 705.
7.   Gu, B.; Rosen, M. *J. Colloid Interface Sci.* **1989**, *129,* 537.
8.   Hua, X.Y.; Rosen, M.J. *J. Colloid Interface Sci.* **1982**, *90*, 212.
9.   Rosen, M.J.; Gu, B. *Colloids and Surfaces* **1987**, *23*, 119.
10. Rosen, M.J.; Murphy, D.S. *J. Colloid Interface Sci.* **1986**, *110*, 224.
11. Rosen, M.J.; Gu,B.; Murphy, D.S.; Zhu, Z.H. *J. Colloid Interface Sci.* **1989**, *129*, 468.
12. Hua, X.Y.; Rosen, M.J. *J. Colloid Interface Sci.* **1988**, *125,* 730.
13. Rosen, M.J.; Murphy, D.S. *J. Colloid Interface Sci.* **1989**, *129*, 208.
14. Zhao, G.-X.; Zhu, B.-Y. *In Phenomena in Mixed Surfactant Systems,* Scamehorn, J.F., Ed.; ACS Symp. Series 311; Amer. Chem. Soc., Washington, D.C., 1986, pp. 184-198.
15. Rosen, M.J.; Zhu, Z.H. *J. Colloid Interface Sci.* **1989**, *133,* 473.
16. Rosen, M.J.; Zhu, Z.H. *J. Amer. Oil Chem. Soc.* **1988**, *65*, 663.

RECEIVED January 22, 1992

Chapter 22

Nonideality at Interfaces in Mixed Surfactant Systems

Paul M. Holland

General Research Corporation, Santa Barbara, CA 93111

Nonideality at various solution interfaces is examined using a model based on the pseudophase separation approach, regular solution approximation for treating nonideality at solution interfaces, and a simplified method for taking changes in molar areas on mixing into account. Comparison with experimental results for mixed anionic-cationic surfactant systems show this approach to provide useful predictions for surface tensions both above and below the critical micelle concentration, and useful predictions of contact angles on Teflon and Parafilm.

The adsorption of surfactants from solution onto interfaces leads to most of the practical benefits derived from mixed surfactant systems. Among the major effects observed are interfacial tension lowering and changes in the wettability of surfaces as reflected in contact angle changes. Since the chemical potential of surfactant molecules in bulk solution effectively controls the composition at solution interfaces at equilibrium, a mixed surfactant model for interfaces can be viewed as a natural extension of solution models for mixed surfactant systems.

Theory

At concentrations far below the mixed CMC, the interface is only sparsely covered by surfactant and a significant amount of "bulk-like" water is present at the interface. As the concentration increases, the surface becomes "saturated" with an adsorbed mixed surfactant monolayer. This regime is indicated by a constant slope in the plot of surface tension versus the logarithm of activity or total surfactant concentration. Here, the Gibbs equation

$$\omega_i = -RT \frac{d \ln C_i^m}{d\gamma_i} \tag{1}$$

0097–6156/92/0501–0327$06.00/0

applies, and can be used to determine the surface area per mole at the interface. In the case of mixed surfactant systems the average area per mole at the interface is obtained. Eventually as the surfactant concentration continues to increase, the CMC or a phase boundary is surpassed and the surfactant activity tends to level off sharply.

At or above concentrations where the Gibbs equation holds, a "surface pseudophase" consisting of a "saturated" monolayer of adsorbed surfactant aggregate can be defined. This approach diverges from the standard surface solution approach (see *ref 1*) because the presence of water at the interface is not explicitly included and the sum of mole fractions of surfactant at the interface is therfore assumed to be unity. In this treatment, any "residual" solvent effects at the interface are now either accounted for in the standard state chemical potentials for the pure components, or in a surface interaction parameter accounting for nonideality in mixed systems.

This provides the basis for developing a tractable and generalized nonideal mixed surfactant model for interfaces (2-4). The mixed monolayer model can be designed for use above the CMC, below the CMC where the Gibbs equation holds, for contact angles, and for extension to multicomponent systems (at least in principle). This can be developed as follows. For concentrations at or above the CMC in a pure surfactant system, the chemical potential of surfactant i in the adsorbed monolayer can be expressed as

$$\mu_i^s = \mu_i^{os} + \pi_i^{max} \omega_i \tag{2}$$

where μ_i^{os} is a standard state chemical potentialat the surface and $\pi_i^{max} \omega_i$ a force field term containing the maximum (constant) surface pressure at or above the CMC and the molar area at the interface. The chemical potential of surfactant component i in solution is given by

$$\mu_i = \mu_i^o + RT \ln C_i^m \tag{3}$$

where μ^o is a standard state chemical potential and C_i^m the monomer concentration. At or above the CMC in the pure system a similar expression results

$$\mu_i = \mu_i^{Mo} = \mu_i^o + RT \ln C_i^* \tag{4}$$

where C_i^* is the CMC of the pure component. At equilibrium the chemical potential of the monomeric species equals that in the mixed monolayer and these can be combined to yield the expression

$$\frac{\mu_i^{os} - \mu_i^o}{RT} + \frac{\pi_i^{max} \omega_i}{RT} = \ln C_i^* \tag{5}$$

where the term with the standard state chemical potentials is in the form of a bulk-surface distribution coefficient.

In a mixed micellar system, the chemical potential can be expressed by

$$\mu_i = \mu_i^o + RT \ln f_i x_i C_i^* \tag{6}$$

where f_i is an activity coefficient and x_i the mole fraction of the ith component in the micelle which account for nonideal mixing in the micelle. At the interface in the mixed system the chemical potential of the ith component becomes

$$\mu_i^s = \mu_i^{os} + RT \ln f_i^s x_i^s + \pi \omega_i'$$ (7)

where π is the total surface pressure and ω_i' its area per mole after mixing. Nonideality in the pseudophase at the interface is accounted for by the term containing an activity coefficient (f_i^s) and mole fraction (x_i^s). Again at equilibrium, chemical potentials of the ith component are equal and equations 6 and 7 can be combined to yield

$$\frac{\mu_i^{os} - \mu_i^o}{RT} + \frac{\pi \omega_i'}{RT} = \ln \left[\frac{f_i x_i C_i^*}{f_i^s x_i^s} \right]$$ (8)

This expression includes a term with standard state chemical potentials in the form of a bulk-surface distribution coefficient, a force field term, and a ratio of activity coefficients and mole fractions in the micelle and monolayer.

Eliminating the bulk-surface distribution coefficient term between equations 5 and 8 allows a generalized nonideal analog of Butler's equation be obtained (2-5)

$$\pi \omega_i' = RT \ln \left[\frac{f_i x_i}{f_i^s x_i^s} \right] + \pi_i^{max} \omega_i$$ (9)

This directly relates properties of the mixed surfactant system to the composition and activity coefficients in the micellar and adsorbed monolayer pseudophases and properties of pure surfactant systems.

For concentrations below the CMC, one can apply the generalized result $C_i^m = f_i x_i C_i^*$ (see chapter 2, this volume) and $C_i^m = \alpha_i C$, which is valid below the CMC, to equation 9 obtaining

$$\pi \omega_i' = RT \ln \left[\frac{\alpha_i C}{f_i^s x_i^s C_i^*} \right] + \pi_i^{max} \omega_i$$ (10)

Together, Equations 9 and 10 allow nonideal mixed monolayers to be treated above and below the mixed CMC by using experimentally measured properties of the pure systems and solving for mole fractions and activity coefficients in the mixed system. Here, a simple regular solution approximation can be used for the activity coefficients with f_i^s (for binary surfactant mixtures) defined as

$$f_i^s = \exp \beta^s (x_i^s)^2$$ (11)

where β^s is a dimensionless interaction parameter analogous to that used sucessfully for treating nonideal mixing in micelles. To solve the model at a solution interface, either Equation 9 or 10 can be applied to each component in the system along with the constraint that mole fractions at the interface sum to unity. This allows either surface interaction parameters to be calculated from experimentally derived values of the surface pressure, or given the interaction parameters, surface pressures in the mixed system to be calculated.

The importance of allowing for changes in the molar areas of surfactant at the interface on mixing is beginning to be recognized (3,4,6,7). While there are a number of ways to approach this, no clearly preferred method has emerged (4). One of the simplest and most tractable of these is to define a ratio for the average change in area per mole for a surfactant on mixing

$$\delta = \frac{\omega_i'}{\omega_i} \tag{12}$$

which allows substitution of the quantity $\delta\omega_i$ for ω_i'. In the case of many anionic and cationic surfactant mixtures, inspection of surface tension versus concentration plots shows that a nearly constant value for the slope (or average area per mole as calculated from Equation 1) is maintained even when the composition of the surfactant mixture is varied. This observation suggests that the use of a single fixed value of δ for a surfactant mixture may be adequate.

The present nonideal mixed monolayer approach can also be extended to contact angles where three different interfaces meet (2-4). This represents a particularly interesting test case. For a surfactant solution in contact with air and a solid substrate, the contact angle is related to interfacial tensions by the Young equation

$$\gamma_{air/soln} \cos\theta = \gamma_{air/solid} - \gamma_{soln/solid} \tag{13}$$

While the contact angle and $\gamma_{air/soln}$ can be directly measured experimentally, the other interfacial tensions cannot. Fortunately, $\gamma_{air/solid}$ values in either mixed or pure systems are virtually identical and constant, particularly in the case of advancing contact angles (8), and these can be backed out of experimental data on pure systems. With these in hand, experimental values of the contact angle and surface tension for both pure water and pure surfactant solutions can be applied in Equation 13 to determine $\gamma_{soln/solid}$. This procedure allows all the required model parameters for the pure surfactant systems to be tabulated.

To solve the model for mixed systems, Equations 9 or 10 can be applied by converting between interfacial tensions and surface pressures using the relationship

$$\pi = \gamma_{H_2O}^o - \gamma \tag{14}$$

where the surface pressure is defined as the difference between the interfacial tension in pure water and surfactant solution. This allows selected experimental results in mixed systems to be used to determine interaction parameters and δ for the solution/solid interface. Once these have been determined, contact angles in mixed systems can be predicted by solving the model (Equations 9 or 10) at the air/solution and solution/solid interfaces, and using

$$\cos\theta = \frac{\gamma_{air/solid} - \gamma_{soln/solid}}{\gamma_{air/soln}} \tag{15}$$

Results and Discussion

To provide a test for the above approach for treating nonideality at interfaces, experimental data from two different anionic/cationic surfactant systems were compared with theory. These mixtures were sodium decyl sulfate ($C_{10}SO_4Na$) with decyltrimethylammonium bromide ($C_{10}TABr$), and sodium decanesulfonate ($C_{10}SO_3Na$) with dodecyl pyridinium bromide ($C_{12}NBr$). For the first comparision, previously unpublished surface tension data over a wide range of composition and concentration above and below the CMC in the $C_{10}SO_4Na/C_{10}TABr$ mixed system was used. These experiments, carried out at 23°C in 0.05 M NaBr, are described elsewhere (2). For the second comparison, previously published results of experimental measurements below the CMC by Gu and Rosen (7) on surface tensions and advancing contact angles on Teflon (PTFE, E.I. duPont de Nemours and Co.) and Parafilm M (paraffin wax, American National Can Co.) in the $C_{10}SO_3Na/C_{12}NBr$ mixed system were used. These results at 25°C in 0.1 M NaBr, were generously provided by Prof. Rosen to the author. Anionic/cationic systems were chosen for these comparisions because they exhibit both large deviations from ideality and significant changes in molar areas at interfaces on mixing.

Surface Tension Above and Below CMC. Results comparing theory and experiment in the $C_{10}SO_4Na/C_{10}TABr$ mixed surfactant system over a wide range of compositions are shown in Figures 1-8, where α represents the fractional composition of $C_{10}SO_4Na$ in the binary mixture. The plotted points are experimentally measured values of the surface tension, the solid line is the prediction of the nonideal mixed surfactant model with $\beta = -13.2$, $\beta^s = -14.7$, and $\delta = 0.741$, and the dashed line the prediction for ideal mixing ($\beta = 0$, $\beta^s = 0$, and $\delta = 1$). It is seen that both the surface tensions and CMCs strongly deviate from ideality, and that the experimental results are well described by the theory using only three parameters. It is only at extreme anionic rich compositions that any significant deviations between theory and experiment become apparent. These deviations may be correlated with the higher surface activity of $C_{10}SO_4Na$ compared to $C_{10}TABr$.

The observation that the model provides good descriptions using a single factor for molar area changes in the $C_{10}SO_4Na/C_{10}TABr$ mixed system suggests that more detailed modeling of molar area changes on mixing may be unnecessary. The observed similarity in the surface and micellar interaction parameters is not suprising. Physical insight would suggest that interaction energies at either a planar (monlayer) or curved (micellar) surface ought to be of similar magnitude. The somewhat weaker interaction indicated by the less negative micellar parameter may be due to curvature effects.

Contact Angles on Teflon and Parafilm. Results for surface tensions and contact angles on Teflon and Parafilm below the CMC in the $C_{10}SO_3Na/C_{12}NBr$ mixed

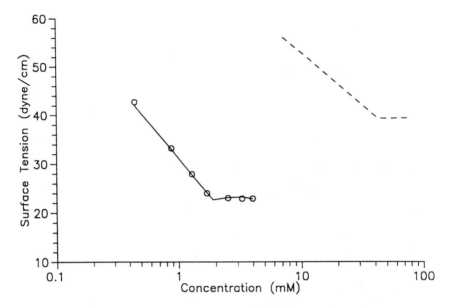

Figure 1. Surface tension in the $C_{10}SO_4Na/C_{10}TABr$ mixed surfactant system for $\alpha = 0.0823$, showing experimental data compared with models (see text).

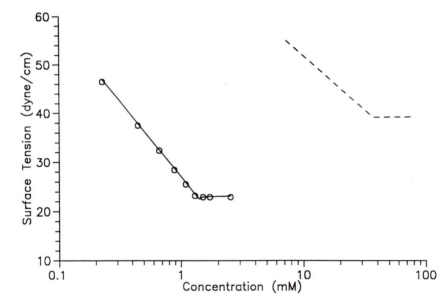

Figure 2. Surface tension in the $C_{10}SO_4Na/C_{10}TABr$ mixed surfactant system for $\alpha = 0.1772$, showing experimental data compared with models (see text).

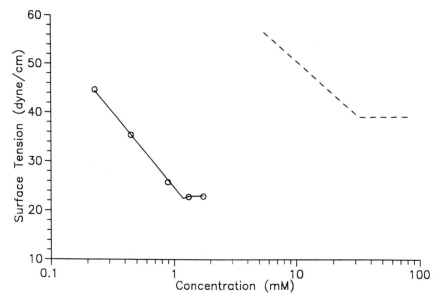

Figure 3. Surface tension in the $C_{10}SO_4Na/C_{10}TABr$ mixed surfactant system for $\alpha = 0.3011$, showing experimental data compared with models (see text).

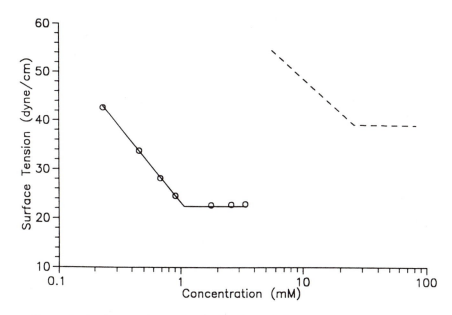

Figure 4. Surface tension in the $C_{10}SO_4Na/C_{10}TABr$ mixed surfactant system for $\alpha = 0.5185$, showing experimental data compared with models (see text).

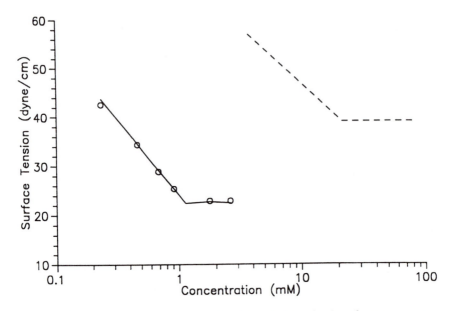

Figure 5. Surface tension in the $C_{10}SO_4Na/C_{10}TABr$ mixed surfactant system for $\alpha = 0.7292$, showing experimental data compared with models (see text).

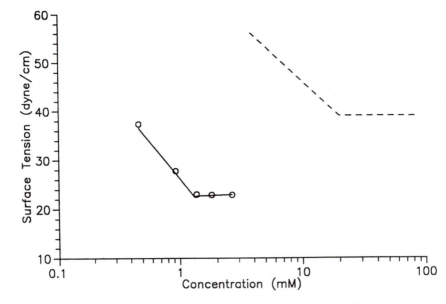

Figure 6. Surface tension in the $C_{10}SO_4Na/C_{10}TABr$ mixed surfactant system for $\alpha = 0.8434$, showing experimental data compared with models (see text).

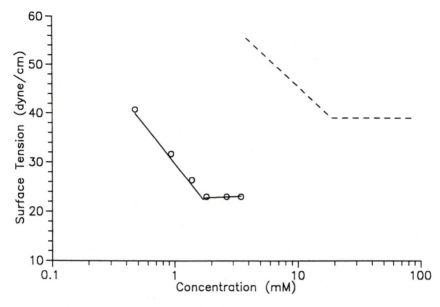

Figure 7. Surface tension in the $C_{10}SO_4Na/C_{10}TABr$ mixed surfactant system for $\alpha = 0.9282$, showing experimental data compared with models (see text).

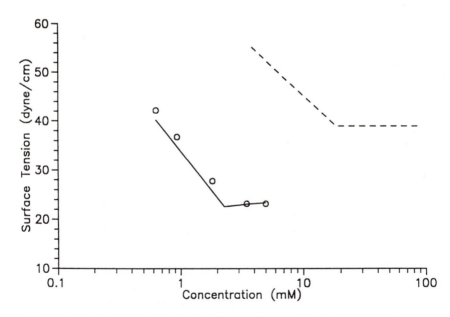

Figure 8. Surface tension in the $C_{10}SO_4Na/C_{10}TABr$ mixed surfactant system for $\alpha = 0.9700$, showing experimental data compared with models (see text).

surfactant system are shown in Figures 9-11, where α represents the fractional composition of $C_{12}NBr$ in the binary mixture. In Figure 9, the plotted points are experimentally measured values of the surface tension at the air/solution interface, the solid line is the prediction of the nonideal mixed surfactant model with β^s = -8.9 and δ = 0.71, and the dashed line the prediction for ideal mixing (β^s = 0, and δ = 1). The results show the surface tensions to deviate significantly from ideality, and that the experimental values are reasonably well described by the theory using these two parameters. The compositions range from anionic rich (α = 0.2286) to cationic rich (α = 0.7713), where the deviations between theory and experiment are seen to be somewhat larger.

For contact angles of surfactant solution on a solid substrate, three interfacial tensions are involved. Of these, only the air/solution surface tension can be directly measured, requiring the other interfacial tensions to be inferred from experimental results and the Young equation. Calculating values for $\gamma_{air/solid}$ for Teflon and Parafilm gives 18.0 and 24.0 dyne/cm, respectively. Together with experimental data on pure surfactant systems, these allow the constants π_i^{max} and ω_i to be determined for the solution/solid interface. Given these parameters the model can be solved. Here, an additional surface interaction parameter, β^σ, for describing nonideality at the solution/solid interface has been introduced.

Results for contact angles on Teflon in the $C_{10}SO_3Na/C_{12}NBr$ system are shown in Figure 10. The plotted points are experimentally measured values of the contact angle on Teflon, the solid line is the prediction of the nonideal mixed surfactant model with β^s = -8.9, β^σ = -6.2, and δ = 0.71, and the dashed line the prediction for ideal mixing (β^s = β^σ = 0, and δ = 1). Results show that contact angles on Teflon deviate significantly from ideality, and that the deviations increase with increasing concentration. The experimental values are reasonably well described by the nonideal theory using these three parameters. As compositions change from anionic rich to cationic rich, the deviations between theory and experiment are seen to increase larger (Figure 10C).

Contact angle results on Parafilm are shown in Figure 11. This shows experimentally measured values of the contact angle for $C_{10}SO_3Na/C_{12}NBr$ solutions on Parafilm (points) with the solid line showing the prediction of the nonideal mixed surfactant model using the same parameters used for Teflon, mainly, β^s = -8.9, β^σ = -6.2, and δ = 0.71. The dashed line shows the prediction for ideal mixing (β^s = β^σ = 0, and δ = 1). It is seen that the contact angles on Parafilm deviate from ideality even more than those on Teflon with the deviations also increasing with increasing concentration. As for Teflon, the experimental values are reasonably well described by the nonideal theory using only three parameters, although as compositions change from anionic rich to cationic rich, the deviations between theory and experiment also increase (Figure 11C). The larger observed deviations between theory and experiment in cationic rich mixtures may be correlated with the much higher surface activity of $C_{12}NBr$ compared to $C_{10}SO_3Na$.

It is especially interesting to note that interfacial tensions and contact angles on both Teflon and Parafilm in the $C_{10}SO_3Na/C_{12}NBr$ system can be modeled over wide range of composition and concentration using the same three parameters. That is, a single factor for the relative change in molar area after mixing (δ = 0.71) can be used for the solution interfaces with air, Teflon and Parafilm, a single value of the surface

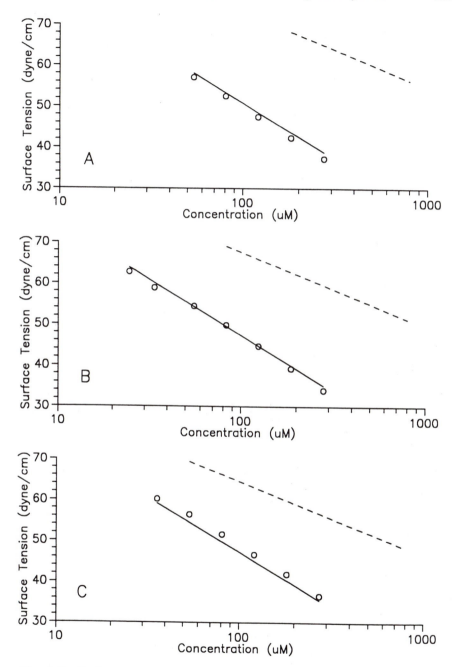

Figure 9. Surface tension versus concentration in the $C_{10}SO_3Na/C_{12}NBr$ mixed surfactant system for (A) $\alpha = 0.2286$, (B) $\alpha = 0.5000$, and (C) $\alpha = 0.7714$, showing experimental data compared with models (data from ref 7, see text).

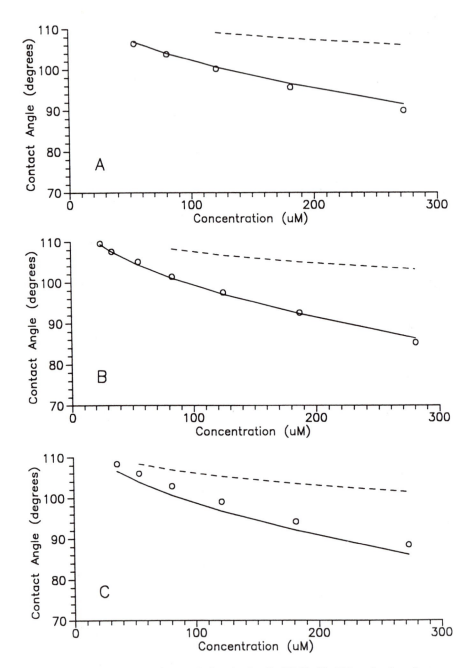

Figure 10. Contact angles on Teflon in the $C_{10}SO_3Na/C_{12}NBr$ mixed surfactant system for (A) $\alpha = 0.2286$, (B) $\alpha = 0.5000$, and (C) $\alpha = 0.7714$, showing experimental data points compared with models (data from ref 7, see text).

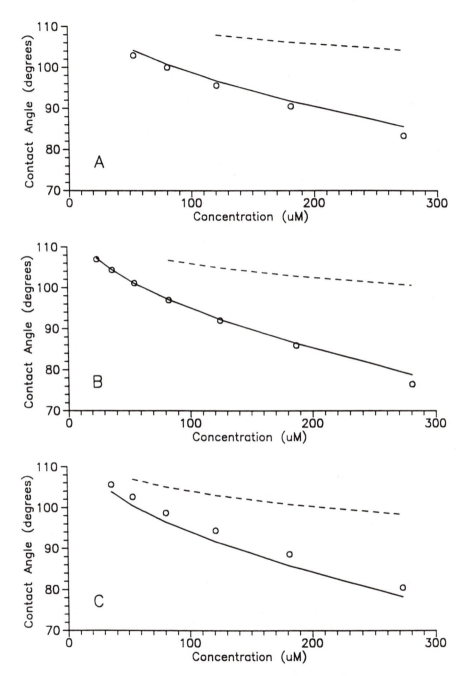

Figure 11. Contact angles on Parafilm in the $C_{10}SO_3Na/C_{12}NBr$ mixed surfactant system for (A) $\alpha = 0.2286$, (B) $\alpha = 0.5000$, and (C) $\alpha = 0.7714$, showing experimental data points compared with models (data from ref 7, see text).

interaction parameter (β^σ = -6.2) can be used at the solid/solution interface for both Teflon and Parafilm, and a surface interaction parameter β^s of -8.9 can be used for surface tensions at the air/solution interface. The observation that a single factor can account for molar area changes on mixing is consistent with interactions in the $C_{10}SO_3Na/C_{12}NBr$ system being dominated by electrostatics, and suggests that a more detailed accounting for molar area changes at interfaces may be unnecessary. The significantly stronger surface interaction parameter at the air/solution interface is likely due to more efficient headgroup/headgroup interaction at a fluid compared to a solid interface. The cause of the striking similarity of the parameter interaction β^σ on Teflon and Parafilm is not clear, but may be related to similar constraining effects of the solid surface on headgroup interactions.

Conclusions

A generalized model for treating nonideal mixed surfactant systems at various solution interfaces and for contact angles has been developed with tractability in mind. This model is based on the pseudophase separation approach, regular solution approximation for treating nonideality at solution interfaces, and a simplified approach for taking changes in molar areas on mixing into account. Comparison with experimental results for the binary mixed anionic-cationic surfactant systems $C_{10}SO_4Na/C_{10}TABr$ and $C_{10}SO_3Na/C_{12}NBr$ shows generally good agreement between the model and experiment both above and below the mixed CMC, and for contact angles on Teflon and Parafilm. It is also observed that a single parameter seems to adequately account for changes in molar areas after mixing in each system.

Legend of Symbols

C	total surfactant concentration
C_i^*	CMC of pure surfactant i
C_i^m	monomer concentration of surfactant i
f_i	activity coefficient of surfactant i in mixed micelles
f_i^s	activity coefficient of surfactant i in mixed monolayer
R	gas constant
T	absolute temperature
x_i	mole fraction of surfactant i in mixed micelles
x_i^s	mole fraction of surfactant i in mixed monolayer at interface
α_i	mole fraction of surfactant i in total surfactant
β	dimensionless interaction parameter in mixed micelle
β^s	dimensionless interaction parameter in monolayer at air/solution interface
β^σ	dimensionless net interaction parameter in monolayer at solution/solid interface
γ	surface tension at interface
δ	molar area of surfactant at interface in mixed system over that in pure system
θ	contact angle
μ_i	chemical potential of surfactant i
μ_i°	standard state chemical potential of surfactant i in solution
μ_i^{Mo}	chemical potential of surfactant i in pure micelles
μ_i^s	chemical potential of surfactant i at surface

μ_i^{os}	standard state chemical potential of surfactant i at surface
π	surface pressure in mixed surfactant system
π_i^{max}	maximum surface pressure of surfactant i at or above CMC in pure system
ω_i	area per mole of surfactant i at interface in pure system
ω'_i	area per mole of surfactant i at interface in mixed system

Acknowledgement

The author wishes to thank Prof. Milton J. Rosen for generously providing a copy of experimental results from his previously published work on contact angles in the $C_{10}SO_3Na/C_{12}NBr$ system.

Literature Cited

1. Lucassen-Reynders, E. H. In *Anionic Surfactants: Physical Chemistry of Surfactant Action,* Lucassen-Reynders, E. H., Ed.; Surfactant Science Series 11; Marcel Dekker, Inc.: New York, NY, 1981; pp 1-54.
2. Holland, P. M. In *Phenomena in Mixed Surfactant Systems;* Scamehorn, J. F, Ed.; ACS Symposium Series 311, American Chemical Society: Washington, DC, 1986; pp 102-115.
3. Holland, P. M. *Colloids Surf.* **1986,** *19,* 171.
4. Holland, P. M.; Rubingh, D. N. In *Cationic Surfactants: Physical Chemistry;* Rubingh, D. N.; Holland, P. M., Eds.; Surfactant Science Series 37; Marcel Dekker, Inc.: New York, NY, 1990; pp 141-187.
5. Defay, R.; Prigogine, I.; Bellemans, A. *Surface Tension and Adsorption,* Longmans: London, 1966; p 167.
6. Rosen, M. J.; Gu, B. *Colloids Surf.* **1987,** *23,* 119.
7. Gu, B.; Rosen, M. J. *J. Colloid Interface Sci.* **1989,** *129,* 537.
8. Neuman, A. W. *Adv. Colloid Interface Sci.* **1974,** *4,* 105.

RECEIVED January 6, 1992

Chapter 23

Compositions of Langmuir Monolayers and Langmuir–Blodgett Films with Mixed Counterions

Dong June Ahn and Elias I. Franses

School of Chemical Engineering, Purdue University, West Lafayette, IN 47907

A new model by the authors (*J. Chem. Phys.,* 1991) has been applied to interpret binary ionic composition data obtained by Deamer et al. (*J. Lipid Res.,* 1967) for stearic acid monolayers with group IIA ions (Be^{2+}, Mg^{2+}, Ca^{2+}, Sr^{2+}, and Ba^{2+}). Comparisons of the model predictions with those by previous electrochemical or thermodynamic models are made. Ion association equilibrium constants and Flory-Huggins mixing parameters have been determined. The model is then extended to ternary ionic systems, such as with H^{+}, Ca^{2+}, and Cd^{2+}. Results for both binary and ternary systems show that the mixing of adsorbed ions in the monolayer plays a critical role in controlling the ionic compositions of Langmuir monolayers and subsequently produced Langmuir-Blodgett films.

Fundamental knowledge of compositions of insoluble Langmuir monolayers is important for controlling their Langmuir-Blodgett (LB) deposition characteristics and the properties of the subsequent LB films formed from these monolayers. Such films are potentially useful in developing ultrathin devices for microelectronic, photonic, or sensor applications *(1)*. The ionic compositions of Langmuir monolayers and their LB films are determined by the competitive adsorption of ions from the bulk subphase. Various experimental data on compositions of LB films have been reported *(2-4)*. For binary ionic systems containing protons and bivalent ions, the composition changes with pH quite abruptly for Pb^{2+}, Be^{2+}, and Sr^{2+} ions, less abruptly for Cd^{2+} and Ca^{2+} ions, and least abruptly for Ba^{2+} and Mg^{2+} ions.

In this paper, various theoretical models *(4-10)* for the Langmuir or LB films ionic compositions are reviewed, and certain limitations in fitting binary composition data are discussed. A new model *(11)* which has been developed to overcome some of these limitations is described further and applied to certain data which have previously not been compared to models. The new model is then applied to predict yet unreported behavior in ternary ionic systems (e.g. H^{+}, Ca^{2+}, and Pb^{2+} ions).

0097–6156/92/0501–0342$06.00/0

These results may provide insight and experimental guidance in ionic compositions and ion exchange processes in Langmuir and LB films interacting with multicomponent ionic solutions.

Review of Previous Models

When a monolayer of a fatty acid is spread on an aqueous subphase, the protons of the carboxylic groups dissociate partly or completely, depending on the ionic content of the water subphase. Counterions in the subphase adsorb on the charged monolayer, depending on the pH, the concentrations of the ions in the subphase, the adsorption capacities of the ions, and the interactions of these ions in the monolayer with the subphase, the fatty acid monolayer, and each other.

One model for describing the equilibrium composition of such Langmuir monolayers has been developed by Matsubara et al. *(5)* and Yamauchi et al. *(6)*. They have used the simplifying assumption that the monolayers are electrically neutral. A chemical reaction equilibrium among ions and surfactant molecules is considered using thermodynamic activities. The Bragg-Williams equation is exploited to obtain the activities of the surfactant complexes. Petrov et al. *(4)* have added in this model the additional assumption that the association constant of protons with the carboxylic groups of the monolayer is the same as those with soluble carboxylic acids. This constant determines the overall equilibrium constant when combined with the equilibrium constant of the other kind of ions with the polar groups. Even though these thermodynamic models can describe certain data well *(4)*, they are not sufficiently rigorous, because they ignore the fact that the monolayers can be charged, thus ignoring that electrostatic effects can significantly affect ions spatial distributions.

Among the electrochemical models which consider these effects *(7-10)*, one which has been recently suggested by Bloch and Yun *(10)* focuses on the monolayer compositions. Others focus on the zeta potentials of bilayer membranes *(7)* or the surface-pressure/surface-area isotherms of monolayers *(8,9)*. The Langmuir isotherm has been used by all these authors to determine the surface densities of surfactant complexes and the surface charges inducing the diffuse layer in the subphase. Bloch and Yun's model is quite successful in describing compositions of fatty acid monolayers interacting with H^+ and Ca^{2+} ions or with H^+ and Cd^{2+} ions *(10)*. However, other experimental data have shown that the above model cannot describe the compositions for $H^+ - Pb^{2+}$, $H^+ - Ba^{2+}$, and other systems *(2,3)*. A more general model is needed.

The 1:1 positively charged complex (RA^+, where R^- is the dissociated fatty acid and A^{2+} is the bivalent ion) has been assumed by Bloch and Yun as the only form of complex between surfactant molecules and bivalent ions, rather than the 2:1 neutral complex (R_2A). The model studied by Lösche et al. *(8)* applies to the 2:1 neutral complex system. The stoichiometry of the complex is not clearly established, however. Hasmonay et al. *(12)* have suggested the coexistence of both types of surface complexes. Irrespective of the stoichiometry, the previous electrochemical models are incapable of describing the composition (slope of composition versus pH) of Langmuir monolayers for $H^+ - Ba^{2+}$ and $H^+ - Pb^{2+}$ systems, which show much lower slopes and much higher slopes respectively than monolayers with H^+ and Ca^{2+} or H^+ and Cd^{2+} *(11)*.

As detailed in the next section, these limitations result from using the Langmuir isotherm which accounts for no interactions among adsorbed ions. The previous electrochemical models are primarily appropriate for the noninteractive competitive adsorption.

Interactive Competitive Adsorption

Recently, a new model has been developed to describe the interactive competitive adsorption of ions *(11)*. The model is briefly as follows. It employs a modified Langmuir isotherm which is expressed with activities of the reaction components instead of their concentrations. The array of surfactant molecules at the air/water interface defines a two-dimensional lattice which is assumed in the Langmuir isotherm. The Flory-Huggins equation, which describes a simple lattice model, is then applied to this lattice containing dimers (R_2A) and monomers (RH and R^-). The dimers simply indicate the association of two surfactant molecules per bivalent ion. The monolayer charge is of course due to R^-. By contrast, previous models assume either neutral monolayers *(4-6)* or monolayers containing RA^+, RH, and R^- *(10)*, i.e. one surfactant molecule associated with one bivalent ion. Whereas charge inversion from negative to positive is possible with this model, with our model the interface is always negatively charged. No clear evidence of charge inversion of the monolayer with increasing bivalent ion concentration is available, unlike the inversion of the zeta potential of bilayer membranes *(7)*. The activities of surface complexes are then derived in terms of a Flory-Huggins binary interaction parameter χ_{HA}. The new model enables one to predict not only the electrochemical properties of Langmuir monolayers but also the effects of the interactions of the adsorbed ions on the compositions of Langmuir monolayers and LB films.

The Flory-Huggins binary interaction parameter plays the critical role in determining slopes in the composition of binary ionic systems versus pH (Figure 1), whether one considers the R_2A or the RA^+ complex. Here, $\phi_{A,LB}$ is the salt composition in LB films and is defined as the ratio of the number of moles of surfactant molecules associated with A^{2+} ions over the total number of moles of surfactant in LB films; C_A is the concentration of the ion A^{2+} in the bulk subphase; K_H and K_A are the surface association constants of protons and A^{2+} ions with the monolayer, respectively. When the interaction parameter χ_{HA} equals zero, the adsorbed ions' interactions have no influence on the adsorption of ions and hence on the monolayer composition. In this case, the adsorption can be regarded as "ideal," or effectively athermal. The previous electrochemical models describe such a case. When $\chi_{HA} < 0$, the mixing of adsorbed ions is favored compared to the ideal mixing case, because the free energy of the monolayer decreases upon mixing. The change of $\phi_{A,LB}$ with the bulk counterion concentration (pH or $p[A^{2+}]$) becomes less abrupt. By contrast, a positive deviation ($\chi_{HA} > 0$) leads to a more abrupt change in composition. Moreover, demixing in the mixed ion monolayer becomes possible for $\chi_{HA} \gtrsim 1.5$ *(13)*. The composition data obtained by Deamer et al. *(2)* for stearic acid monolayers interacting with group IIA ions are quantitatively interpreted here (Figure 2). The ions adsorb more strongly in the order Ba^{2+}, Sr^{2+} or Mg^{2+} (distinction unclear), Ca^{2+}, and Be^{2+}. This order does *not* follow the order of increasing covalency or ionic radius, which suggests that energetic interactions, rather than packing considerations, are important for the individual ions equilibrium constants. According to the interpretations given above, $H^+ - Be^{2+}$ and $H^+ - Sr^{2+}$ systems show positive-nonideal competitive adsorption, whereas the $H^+ - Mg^{2+}$ system shows negative-nonideal adsorption (Table I). Bloch and Yun's model agrees well with data for H^+ and Ca^{2+}, for which competitive adsorption is nearly ideal ($\chi_{HA} \approx 0$). Such a limitation results from the following reason. The conventional Langmuir isotherm, which includes no effects of adsorbed ions interactions, predicts parallel composition curves and cannot describe the nonideal adsorption data. Moreover, the ion association constants for soluble carboxylic acids *(14)* have been assumed to apply to the Langmuir monolayer system in Bloch and Yun's model *(10)*. Discrepancies of their model descriptions from the composition data indicate, however, that the surface association constants of ions with the monolayer may differ from the bulk association constants of individual ions. Table I shows that the surface association constant for Be^{2+} is about the same as its bulk association constant. For the other ions, such a comparison is not made here, since the bulk association constants for R_2A salts are not available in the literature. As

Table I. Parameter Values Used in Figure 2

| Ions | Present Model [a] | | Literature Value for Soluble Carboxylic Acids [b] | |
	χ_{HA}	K_A,[c] dimensionless	K_A,[c,d] dimensionless	K_A^{+},[e] M^{-1}
Be^{2+}	1.40(0.20)	$9.1 \times 10^5 (1.3 \times 10^5)$	7.1×10^5	41.7
Mg^{2+}	-1.81(0.58)	$4.9 \times 10^0 (3.3 \times 10^0)$	— [d]	3.55
Ca^{2+}	-0.52(0.25)	$3.9 \times 10^2 (9.0 \times 10^1)$	—	3.72
Sr^{2+}	1.40(0.38)	$8.1 \times 10^0 (8.7 \times 10^{-1})$	—	3.09
Ba^{2+}	— [f]	$1.7 \times 10^{-4} (1.6 \times 10^{-4})$	—	2.82

[a] Best fits were obtained for data by Deamer et al. *(2)* with $K_H = 3.63 \times 10^6$. The values in parentheses are the standard deviations.

[b] Values obtained for the ionic strength of 0.1M were cited from Smith and Martell *(14)*.

[c] Association constant to form the 2:1 R_2A complex.

[d] Value determined from literature value, using mole fractions instead of concentrations. For the other ions, values were not found.

[e] Association constant to form the 1:1 RA^+ complex, used with Bloch and Yun's model.

[f] Value too imprecise, since the corresponding K_A is too small ($K_A \ll 1$).

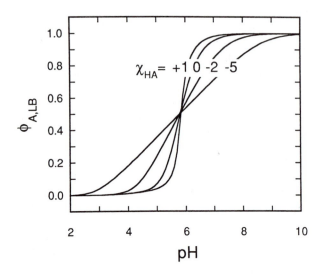

Figure 1. Dependence of LB composition on pH for different values of χ_{HA}; in this example, $C_A = 10^{-4}$ M and $K_H/K_A = 10^3$.

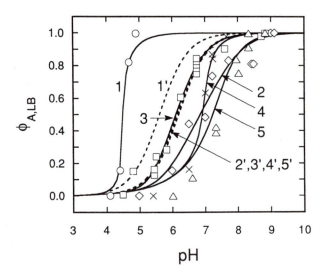

Figure 2. Comparison of models to LB composition data for stearic acid (IR analysis; $C_A = 2 \times 10^{-4}$ M; $T = 300$ K) by Deamer et al. *(2)* : $\bigcirc$, 1, and 1': Be^{2+}; $\Diamond$, 2, and 2': Mg^{2+}; $\square$, 3, and 3': Ca^{2+}; $\times$, 4, and 4': Sr^{2+}; $\triangle$, 5, and 5': Ba^{2+}; solid lines (1, 2, 3, 4, and 5): best fit predictions of present model *(11)* ; dotted lines (1', 2', 3', 4', and 5'): predictions from Bloch and Yun's model *(10)*.

found previously *(11)*, however, the surface pK's of other ions (Pb^{2+} and Cd^{2+}) can differ substantially from the bulk pK's, since surface tension and surface charge can influence the energetics and entropy of association of ions.

In principle, our model can be used to estimate K_H, K_A, and χ_{HA}. Because of insufficiently precise data, only if K_H is fixed can K_A and χ_{HA} be reasonably estimated. In the absence of a better alternative, we have set the value of K_H as equal to 3.63×10^6 *(11)*, which has been estimated from soluble carboxylic acid data, as done by Petrov et al. *(4)*. There is room for improvement in the precision of the model fitted parameters, if independently obtained estimate of K_H become available. The parameter values estimated in the present study for the $H^+ - Ca^{2+}$ system are in fair agreement with those from Reference 11: $K_A = 390 \pm 90$ for stearic acid at the surface pressure (Π) where the area-per-molecule is about 0.2 nm^2, versus $K_A = 290 \pm 81$ for arachidic acid at $\Pi = 29.5$ mN/m, and $\chi_{HA} = -0.52 \pm 0.25$ versus -0.76 ± 0.51. However, there is a substantial difference for the $H^+ - Ba^{2+}$ system: $K_A = 1.7 \times 10^{-4}$ versus 4.4. Such a system is shown elsewhere *(11)* to have a large difference between the compositions $\phi_{A,LB}$ and ϕ_A. We consider the first value less reliable, possibly because (i) the data by Kobayashi et al. *(3)* are of higher quality, since they used the LB deposition process rather than a monolayer "skimming" used by Deamer et al. *(2)* ; (ii) when $\phi_A << \phi_{A,LB}$, as in this case, the monolayer transfer process introduces a large uncertainty in the amount of entrained ions with the monolayer. More work is needed to clearly establish the reasons for this discrepancy.

In Figure 3, we compare the present model to a thermodynamic model and to the present model modified with the additional assumption that the Langmuir monolayer is neutral ($\phi_0 = 0$, where ϕ_0 is the surface fraction of the dissociated surfactant molecules R^-). This assumption makes little difference on the compositions $\phi_{A,LB}$ for the $H^+ - Pb^{2+}$ and the $H^+ - Cd^{2+}$ systems, but it makes a significant difference for the $H^+ - Ba^{2+}$ system. The entropy of mixing of surfactant complexes is still taken into account in the modified model, because the Flory-Huggins equation is applied rather than the Bragg-Williams equation, which is employed in the thermodynamic models *(4-6)*. As shown in Figure 3, using the latter model causes a larger discrepancy from the present model. In fitting values of key parameters for a given data set, the thermodynamic model leads to overestimating K_A and to underestimating the magnitude of χ_{HA}, especially for the negative-nonideal adsorption case.

Extension of New Model to Ternary Ionic Systems

Systems containing two kinds of bivalent ions and protons are considered next with the following competitive adsorption equilibria:

$$R^- + H^+ \overset{K_H}{\longleftrightarrow} RH \tag{1}$$

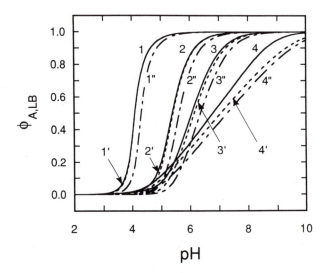

Figure 3. Comparison of present model (solid lines; 1, 2, 3, and 4) to modified present model with $\phi_0 = 0$ (dotted lines; 1', 2', 3', and 4') and thermodynamic model *(4-6)* (broken lines; 1", 2", 3", and 4"): 1, 1', and 1": Pb^{2+} ($\chi_{HA} = 0.78$, $K_A = 3.7 \times 10^6$); 2, 2', and 2": Cd^{2+} ($\chi_{HA} = 0.13$, $K_A = 8.6 \times 10^3$); 3, 3', and 3": Ca^{2+} ($\chi_{HA} = -0.76$, $K_A = 2.9 \times 10^2$); 4, 4', and 4": Ba^{2+} ($\chi_{HA} = -4.1$, $K_A = 4.4$); $C_A = 3 \times 10^{-4}$ M and T = 293 K. Parameter values fitted to data by Kobayashi et al. *(3)* were used *(11)*.

$$2R^- + A^{2+} \overset{K_A}{\longleftrightarrow} R_2A \tag{2}$$

$$2R^- + B^{2+} \overset{K_B}{\longleftrightarrow} R_2B \tag{3}$$

Since four components are present in the monolayer (R^-, RH, R_2A, and R_2B), a total of six binary interaction parameters can be defined. As done before *(11)*, it is assumed that interactions of adsorbed ions with unoccupied sites (i.e. dissociated surfactant molecules R^-) are negligible compared with the other interactions. Hence, the following binary parameters are important: χ_{HA}, χ_{HB}, and χ_{AB}. The Flory-Huggins equation then leads to the following expressions for activities a_i of surface components (i = 0 for R^-, H for RH, A for R_2A, and B for R_2B):

$$\ln a_0 = \ln \phi_0 + \frac{1}{2}(\phi_A + \phi_B) - \chi_{HA}\phi_H\phi_A - \chi_{HB}\phi_H\phi_B - \chi_{AB}\phi_A\phi_B \tag{4}$$

$$\ln a_H = \ln\phi_H + \frac{1}{2}(\phi_A + \phi_B) + \chi_{HA}(1-\phi_H)\phi_A + \chi_{HB}(1-\phi_H)\phi_B - \chi_{AB}\phi_A\phi_B \tag{5}$$

$$\ln a_A = \ln\phi_A - (\phi_0+\phi_H) + 2[\chi_{HA}\phi_H(1-\phi_A) - \chi_{HB}\phi_H\phi_B + \chi_{AB}(1-\phi_A)\phi_B] \tag{6}$$

$$\ln a_B = \ln\phi_B - (\phi_0+\phi_H) + 2[-\chi_{HA}\phi_H\phi_A + \chi_{HB}\phi_H(1-\phi_B) + \chi_{AB}\phi_A(1-\phi_B)] \tag{7}$$

where ϕ_i is the surface (or area) fraction of the i-component *(11)*. Here, a ternary interaction among components (the term $\chi_{HAB}\phi_H\phi_A\phi_B$) is neglected. Then, one gets the following coupled equations for the above competitive adsorption equilibria:

$$\phi_0 = 1/D \tag{8}$$

$$\phi_H = K_H a_H^I \exp(-\chi_{HA}\phi_A - \chi_{HB}\phi_B)/D \tag{9}$$

$$\phi_A = K_A a_A^I \phi_0 \exp[1 - 2(\chi_{HA}\phi_H + \chi_{AB}\phi_B)]/D \tag{10}$$

$$\phi_B = K_B a_B^I \phi_0 \exp[1-2(\chi_{HB}\phi_H+\chi_{AB}\phi_A)]/D \tag{11}$$

where

$$D \equiv 1 + K_H a_H^I \exp(-\chi_{HA}\phi_A-\chi_{HB}\phi_B) + K_A a_A^I \phi_0 \exp[1-2(\chi_{HA}\phi_H+\chi_{AB}\phi_B)]$$

$$+ K_B a_B^I \phi_0 \exp[1-2(\chi_{HB}\phi_H+\chi_{AB}\phi_A)] \tag{12}$$

and a_i^I is the solution ion activity at the surface of the subphase. This system of four coupled equations is the effective overall multicomponent adsorption isotherm, which considers effects of nonideal binary mixing of bound ions in the monolayer. The diffuse layer is then induced by the surface charge due to the surfactant molecules R^-. The spatial distribution of the ions in the diffuse layer can be analytically determined from the Poisson-Boltzmann distribution *(11,15)*.

The symbols ϕ_A and ϕ_B represent the surface fractions of the salts with bivalent ions A^{2+} and B^{2+} in the Langmuir monolayer, respectively. To obtain the composition of LB films, one needs appropriate corrections, because the Langmuir monolayers are neutralized during the transfer process to make the LB films *(11)*. The number of ions incorporated into the monolayer at this stage is assumed to be proportional to their relative populations in the interfacial diffuse layer. The compositions of salts in LB films $\phi_{A,LB}$ and $\phi_{B,LB}$ are then obtained as follows:

$$\phi_{A,LB} = \phi_A + \frac{2\Gamma_A}{\Gamma_H+2\Gamma_A+2\Gamma_B} \phi_0 \tag{13}$$

$$\phi_{B,LB} = \phi_B + \frac{2\Gamma_B}{\Gamma_H+2\Gamma_A+2\Gamma_B} \phi_0 \tag{14}$$

where Γ_H, Γ_A, and Γ_B are the densities of protons and bivalent ions A^{2+} and B^{2+} in the interfacial layer, respectively. As usually done in composition studies *(4,11)*, it is assumed that no hydrodynamic effects are involved that could disturb the equilibrium composition of the monolayer.

Certain predictions of the model for compositions of LB films for $H^+ - Ca^{2+} - Cd^{2+}$ and $H^+ - Ca^{2+} - Pb^{2+}$ systems are shown in Figures 4 and 5. These results describe how the monolayer composition changes as another kind of bivalent ions (Cd^{2+} or Pb^{2+}) is introduced into the binary ionic system of H^+ and Ca^{2+}. No information is yet available on χ_{Ca-Cd} or χ_{Ca-Pb}. No comparison with experimental data is available at present, either. A pH of 11, at which no protons are adsorbed, has been chosen for these calculations, because the effect of interaction between two kinds of bivalent ions on the composition can be seen more clearly by having a minimal effect of interactions of these ions with protons. At this

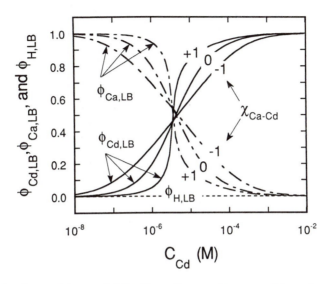

Figure 4. Compositions of LB films ($\phi_{Cd,LB}$, $\phi_{Ca,LB}$, and $\phi_{H,LB}$) containing H^+, Ca^{2+}, and Cd^{2+} ions; pH = 11, $C_{Ca} = 10^{-4}$ M; the parameter values used in Figure 3 were used here for each ion and for χ_{H-Ca} and χ_{H-Cd}. Results are presented for three values of χ_{Ca-Cd} which is presently unknown.

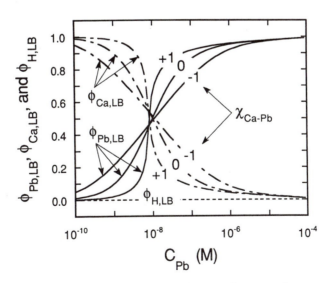

Figure 5. Same as Figure 4, but for H^+, Ca^{2+}, and Pb^{2+} ions.

pH, $\phi_{H,LB}$ is essentially zero. Figure 4 shows that Cd^{2+} ions are more favored by Langmuir monolayers than Ca^{2+} ions, because of the Cd^{2+} ions' larger association constant. Monolayers are even more selective to adsorbing Pb^{2+} ions than Cd^{2+} or Ca^{2+} in Figure 5. The calculations predict a substantial selectivity of Langmuir monolayers to different bivalent ions. A transition of the composition (defined as the midpoint of the S-shaped curve, $\phi_{A,LB} \approx \phi_{B,LB}$) occur roughly at such a condition that a product of the ion concentration and the ion association constant is about the same for both kinds of bivalent ions. This is because the denominator D in the isotherm is reduced at the transition to $(1+K_H C_H+K_A C_A+K_B C_B)$, where C_i is the bulk ion concentration, as an order-of-magnitude analysis clearly shows. The width of the transition region (concentration range between $\phi_{A,LB} = 0.05$ to 0.95) depends a lot on the value of the Flory-Huggins interaction parameter χ_{AB}. As in the case of $\phi_{A,LB}$ versus pH for binary ionic systems (Figure 1), where the transition width depends on the nonideality in mixing of adsorbed H^+ and A^{2+} ions, the nonideality in mixing of adsorbed A^{2+} and B^{2+} ions is critical in controlling that width of $\phi_{A,LB}$ and $\phi_{B,LB}$ versus C_B in ternary ionic systems at fixed pH (Figures 4 and 5).

Conclusions

The new electrochemical interactive model (11) has been applied successfully to literature binary ionic composition data for LB films with group IIA ions. Positive-nonideal mixing ($\chi_{HA} = 1.4$) of adsorbed ions appears to occur for the $H^+ - Be^{2+}$ and the $H^+ - Sr^{2+}$ systems whereas negative nonideal mixing occurs for the $H^+ - Mg^{2+}$ system ($\chi_{HA} = -1.8$). The $H^+ - Ca^{2+}$ system ($\chi_{HA} = -0.5$) is close to ideal mixing. The Flory-Huggins parameter could not be determined for the $H^+ - Ba^{2+}$ system. The model has then been extended to ternary ionic systems, for which theoretical predictions show highly selective ion exchange capabilities of Langmuir monolayers and LB films.

Notation

a_i	activity of i-component in the monolàyer.
a_i^I	solution i-ion activity at the surface of the subphase
C_i	bulk ion concentration
D	denominator of the terms of the isotherm as defined in equation 12
K_i	equilibrium constants governing adsorption of counterions
R	$CH_3(CH_2)_n COO$
T	absolute temperature
Γ_i	density of i-ion in the interfacial layer
Π	surface pressure

ϕ_i	surface mole fraction of surfactant associated with the i-ion (i = A or B or H) in the Langmuir monolayer
$\phi_{i,LB}$	surface mole fraction in the LB film
χ_{ij}	Flory-Huggins binary interaction parameter

Subscripts

A	A^{2+} ions
B	B^{2+} ions
H	protons
0	R^- in the monolayer

Acknowledgments

This research was supported in part by NSF grant #8604904 and by a David Ross fellowship to Mr. D.J. Ahn.

Literature Cited

1. Roberts, G.G. In *Langmuir-Blodgett Films;* Roberts, G.G., Ed.; Plenum: New York, NY, 1990; pp. 317-411.
2. Deamer, D.W.; Meek, D.W.; Cornwell, D.G. *J. Lipid Res.* **1967**, *8*, 255.
3. Kobayashi, K.; Takaoka, K.; Ochiai, S. *Thin Solid Films* **1988**, *159*, 267.
4. Petrov, J.G.; Kuleff, I.; Platikanov, D. *J. Colloid Interface Sci.* **1982**, *88*, 29.
5. Matsubara, A.; Matuura, R.; Kimizuka, H. *Bull. Chem. Soc. Jpn.* **1965**, *38*, 369.
6. Yamauchi, A.; Matsubara, A.; Kimizuka, H.; Abood, L.G. *Biochim. Biophys. Acta* **1968**, *150*, 181.
7. McLaughlin, S.; Mulrine, N.; Gresalfi, T.; Vaio, G.; McLaughlin, A. *J. Gen. Physiol.* **1981**, *77*, 445.
8. Lösche, M.; Helm, C.; Mattes, H.D.; Möhwald, H. *Thin Solid Films* **1985**, *133*, 51.
9. Pezron, E.; Claesson, P.M.; Berg, J.M.; Vollhardt, D. *J. Colloid Interface Sci.* **1990**, *138*, 245.
10. Bloch, J.M.; Yun, W. *Phys. Rev. A* **1990**, *41*, 844.
11. Ahn, D.J.; Franses, E.I. *J. Chem. Phys.* **1991**, *95*, 8486.
12. Hasmonay, H.; Vincent, M.; Dupeyrat, M. *Thin Solid Films* **1980**, *68*, 21.
13. Prausnitz, J.M.; Lichtenthaler, R.N.; de Azevedo, E.G. *Molecular Thermodynamics of Fluid-Phase Equilibria;* Prentice-Hall: Englewood Cliffs, NJ, 1986; pp. 274-370.
14. Smith, R.M.; Martell, A.E. *Critical Stability Constants;* Plenum: New York, NY, 1989; Vol. 6.
15. Abraham-Shrauner, B. *J. Math. Biology* **1975**, *2*, 333; errata in **1977**, *4*, 201.

RECEIVED January 6, 1992

Chapter 24

Adsorption from Aqueous Binary Surfactant Mixtures onto the Solid–Liquid Interface

A Kinetic Study with Attenuated Total Reflection and Fourier Transform Infrared Spectroscopies

Alexander Couzis and Erdogan Gulari[1]

Department of Chemical Engineering, University of Michigan,
Ann Arbor, MI 48109

The adsorption kinetics of a mixture of sodium dodecyl sulfate and sodium laurate onto Al_2O_3 in contact with the aqueous solution of the surfactants was studied using Attenuated Total Reflection IR spectroscopy. In addition the displacement kinetics of SDS by SLA were studied. ATR provides us the unique ability to study the adsorption process *in situ* on surfaces that are not IR transparent. The results obtained from this study suggest that adsorption involves a complicated mechanism, which strongly depends on the bulk solution concentration. The data also suggest that the adsorption process involves three stages; diffusion to the solid-liquid interface, adsorption with an initial conformation, and conformational changes that lead to an irreversibly adsorbed surfactant layer. The exchange experiments show that an irreversibly adsorbed surfactant layer can be desorbed and substituted by another surfactant.

The adsorption of surfactants from solution on to a solid surface is of great technological and scientific interest, because of its application in commercial processes such as enhanced oil recovery, floatation, lubrication and adhesion. In addition, the adsorption phenomenon is of fundamental importance in understanding the solution and interfacial behavior of surfactants.

Most studies in adsorption, reported in the literature today, focus on the equilibrium properties and behavior of single component and mixtures of surfactants *(1-8)*. These types of studies are of great importance in understanding the thermodynamics involved, and providing means of determining the amount of material adsorbed, although they do not provide any information on the dynamics or kinetics of the adsorption process. The small number of kinetic studies*(9,10)* can be attributed to the lack of reliable methods that can monitor the kinetics of adsorption in solid-liquid systems. The traditional bulk methods, generally

[1]Corresponding author

0097–6156/92/0501–0354$06.00/0

involve bringing the solid, in powder form, and the solution in contact, letting the system equilibrate, separating the slurry and analyzing the residual solution *(1)*. These techniques lack the speed to monitor the kinetics, cannot be conducted *in situ*, and do not provide any information about the surface-surfactant interaction. In total contrast with the bulk techniques, employing spectroscopic techniques provides solutions to most of the above problems. Two such techniques have appeared recently in the literature, Electron Spin Resonance (ESR) *(10)* and Attenuated Total Reflection Fourier Transform Infrared Spectroscopy (ATR-FTIR) *(9)*. The latter is the technique employed in this study. Attenuated Total Reflection Spectroscopy (ATR) or Internal Reflection Spectroscopy (IRS) is a method of obtaining an infrared spectrum of species located near the surface of an IR transparent crystal. It was first developed by Harrick *(11)*, and its capability is attributed to the presence of an evanescent wave of light when total reflection occurs at the interface between two materials with different indices of refraction *(11)*.

Although ATR infrared spectroscopy has been familiar as a convenient surface analytical method, and has been applied for qualitative purposes, its capability to analyze very thin films has not been extensively examined. Compared to ellipsometry, viscometric and magnetic resonance techniques used in studying the solid-liquid interface, ATR can provide information about the bond between the adsorbed species and the substrate, and distinguish between the chains that are actually adsorbed from those that are in the vicinity of the interface. Furthermore, it provides the ability to conduct measurements *in situ*, thus allowing kinetic studies in the presence of the solvent. Also, by borrowing thin film deposition techniques from the microelectronic's area, we can study the adsorption on solid surfaces that are not IR transparent, and still achieve sensitivity that allows a detailed study of the adsorption kinetics and the interaction of the surfactants with the solid surface.

In this study we utilize these advantages of ATR, for the first time to our knowledge, to monitor, in detail, the kinetics of adsorption of a mixture of sodium dodecyl sulphate and sodium laurate in an aqueous solution on an alumina surface, that is in contact with the solution. The main difference from previously reported studies *(9,12)* is that in this study, the solid surface studied is an oxide, Al_2O_3, that is not transparent in the mid IR range.

Experimental

Materials. High purity (99-100%) sodium lauric acid (SLA) was purchased from Sigma Chemical and was used without further purification. High purity (99%) sodium dodecyl sulfate (SDS) was purchased from BDH Chemical Ltd and was used as is. The H_2O used for the aqueous solutions was deionized and doubly distilled. Germanium Internal Reflection Elements (IRE) were purchased from Harrick Scientific Co. and were 50x10x3mm and 50x10x2mm in size and with 45 or 60 degrees angle of incidence. ZnSe IRE was purchased from Spectra Tech and Harrick Scientific Co, and had dimensions of 50x10x3mm and 45 degrees angle of incidence. The germanium IREs, before every use, were polished using $0.3\mu m$ and $0.05\mu m$ Al_2O_3 suspensions, purchased from Mager Scientific, and then rinsed with acetone, water, and methanol. Following drying of the IRE, a 300Å thick Si_3N_4 film, was deposited using sputtering, followed by evaporation of a 600Å thick Al_2O_3 film on the element. The Si_3N_4 was used as a passivation layer because we found that direct deposition of Al_2O_3 on Ge would produce a germanium oxide layer which is water soluble, leading to liftoff of the Al_2O_3 layer when the IRE was wetted by an aqueous solution. For the ZnSe IRE polishing and rinsing was done in a similar fashion and was followed by evaporation of Al_2O_3 directly on the element. During the evaporation the bevel surfaces of the IRE were protected with a coating of photoresist, which was removed using acetone, after the deposition.

Apparatus. IR spectra were taken using a single beam Mattson Cygnus 100 spectrometer. The sample chamber is purged with nitrogen. A wide band, liquid nitrogen cooled Mercury-Cadmium-Telluride (MCT) detector with a 1x1mm active area was used. Spectra were taken with 4cm^{-1} resolution and a mirror scanning speed of 2.53 cm/s and were Fourier transformed with triangular apodization. The IRE's were checked for hydrocarbon contamination before each experiment was conducted. The special cell used, constructed in our laboratory, and the flow system used for the introduction of the solution can be seen in figures 1a and 1b. The liquid cell consisted of two teflon blocks, with surfaces machined parallel and flat. In each block, a channel was milled, having dimensions of approximately 30x2x2mm. The cell was then sealed by sandwiching the IRE between the blocks and applying pressure through a clamping device. The flow system consisted of teflon tubing and teflon Swagelok fittings. Following this scheme, aside from the coated IRE, the solution came in contact with relatively inert materials. Measurements were taken with the surfactant solution flowing over the solid surface. The flow rate was adjusted so that the total volume cell was exchanged once every hour. The extinction coefficients for the surfactants used were calculated from transmission spectra of their solutions, using a liquid cell with variable pathlengths, purchased from Harrick Scientific Co. Spectra of the surfactants, in solid state, were obtained using transmission spectra of KBr pressed disks.

Data Analysis. The total amount of adsorbed surfactant on the solid surface is monitored by the increase of the absorbance peak due to the CH_2 stretch at 2850 and 2920cm^{-1}. A typical time series of the increasing absorbance can be seen in figure 2. The individual surfactant adsorption is monitored by the increase of the absorbance of the sulfonate group at 1060cm^{-1} for SDS and the symmetric carboxyl stretch at 1410cm^{-1}. Using extinction coefficients we calculated from transmission FT-IR, we were able to calculate the actual adsorbed amounts from the absorbance values for the CH_2 stretching. This can be reliably done only if the cross section of the molecule, for the particular vibration mode, does not change significantly upon adsorption, as is the case for the CH_2 stretching. For the sulfonate and the carboxyl bands this assumption cannot be validated, so we are unable to calculate the amounts of the individual surfactant adsorbed. However, because the absorbance is proportional to the adsorbed amount, we can still draw conclusions about the relative behavior of the surfactants.

The measured total absorbance using FTIR-ATR, A_{total}, is the sum of contributions from the bulk solution, A_{bulk}, as well as from the interface, $A_{interface}$. The contribution of the bulk solution extends as far as the effective depth of penetration of the evanescent wave, and the contribution of the interface extends to a distance equal to the thickness of the adsorbed surfactant layer. Therefore, we have:

$$A_{total} = A_{interface} + A_{bulk} \tag{1}$$

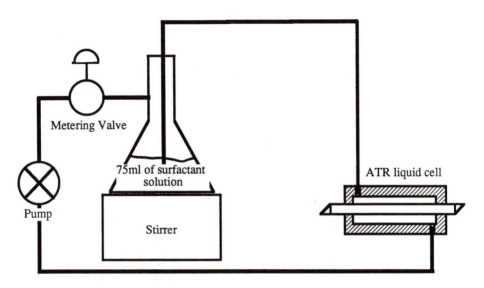

Figure 1a: Experimental Flow System.

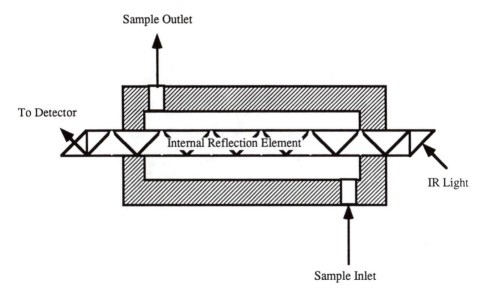

Figure1b: Schematic of the Internal Reflection Liquid Cell

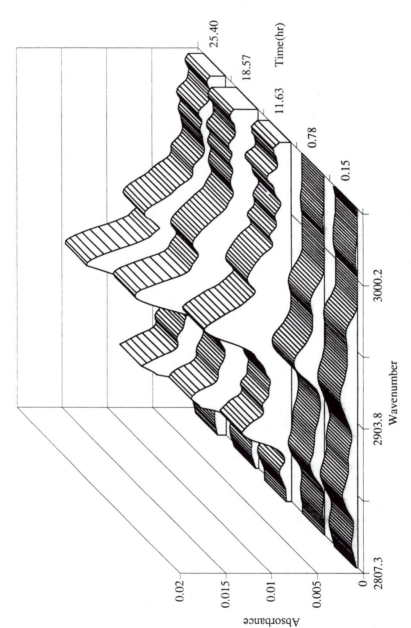

Figure 2: Time series of spectra for adsorbing surfactants. The C-H stretching region is shown. The two main peaks at 2850cm⁻¹ and 2920cm⁻¹ correspond to the symmetric and asymmetric C-H₂ stretching respectively.

Equation 1 can be rewritten *(13)* as:

$$\frac{A_{total}}{N} = \alpha\, C_b\, d_e + \alpha\, (2\,\frac{d_e}{d_p})\, \Gamma_i \qquad (2)$$

Where: α is the extinction coefficient
C_b is the concentration in the bulk
Γ_i is the surface excess
d_e is the effective thickness
d_p is the depth of penetration of the evanescent wave
N is the total number of reflections in the ATR element
($N=\frac{1}{d}$ cotθ, θ=angle of incidence , l=IRE length, d=IRE thickness)

In the derivation of equation 2, it is assumed that the surfactant concentration near the interface assumes a step profile: $C(z)=C_i+C_b$ for $0<z<t$ and $C(z)=C_b$ for $t<z<\infty$. C_i is the interfacial surfactant concentration and is equal to $\frac{\Gamma_i}{t}$, and t is the thickness of the adsorbed surfactant layer. Any other more realistic concentration profile can also be used. The penetration depth, d_p, is calculated using the following equation *(11)*:

$$\frac{d_p}{\lambda_1} = \frac{1}{2\pi(\sin^2\theta - n_{21}{}^2)^{1/2}} \qquad (3)$$

With n_{21} is the ratio of refractive indices of Al_2O_3 to the IRE material, and λ_1 is the wavelength in the denser medium, λ/n_1.

When the thickness of the deposited films is small compared to the penetration depth, the effective depth of sampling varies with the angle of incidence, and polarization of the incident beam. Under this assumption, the effective thickness, d_e is the linear average of $d_e\|$ and $d_{e\perp}$ *(11,13)*, which are calculated by the following equations *(11)*:

$$\frac{d_{e\perp}}{d} = \frac{4\,n_{21}\cos\theta}{(1-n_{31}{}^2)} \qquad (4)$$

$$\frac{d_e\|}{d} = \frac{4n_{21}\cos\theta[(1+n_{32}{}^4)\sin^2\theta - n_{31}{}^2]}{(1-n_{31}{}^2)[(1+n_{31}{}^2)\sin^2\theta - n_{31}{}^2]} \qquad (5)$$

where : $d_e\|$ and $d_{e\perp}$ are the effective depth of penetration for parallel and perpendicularly polarized light.
n_{21} is the ratio of refractive indices of Al_2O_3 to the IRE material.
n_{31} is the ratio of refractive indices of the solution to the IRE material.
n_{32} is the ratio of the refractive indices of the solution to Al_2O_3.
θ is the angle of incidence of the light.

Results and Discussion

Adsorption Kinetics. Adsorption experiments using the Al_2O_3 coated IRE were performed for binary aqueous solutions of SDS and SLA for 4 different solution concentrations and compositions. In figure 3 we can see the results obtained for different solutions. In all the samples the adsorbed amount shows a rapid increase. Within the first hour the rate decreases considerably. For the highest concentration sample, within that time the adsorbed amount stops increasing. For the solution with concentration of 0.01M SDS and 0.05M SLA, the rate of adsorption decreases rapidly after approximately one hour. The amount of surfactant adsorbed reaches a constant value within 3 hours of the initial contact of the solution and the solid surface. Both, samples have total concentrations above the CMC of the solution. The CMCs for these solutions were determined using ideal solution theory. The CMC for the equimolar solutions was calculated to be 15mM, and for the solution of 1:5 ratio SDS:SLA, it was 22mM. The other two solutions have a total concentration below the CMC. For these samples, the adsorbed amount increases, at the early stages, very rapidly. As time progresses, the adsorbed amount grows through a series of plateaus. These plateaus are are of a longer duration for the lower concentration sample. This result clearly shows that the adsorption process cannot be seen as an elementary surface "reaction", but is a complicated process involving various mechanisms, taking place with different timescales. Similar results have been reported for the adsorption of SDS on kaolinite surfaces *(6)*, and for adsorption of polymers on a silver surface *(16)*. One can argue that adsorption involves three stages *(9,14)*. First diffusion of the surfactant molecule through a stagnant boundary layer to the solid-liquid interface, which is a very rapid process. During the initial stages of adsorption the increase in the adsorbed amount is related to the diffusion coefficient and time by the equation *(17,18)*

$$\Gamma_i = 2\,C_b \sqrt{\frac{D_{app}t}{\pi}} \tag{6}$$

where: Γ_i = total surface excess of the surfactants on the solid surface
             C_b = total bulk concentration of the surfactants A and B
             D_{app}=apparent diffusion coefficient
             t = elapsed time

In figure 4 , the surface excess is plotted against the square root of the elapsed time, for the first hour. It clearly shows that the data lies on a straight line, supporting the statement that the first step in the adsorption process is the diffusion to the solid surface.

The second stage is the actual adsorption of the surfactant on the interface, which at this point is reversible *(14)*. The third stage involves a reconformation of the adsorbed molecules, possibly of the hydrophobic chain that would lead to additional adsorption or desorption of the surfactant. This last stage is a very slow process and leads to a state where the surfactant is irreversibly adsorbed on the surface *(14)*. It also seems likely that during this reconformation process there is no surface accessible to the surfactant for additional adsorption, explaining the very slow rate of adsorption, or the presence of plateaus for the adsorbed amount. The key point here is to understand that the adsorption of

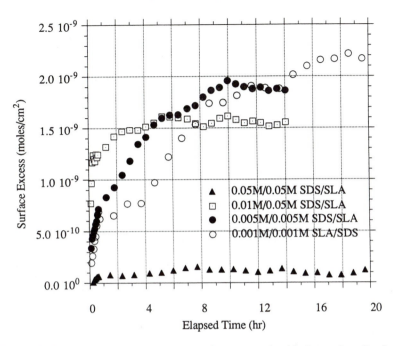

Figure 3: Total surface excess of the surfactant on the Al_2O_3 surface for four different solution concentrations.

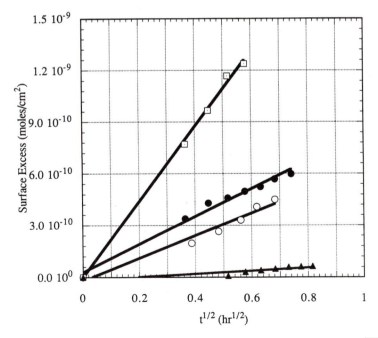

Figure 4: Total surface excess of the surfactant on the Al_2O_3 surface vs. $\sqrt{time}$. for the data in the first hour. Symbols as in figure 3.

surfactants, and especially of mixtures, is not a elementary process. To further emphasize this point, we compare the data with the first order kinetic model:

$$\Gamma_i = C_{bA} K_{eqA} [1 - \exp(\frac{k_{1A} t}{K_{eqA}})] + C_{bB} K_{eqB} [1 - \exp(\frac{k_{1B} t}{K_{eqB}})] \qquad (7)$$

where: Γ_i = total surface excess of the surfactants on the solid surface
C_b = total bulk concentration of the surfactants A and B
K_{eq} = adsorption equilibrium constant for the surfactant A and B
k_1 = rate constant for the adsorption of surfactant A and B

As shown in figure 5, there is a systematic deviation from the first order line, illustrating the complexity of the adsorption mechanism.

Comparing the total amounts of surfactant adsorbed after 16 hours, we can see that for the high concentration sample, which is above the mixture's CMC, it is approximately one order of magnitude less than all the other samples. The adsorbed amount for the sample 0.01M/0.05M SDS/SLA is of comparable value to the dilute samples, even though its concentration is above the solution's CMC. This is an expected result because it is known from equilibrium data that for some surfactant systems the adsorption isotherm exhibits a maximum *(6,16)* , usually in the vicinity of the CMC. These result along with the fact that the surface excess calculated from our experiments agrees with equilibrium data, strengthens the argument that ATR-FTIR is a reliable method for the *in situ* study of the solid-liquid interface.

Displacement Kinetics. We also conducted experiments involving the sequential adsorption of the two surfactants. A ZnSe IRE was used, which did not allow us to quantify our results, because the 45° angle of incidence used is very close to the critical angle of incidence for this system. However, we can monitor the relative adsorption kinetics because the integrated absorbance of the bands monitored is proportional to the amount adsorbed, as already stated. The bands monitored for this experiment are the the two modes of CH_2 stretching at $2850cm^{-1}$ and $2920cm^{-1}$ and the two sulfonate stretching bands at $1060cm^{-1}$ and $1200cm^{-1}$. As shown in figures 6 and 7, the absorbances for the symmetric and asymmetric stretching for these bands compare very well, which is a reliable internal check of our results.

Initially, a SDS solution, of concentration 0.01M and pH of 5.2, was initially brought in contact with the Al_2O_3 surface. After 42 hours, at which time, equilibrium had not yet been attained, the solution was switched to pure water for 4 hours. During this time, a very small reduction of the integrated absorbance is observed and can be attributed to the loss of the bulk solution contribution. This results also leads to the conclusion that the adsorption of SDS on Al_2O_3, is irreversible. Following the flushing with pure water, a solution of 0.01M SLA and pH of 9.2 was introduced. Again the total adsorbed amount, and the SDS adsorbed amount was monitored. As can be seen in figure 6, the integrated absorbance of the CH_2 stretching increases with time after the introduction of the SLA solution. This means that the total amount of surfactant on the interface is increasing. The rate of increase is initially high but as time progresses, it decreases, leading to a plateau. The behavior of SDS, which is already adsorbed on the interface, is shown in figure 7. After the introduction of the SLA solution

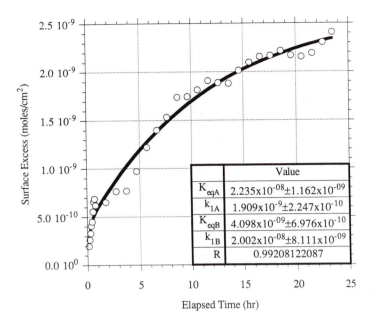

Figure 5. Total surface excess vs. time for the most dilute solution, 0.001M SDS and 0.001M SLA. The line is the fit of equation 7, as described in the text, and the calculated parameters are shown in the plot.

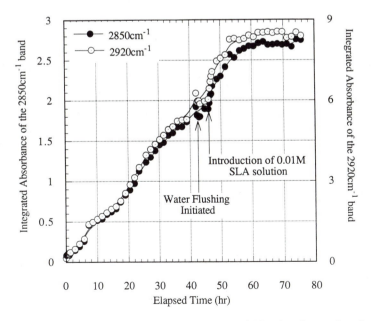

Figure 6: Integrated absorbance of the two CH_2 stretching bands as a function of time.

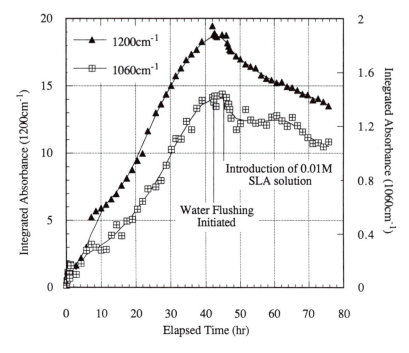

<u>Figure 7:</u> Integrated absorbance of the two S=O stretching bands of SDS as a function of time.

the amount adsorbed on the interface decreases with time, indicating that SDS is being displaced by the adsorbing SLA. This effect is very interesting because it clearly shows that a surfactant that is irreversibly adsorbed on a solid surface, can be desorbed if the proper conditions are met. In this case the pH is raised, which is known from equilibrium studies *(15)* to decrease the amount of SDS adsorbed on Al_2O_3.

Comparing the ratio of the integrated absorbance between the S=O stretching and the CH_2 stretching we find that the amount of SDS displaced is directly proportional to the additional amount of SLA adsorbed on the interface. This observation leads us to believe that the displacement is on a one to one basis. To our knowledge, this was the first time that displacement experiments were conducted with adsorbing surfactants, and they clearly show that by properly selecting the materials involved, a surface can be modified at will.

Conclusions

In this paper we have shown that attenuated total reflection IR spectroscopy can be successfully used to study in detail, the kinetics of adsorption of mixtures of surfactants on an Al_2O_3 surface from solution *in situ*. We have demonstrated that ATR is a powerful technique that allows us to study the adsorption onto solid-liquid interfaces, even if the solid is not IR transparent.

The results show that the rate of adsorption as well as the amount adsorbed at any given time is a strong function of the solution's concentration. From the same data we can also conclude that adsorption of surfactants is a very complicated process, that involves, at least, three steps; diffusion, adsorption and reconformation. The adsorption process of SDS on Al_2O_3 for long contact times of the solution and the solid is irreversible, but the SDS film can be displaced to a point by SLA , when its solution is brought in contact with the solid surface.

Literature Cited

1. Scamehorn, J.F.; Schechter, R.C.; Wade, W.H.*J. Colloid interface Sci,* **1982**, 85, 463.
2. Scamehorn, J.F.; Schechter, R.C.; Wade, W.H.*J. Colloid interface Sci,* **1982**, 85, 479.
3. Scamehorn, J.F.; Schechter, R.C.; Wade, W.H.*J. Colloid interface Sci,* **1982**, 85, 494.
4. Roberts, B.L.; Scamehorn, J.F.; Harwell, J.H. in *Phenomena in Mixed Surfactant Systems*; Scamehorn, J.F., Ed.; ACS Symposium Series 311, ACS: Washington, DC, 1986, 200-215.
5. Dick, S.G.; Fuerstenau, D.W.; Healy, T.W. *J. Colloid interface Sci,* **1971**, 37, 595.
6. Hanna, H.S.; Somasundaran, P. *J. Colloid interface Sci,* **1979**, 70, 181.
7. Chandar, P.; Somasundaran, P.; Turro, N.J. *J. Colloid interface Sci,* **1987**, 117, 31.
8. Somasundaran, P.; Fuerstenau, D.W. *J. Phys. Chem.,* **1966**, 70, 90.
9. McKeigue, K.; Gulari, E. in *Surfactants in Solution* ; Mittel, K.L.; Lindman, B., Ed.; Plenum Publishing Corporation, 1984, vol. 2, 1271-1289.
10. Malbrel, C.A; Somasundaran, P.; Turro, N.J. *J. Colloid interface Sci,* **1990**, 137, 600.
11. Harick, N.J. *Internal Reflection Spectroscopy* ; Harrick Scientific Corp., Ossining, New York, 1987, 2nd ed.
12. Kellar, J.J.; Cross, W.M.; Miller, J.D. *Applied Spectroscopy,* **1989**, 43, 1456.
13. Sperline, R.P.; Muralidharan, S.; Freiser, H. *Langmuir* , **1987**, 3, 198.
14. Zawadzki, M.E.; Harel, Y; Adamson, A.W. *Langmuir,* **1987**, 3, 363.
15.Fuesrstenau, D.W.; Wakamatsu, T. *Faraday Discuss. Chem. Soc.* , **1976**, 59, 157.
16.Tassin, J.F.; Siemens, R.L.; Tang, W.T.; Hadziioannou, G.; Swalen, J.D.; Smith, B.A. *J. Phys. Chem.,* **1989**, 93, 2106.
17.Langmuir, I.; Schaefor, V.J. *J. Am. Chem. Soc.,* **1937**, 59, 2400.
18.Petrow, J.G.; Miller, R. *Colloid Polym. Sci.,* **1977**, 255, 669.

RECEIVED January 6, 1992

Chapter 25

Thermodynamic Study of Adsorption of Anionic–Nonionic Surfactant Mixtures at the Alumina–Water Interface

Edward Fu, P. Somasundaran, and Qun Xu

Langmuir Center for Colloids and Interfaces, Henry Krumb School of Mines, Columbia University, New York, NY 10027

Mechanisms of adsorption of an anionic surfactant, sodium octylbenzenesulfonate ($C_8\phi S$), and a nonionic surfactant, dodecyloxyheptaethoxyethylalcohol ($C_{12}EO_8$), and their mixtures on alumina were investigated by adsorption and microcalorimetric studies. Adsorption of anionic surfactant alone on alumina was initially highly exothermic due to the electrostatic interaction with the substrate. Further adsorption leading to solloid (hemimicelle) formation is mainly an entropy driven process. The entropy effect was found to be more pronounced for the adsorption of anionic-nonionic surfactant mixtures than that for anionic $C_8\phi S$ alone. High surface activity of the nonionic $C_{12}EO_8$ and its hydrophobic interaction with adsorbed $C_8\phi S$ is proposed to be the main mechanism for the marked entropy effect for mixture adsorption.

Mixtures of anionic and nonionic surfactants have shown enhanced surface activity and salt tolerance (*1-3*) which are highly advantageous for applications in industrial processes such as enhanced oil recovery, detergency and flotation (*4-5*). However, an overwhelming majority of basic studies of surfactant adsorption have been performed with systems containing single surfactants. Although the solution behavior of ionic/nonionic surfactant mixtures have been extensively studied in the past (*1-3, 6-10*), only very limited amount of work has been done on adsorption at solid/liquid interface (*11-13*). The mechanisms by which ionic/nonionic surfactant mixtures adsorb have not been well understood, in part due to the lack of thermodynamic data for such adsorption. Regular mixing theory, which is usually adequate for fitting CMC data of ionic/nonionic surfactant mixtures, was found to be inadequate for correlating the data for adsorption of anionic/nonionic surfactant mixtures at alumina/water interface (*11*). It is clear that direct measurement of thermodynamic parameters is needed for understanding the mechanisms and for developing better theoretical models for adsorption of surfactant mixtures at solid/liquid interfaces.

0097–6156/92/0501–0366$06.00/0

In this study, adsorption at alumina/water interface was conducted with isomerically pure anionic and nonionic surfactants and their mixtures. Enthalpy of adsorption was measured using a microcalorimetry system. The adsorption mechanism was discussed on the basis of adsorption and microcalorimetric results.

Materials and Methods

Alumina: Linde A alumina was purchased from Union Carbide Co., USA. It had a mean diameter of 0.3 microns and a surface area of 14 m^2/g as measured by N_2 BET adsorption using a Quantasorb system.

Anionic surfactant: Sodium para-octylbenzenesulfonate ($C_8\phi S$) was synthesized in our laboratory. High performance liquid chromatography data of this compound showed it to be more than 97% isomerically pure.

Nonionic surfactant: Dodecyloxyheptaethoxyethylalcohol ($C_{12}EO_8$) was purchased from Nikko Chemicals, Japan. This surfactant was specified to be monodispersed and at least 97% pure.

NaCl, HCl and NaOH used for regulating ionic strength and pH were of A.R. grade. Triply distilled water (conductivity $1-2 \times 10^{-6}$ mhos) was used throughout the experiments.

Adsorption: Adsorption experiments were conducted in capped 50 ml centrifuge tubes at a constant ionic strength of 0.03 M NaCl and a solid/liquid ratio of 0.1 w/w. The samples were kept in a water bath set at 50 °C for three days during which time the pH was adjusted using 0.1 N HCl or NaOH. The sample was centrifuged for 30 minutes at 4500 rpm inside an incubator set at 50 °C and 20 ml of supernatant was pipetted out for analysis. Alcohol concentrations were measured by high performance liquid chromatography using 90:10 v/v solvent mixtures of acetonitrile and water, a reverse phase column, and a refractive index detector. Sulfonate concentrations above 2×10^{-4} kmol/m^3 were measured by a two phase titration technique (*14*). Dilute solutions below 2×10^{-4} kmol/m^3 were analyzed by UV absorbance at 223 nm using a Beckman DU-8 UV-visible spectrophotometer.

Microcalorimetry: Calorimetric experiments were performed using an LKB 2107 differential isoperibol microcalorimetry system. The actual temperature in the air bath was 49.3 ± 0.1 °C for the experiments. Within the calorimeter were two 18 carat gold vessels, one for sample solutions and the other for reference. Heat was generated when the materials in the vessels were mixed by rotating the cylindrical calorimetrical body. Each mixing action consists of two complete revolutions, one in each direction, and was repeated for a sufficient number of times to ensure complete mixing. The heat was passed through a thermopile and was transduced into voltage signal. The area under the voltage-time curve was integrated using a digital read-out system and was given as

$$A_R = \sum_{i=1}^{n} a_i - n a_m$$

Where A_R is the integrated area of the reaction, a_i the integrated area due to each of the n mixing actions, and a_m the integrated area due to heat effects from rotation of the calorimeter. The calibration between the area and the heat was done by passing a known current through the calibration heater in the calorimeter. The heat values thus obtained were converted to enthalpies through mass balance calculations. The accuracy of the calorimeter was tested by measuring the enthalpy of dilution of NaCl solution and comparing with that reported in the literatures, the mean error is about 3%. In order to calculate the heat of adsorption, the heat due to demicellization which resulted from adsorption was subtracted using the heat of demicellization data (ΔH_m) obtained with surfactant solutions (19).

Results and Discussions

Figure 1 shows the surface tension curves for the two surfactants used in this study. It is seen that the nonionic dodecyloxyheptaethoxyethylalcohol ($C_{12}EO_8$) is much more surface active than the anionic sodium octylbenzenesulfonate ($C_8\phi S$) with a critical micellar concentration of 7.0×10^{-5} M compared to 6.5×10^{-3} M. The critical micellar concentration data measured for the mixtures of the two surfactant was found to obey regular solution behavior (15) which assumes complete random mixing (entropy of mixing being zero). The state functions for the regular mixing theory are given as (9)

$$\Delta S^E = 0$$
$$\Delta H_m = X_1 X_2 \beta RT$$
$$\Delta G^E = \Delta H_m$$

Where X_1 and X_2 are the mole fractions of surfactant component 1 and 2 in the mixture, β is the parameter measuring the interaction between the component surfactants. A negative β suggests the interaction to be energetically favorable. For the critical micellar concentration data of the mixtures of $C_8\phi S$ and $C_{12}EO_8$, a β value of -3.5 was obtained (15). For the same mixture system, however, the interaction parameter for excess enthalpy of mixed micellization is $\beta = -8.5$ (Fig. 2). Similar discrepancies were also observed with other anionic/nonionic surfactant mixtures (16), indicating that the assumption of zero mixing entropy in regular mixing theory (9) is not thermodynamically rigorous. For adsorption at solid/liquid interfaces, the mixing of surfactant molecules is even less random due to geometric constraints introduced by such factors as adsorption sites distribution on solid surfaces. This may account for the fact that regular mixing theory was inadequate for modelling adsorption of anionic/nonionic surfactant mixtures at alumina/water interface (11). To clearly understand the mechanisms involved in the adsorption of surfactant mixtures at solid/liquid interfaces, direct measurements of thermodynamic data is highly necessary.

Adsorption isotherms for pure sodium octylbenzenesulfonate ($C_8\phi S$) and dodecyloxyheptaethoxyethylalcohol ($C_{12}EO_8$) and their 1:1 mole mixtures are shown in figure 3. The shape of the isotherm for pure $C_8\phi S$ adsorption is typical of that obtained for adsorption of ionic surfactants on oppositely charged

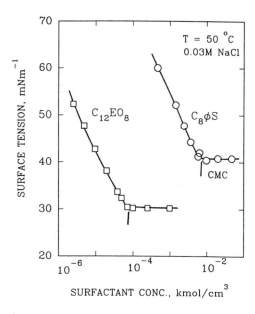

Figure 1. Surface tension vs surfactant concentration for sodium octylbenzenesulfonate ($C_8\phi S$) and dodecyloxyheptaethoxyethylalcohol ($C_{12}EO_8$).

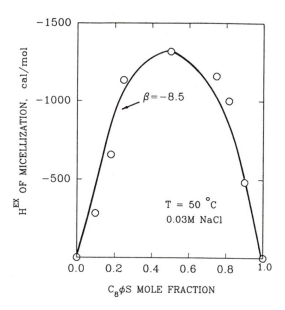

Figure 2. Excess heat of mixed micellization for 1:1 $C_{12}EO_8/C_8\phi S$ mixtures vs sulfonate ($C_8\phi S$) mole fraction.

minerals (*17*). Above a concentration of 4.5×10^{-4} kmol/m^3, solloid formation (hemimicellization) to form solloid (hemimicelles) occurs and this region of enhanced adsorption extends up to a concentration of 6.5×10^{-3} kmol/m^3, which corresponds to the CMC of $C_8 \phi S$. Compared to that of pure sulfonate adsorption, the plateau level of pure ethoxylated alcohol adsorption is three orders of magnitude lower. For adsorption from 1:1 mole mixtures of the two surfactants, however, the ethoxylated alcohol adsorbs to the same extent as the sulfonate. The interaction between surfactants in the mixtures at the solid/liquid interface is synergistic since the isotherm for sulfonate adsorption from the mixtures also shifted to lower concentrations and the slope of the initial hemimicellar region increased from 3.7 to 6.6 for the pure sulfonate adsorption.

The adsorption isotherm for anionic sodium octylbenzenesulfonate ($C_8 \phi S$) on alumina is given in Figure 4 along with the heat of adsorption. The shape of both adsorption and heat curves is similar. The adsorption enthalpy curve, obtained by dividing heat of adsorption with adsorption density, shows different regions (Figure 5) similar to that reported in the literature (*18*). The first region below an adsorption densities of 10^{-12} mol/cm^2 shows the enthalpy to be highly exothermic due to the electrostatic adsorption of the sulfonate on the positive sites on the alumina surface. Above 10^{-12} mol/cm^2, the adsorbed surfactant species associate (*17*) and the adsorption isotherm shows an abrupt increase in the slope which corresponds to an increase in adsorption free energy (Fig. 4). However, in this region adsorption enthalpy shows a marked decrease. Since the adsorption free energy consists of both enthalpy and entropy terms, the aggregation of surfactant molecules at the solid/liquid interface can be considered to be an entropy driven process.

The adsorption isotherm and the corresponding heat of adsorption plot for pure dodecyloxyheptaethoxyethylalcohol ($C_{12}EO_8$) on alumina are given in Figure 6. Similar to the adsorption isotherm, the heat of adsorption curve also shows two regions. The heat measured is very low and shows much scattering as compared to the data for pure sulfonate adsorption (Figure 4). The enthalpy of adsorption for this system was not derived due to the large scattering of the data.

The adsorption isotherm and the heat of adsorption curve for 1:1 $C_8 \phi S / C_{12}EO_8$ mixtures system are shown in Figure 7. The shape of the heat curve mirrors that for adsorption, except for the absence of a maximum. The adsorption enthalpy however was surprisingly constant as a function of adsorption density (Figure 8). As compared to the adsorption of pure sulfonate, adsorption of surfactant mixtures on alumina is less exothermic although adsorption levels are significantly higher before attainment of plateau. These results clearly suggest a stronger entropy effect for mixture adsorption than for single sulfonate adsorption.

To further understand the more pronounced entropy effect for mixture adsorption, the composition of the mixed adsorption layer was determined and the results are given in Figure 9 as a function of the total residual surfactant concentration. It is seen from Figure 9 that the adsorbed layer was always in rich of the nonionic $C_{12}EO_8$ as $C_{12}EO_8 / C_8 \phi S$ ratio in the adsorbed layer was always greater than 1. The enrichment of the adsorbed layer with respect to nonionic $C_{12}EO_8$, especially at low adsorption densities where the $C_{12}EO_8 / C_8 \phi S$ ratio was

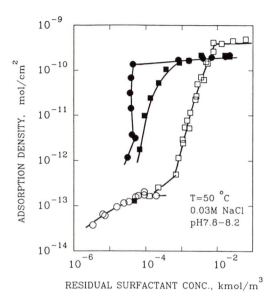

Figure 3. Isotherms for adsorption on alumina vs individual residual surfactant concentration. —□—, pure $C_8\phi S$ adsorption; —■—, $C_8\phi S$ adsorption, from 1:1 $C_{12}EO_8/C_8\phi S$ mixture; —○—, pure $C_{12}EO_8$ adsorption; —●—, $C_{12}EO_8$ adsorption from 1:1 $C_{12}EO_8/C_8\phi S$ mixture, 50 °C, 0.03M NaCl, pH7.8-8.2.

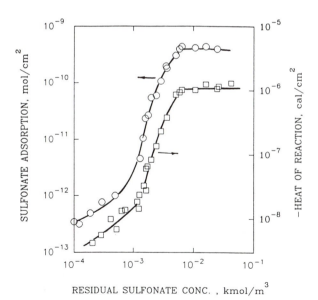

Figure 4. Adsorption isotherms and heat of adsorption of pure octylbenzenesulfonate ($C_8\phi S$) on alumina, 50 °C, 0.03 M NaCl, pH8.2.

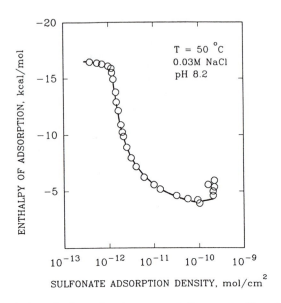

Figure 5 Enthalpy of adsorption for pure octylbenzenesulfonate ($C_8\phi S$) on alumina vs sulfonate adsorption density.

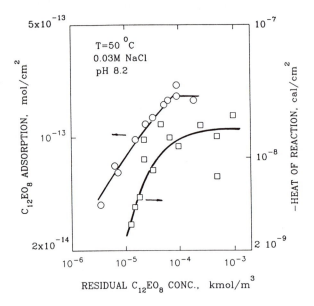

Figure 6   Adsorption isotherms and heat of adsorption of pure dodecyloxyheptaethoxyethylalcohol ($C_{12}EO_8$) on alumina.

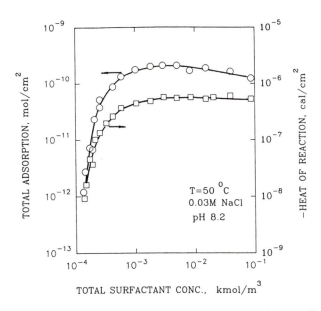

Figure 7 Adsorption isotherms and heat of adsorption of 1:1 $C_{12}EO_8/C_8\phi S$ mixtures on alumina.

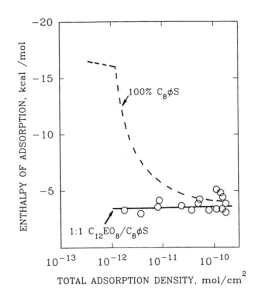

Figure 8 Enthalpy of adsorption of 1:1 $C_{12}EO_8/C_8\phi S$ mixtures on alumina, 50 °C, 0.03M NaCl, pH8.2.

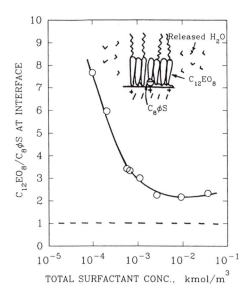

Figure 9 $C_{12}EO_8/C_8\phi S$ ratios at alumina/water interface for the adsorption of 1:1 $C_{12}EO_8/C_8\phi S$ mixtures on alumina.

as high as 7 - 8, is interesting particularly since $C_{12}EO_8$ by itself barely adsorbs on alumina (Fig. 3). It is proposed that the adsorbed anionic sulfonate provides nucleation sites for the adsorption of the nonionic $C_{12}EO_8$ through hydrophobic chain-chain interaction at the interface. The high activity of the nonionic surfactant $C_{12}EO_8$ at the solid/liquid interface is manifested by its higher surface activity in solution as compared to that of the anionic sulfonate (Fig. 1). During the process of adsorption on alumina, water molecules associated with the nonionic surfactant molecules are released to become free water molecules, thus increasing the entropy of the system. The presence of the nonionic surfactant between the sulfonate ions in the hemimicelles should reduce the lateral electrostatic repulsion between the ionic sulfonate head groups so that adsorption of the sulfonate from the mixtures also enhanced (Fig. 3). The co-adsorption of nonionic surfactant was found not to change the zeta potential of alumina particles as a function of sulfonate adsorption density (*15*), further suggesting that entropy effect through hydrophobic chain-chain interaction is the main mechanism for the adsorption of the anionic/nonionic surfactant mixtures at the alumina/water interface.

Conclusions

Adsorption of anionic/nonionic surfactant mixtures at the alumina-water interface is synergistic in nature; adsorption of both surfactants is enhanced by the co-adsorption of the other. The nonionic surfactant, dodecyloxyheptaethoxyethylalcohol ($C_{12}EO_8$), is highly hydrophobic and its adsorption is strongly induced by the co-adsorbed anionic sodium octylbenzenesulfonate species ($C_8\phi S$) through hydrophobic chain-chain interactions at the alumina-water interface. The adsorbed $C_8\phi S$ molecules can be considered to serve as hydrophobic nucleation sites for the abstraction of the nonionic $C_{12}EO_8$ into the adsorbed layer. This adsorption process is strongly entropy driven since such a process is accompanied by the release of many water molecules from around each nonionic surfactant molecule.

Acknowledgements

Financial support of the National Science Foundation (CTS-89-21570), Department of Energy, Unilever Research, and BP America is gratefully acknowledged.

References

1.   Hua, X.Y. and Rosen, M.J., *J. Colloid Interface Sci.*, **90**, 212(1982).
2.   Scamehorn, J.F., Schechter, R.S. and Wade, W.H., *J. Colloid Interface Sci.*, **85**, 494(1982).
3.   Holland, P.M., in **Phenomena in Mixed Surfactant Systems**, J.F. Scamehorn (ed), ACS Symp. Series 311, pp.102-115, 1986.
4.   von Rybinski, W. and Schwuger, M.J., *Langmuir*, **2**, 639(1986).

5.   Rubingh, D.N. and Jones, T., Ind. Eng. Chem. Prod. Res. Dev., **21**, 176(1982).
6.   Scamehorn, J.F., Schechter, R.S. and Wade, W.H., *J. Dispersion Sci. Technology*, **3**, 261(1982).
7.   Moroi, Y. Nishikido, N., Saito, M. and Matuura, R., *J. Colloid Interface Sci.*, **52**, 356(1975).
8.   Holland, P.M. and Rubingh, D.N., *J. Phys. Chem.*, **87**, 1984(1983).
9.   Rubingh, D.N., in **Solution Chemistry of Surfactants**, K.L. Mittal (ed), vol. **1**, pp.337-354, Plenum, New York, 1979.
10.   Bacon, K.J. and Barnes, G.T., *J. Colloid Interface Sci.*, **67**, 70(1978).
11.   Harwell, J.H., Roberts, B.L. and Scamehorn, J.F., *Colloids & Surfaces*, **32**, 1(1988).
12.   Xu, Q., Vasudevan, T.V. and Somasundaran, P., *J. Colloid Interface Sci.*, **142**, 528(1991).
13.   Esumi, K., Sakamoto, Y. and Meguro, K., *J. Colloid Interface Sci.*, **134**, 283(1990).
14.   Reid, V.W., Longman, G.F. and Heinerth, E., *Tenside*, **5**, 90(1968).
15.   Somasundaran, P., Fu, E. and Xu, Q., accepted for publication in *Langmuir*.
16.   Rathman, J.F. and Scamehorn, J.F., *Langmuir*, **4**, 474(1988).
17.   Fuerstenau, D.W., Healy, T.W., and Somasundaran, P., *Tran AIME*, **229**, 321(1964).
18.   Partyka, S., Lindheimer, M., Zaini, S., Keh, E. and Brun, B., *Langmuir*, **2**, 101(1986).
19.   Fu, E., "Adsorption of Anionic-Nonionic Surfactant Mixtures on Oxide Minerals," Ph.D. thesis, Columbia University, 1987.

RECEIVED April 27, 1992

Chapter 26

Irreversible Adsorption of Poly(isobutenyl)-succinimide Dispersants onto Calcium Alkylarylsulfonate Colloids

B. L. Papke

Texaco Research Center, P.O. Box 509, Beacon, NY 12508

The reversibility of poly(isobutenyl)succinimide dispersant adsorption onto alkaline calcium alkylarylsulfonate colloidal dispersions was measured using ultracentrifugation and Fourier transform infrared techniques. Adsorption techniques were used to obtain quantitative information on factors affecting the interaction ratio and strength; dispersant adsorption was shown to occur through interaction of the polar succinimide/amine headgroup with the sulfonate colloid. Poly(isobutenyl)succinimide/calcium sulfonate molar interaction ratios as high as 0.25 were observed. Adsorption of poly-(isobutenyl)succinimide dispersants onto calcium sulfonate colloids is shown to be irreversible in at least some cases. Evidence for the formation of a new mixed surfactant colloid was obtained by light scattering particle size measurements.

Alkaline earth alkylarylsulfonates, the salts of high molecular weight (ca. 450), oil-soluble sulfonic acids, are known to form inverted micellar structures in hydrocarbon media (1). These surfactants can be used to prepare non-aqueous colloidal dispersions of an inorganic base (such as a carbonate or hydroxide). The inorganic base (typically amorphous calcium carbonate) is present as small submicron particulates solubilized in hydrocarbon solvents by an adsorbed layer of the sulfonate surfactant. These alkaline colloidal dispersions form the basis for a widely used class of gasoline and diesel engine lubricant additives (2), providing an inexpensive source of oil-soluble base to aid in the neutralization of corrosive acids generated

0097–6156/92/0501–0377$06.00/0

during engine operation. The manufacturing processes
used to prepare these colloidal dispersions have been
described by Marsh (3); particle sizes and properties
have been measured by small angle neutron scattering (4),
transmission electron microscopy (5), and ultracentrifugation sedimentation rates (6). For simplicity, these
colloidal dispersions are referred to as basic (or
alkaline) calcium sulfonates. Particle sizes vary with
manufacturing conditions and the amount of inorganic base
incorporated, but generally range from 80 to 160
Angstroms in diameter, including the thickness of the
adsorbed surfactant layer. The properties of sterically
stabilized, non-aqueous colloidal dispersions, such as
the basic alkaline earth sulfonates, have been well
documented (7).

Poly(isobutenyl)succinimides, a second widely utilized
class of engine lubricant additives, are oil-soluble
surfactants prepared by reaction of low molecular weight
polyisobutylene (PIB, ca. 1000 - 2500 av. M. W.) with
maleic anhydride and a polyethylene amine. For amines
having more than one primary nitrogen, such as
triethylene tetramine, bisimide structures are usually
prepared (Figure 1). Poly(isobutenyl)succinimides
function by reducing the formation of oil-insoluble
sludge in engine operation, and are believed to function
through a steric stabilization mechanism involving
adsorption of the polar portion of the succinimide
dispersant molecule onto engine sludge particulates (8).
Succinimide dispersants may associate in hydrocarbon
solvents to form weakly bound micelles, but micelle
formation occurs much more readily in the presence of
polar materials (such as acids or water) (9). The
solubilization of acids (organic and inorganic) by
succinimide dispersants in hydrocarbon solvents has been
described (10,11).

Calcium alkylarylsulfonates and poly(isobutenyl)-
succinimide dispersants are both typically present in
commercial engine lubricants, and therefore it is of
practical interest to understand whether physical
interactions occur between these two classes of
surfactant materials. Lubricant additive interactions
may affect engine performance, either synergistically or
antagonistically (12, 13), and a better understanding of
factors affecting engine lubricant additive interactions

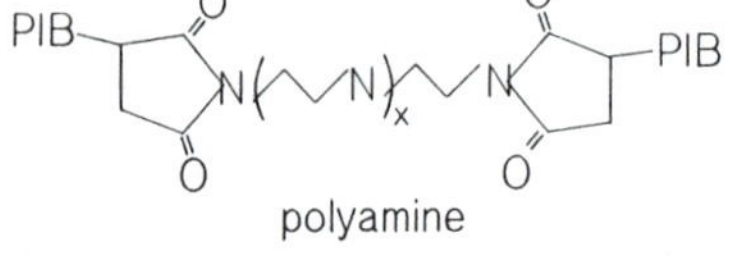

Figure 1. Idealized poly(isobutenyl)bissuccinimide
dispersant structure.

is needed. The present study describes the novel application of an ultracentrifugation/infrared technique to identify direct physical interactions between succinimide dispersants and basic calcium sulfonate colloids.

EXPERIMENTAL

Materials. Three slightly different alkaline calcium alkylarylsulfonate colloids were used in the present study. The sulfonate surfactant in all cases was a mixture of approximately 65 wt% petroleum sulfonate (a largely monoalkaryl sulfonate) and 35 wt% synthetic sulfonate (mostly dialkyl C-12 benzene sulfonates) with a combined average molecular weight of about 1000 g/mole. Calcium alkylarylsulfonate surfactants in this molecular weight range are known to form inverse micellar structures in hydrocarbon media (1). A more detailed physical characterization of the calcium alkylaryl-sulfonate surfactant is available elsewhere (14). The three alkaline calcium sulfonate colloids in this study differed only in the amount of solubilized inorganic base they contained. The first sulfonate, referred to as a "neutral" calcium sulfonate, had a 0.5:1 base:sulfonate molar ratio. The second calcium sulfonate contained a large amount of solubilized inorganic base, and had a base:sulfonate molar ratio of 20:1. This sulfonate contained approximately 38 wt% calcium carbonate, 18 wt% calcium sulfonate and 44 wt% of a non-polar hydrocarbon carrier oil. The third sulfonate was similar to the second, but had a reduced base:sulfonate molar ratio of 12.5:1. Heptane (Aldrich, 99% spectrophotometric grade) was used as a nonpolar solvent in all studies.

Succinimide dispersants used in this study were prepared by the reaction of polyisobutenyl succinic acid anhydride with a polyethylene polyamine (pentaethylene-hexamine (PEHA), Dow Chemical Co., or ethylenediamine (EDA), Aldrich, 99%), in a 2:1 molar ratio. The polyisobutenyl succinic anhydride was prepared from reaction of polyisobutylene (PIB) with maleic anhydride using either 960 or 2060 average molecular weight PIB (H-100 or H-1500; Amoco Chemical Company). The preparation and characterization of succinimide dispersants is described in the literature (8, 15). Succinimide dispersants used in the present study contained approximately 30% unreacted polyisobutylene due to incomplete conversion of the original polyisobutylene to the polyisobutenyl succinic acid anhydride. The succinimide dispersants were present in a non-polar hydrocarbon carrier oil to facilitate handling; the active dispersant concentration (excluding unreacted PIB) was about 37%. Calculated molecular weights for the dispersants in this study are as follows:
(a) H-100 PIB/PEHA/bissuccinimide - 2465 g/mol
(b) H-1500 PIB/EDA/bissuccinimide - 4350 g/mol
(c) H-1500 PIB/PEHA/bissuccinimide - 4750 g/mol

Instrumentation. Ultracentrifuge separations were
conducted using a Beckman L8-55 ultracentrifuge equipped
with a Beckman 50.2 Ti fixed angle rotor. Fourier
transform infrared spectra were obtained using a Nicolet
510 infrared instrument and 0.05 cm KCl fixed pathlength
solution cells. Light scattering studies utilized a
Nicomp 370 submicron particle analyzer supplimented with
a 15 mW HeNe Spectra-Physics laser. This instrument was
supplied with a software package capable of analyzing
Gaussian unimodal and non-Gaussian multi-modal size
distributions. Volume-weighted size measurements are
reported.

Procedures. Direct physical interactions between basic
calcium sulfonate colloidal dispersions and
poly(isobutenyl)succinimides were quantitatively
measuring using adsorption techniques (16).
Ultracentrifugation was used in the present study as a
physical separation technique; this technique was
feasible because a large difference in ultracentrifugation sedimentation rates exist between the calcium
sulfonate colloids and the succinimide dispersants.
Basic calicum sulfonates can be removed from dilute
low-density hydrocarbon solutions within 1-3 hours
through ultracentrifugation (6); succinimide dispersants
used in the present study were essentially unaffected
under similar conditions. However, interactions between
the basic sulfonate and succinimide dispersant will cause
the succinimide dispersant to be physically removed along
with the calcium sulfonate colloid during ultracentrifugation (Figure 2). All adsorption measurements
described in this paper were conducted using the 20:1
(base:sulfonate) calcium sulfonate colloid.
  In a typical adsorption experiment, a weighed amount
of the basic calcium sulfonate colloid was mixed with the
succinimide dispersant in petroleum ether to ensure a
completely homogeneous mixture. The petroleum ether was
removed at 80 C under N2, and the dispersant/detergent
additive mixture was placed in a thermostated oven at 100
C for 4.0 hrs. After cooling, the additive mixture was
diluted 1/6 (wt/wt) with distilled heptane, stirred for
at least 30 minutes to mix throughly, and ultracentrifuged for 3.0 hrs at 18,000 RPM. Translucent
polypropylene tubes were marked to show solvent volumes
prior to centrifugation; the top eighth of the
ultracentrifuged solution was discarded and a cut was
carefully taken of the next quarter of the tube volume
(using a flat-tip syringe). The heptane solvent in this
fraction was stripped under vacuum, and a solution cell
infrared spectrum obtained on the remaining oil solution.
The intensity of the succinimide carbonyl adsorption band
at 1705 cm-1 was recorded and compared with a dispersant

standard calibration curve to quantitatively determine the amount of remaining dispersant.

Light scattering particle size measurements were conducted both on the original calcium sulfonate colloid and on mixtures of the sulfonate colloid with the succinimide dispersant (heated at 100 C for 4 hours). Samples were prepared using filtered heptane (Nucleopore 0.44 micrometer) in concentrations between 3.5 and 0.5 wt%, depending on the scattered light intensity (care was taken to avoid concentrations where multiple scattering occurred). All samples were filtered immediately prior to analysis (Nucleopore 0.22 micrometer gravity filtration). All light scattering studies were conducted using the 12.5:1 (base:sulfonate) calcium sulfonate colloid.

Interaction reversibility was determined using a competitive interaction between a "neutral" calcium sulfonate, a basic calcium sulfonate colloid, and a succinimide dispersant. The "neutral" sulfonate had particle size of 50-60 A (17) and a slow sedimentation rate similar to that observed for the succinimide dispersant. The 20:1 (base:sulfonate) calcium sulfonate colloid had a particle size of ca. 130 A (volume-weighted size measured by light scattering) and a relatively rapid sedimentation rate. An order of addition study was conducted (i.e. mix "neutral sulfonate + succinimide, heat @ 100 C for 4 hrs; followed by addition of the "basic" sulfonate and reheating @ 100 C/4 hrs), the procedure was reversed, and the results were compared after ultracentirfugation to determine the reversibility of the succinimide/sulfonate interaction. For irreversible interactions, premixing the basic sulfonate with the dispersant is expected to result in a much greater removal of dispersant from solution; the mixing order would be irrelevant for completely reversible interactions.

RESULTS and DISCUSSION

Basic Calcium Sulfonate / Polyisobutenyl Succinimide Aggregation Behavior. Quantitative information on interactions between basic calcium sulfonate colloids and polyisobutenyl succinimide dispersants was obtained by considering the sulfonate colloid to be a solid adsorbent. Adsorption curves are shown for the basic calcium sulfonate colloid and two different polyisobutenyl bissuccinimide dispersants in Figure 3. The adsorption curves are all Langmuirian in shape, that is, they followed the equations:

$$n = n_m \frac{bC}{1 + bC} \qquad (1)$$

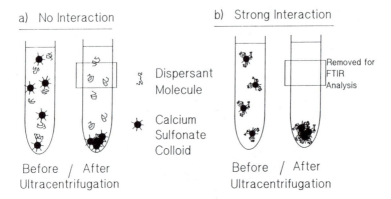

Figure 2. Ultracentrifugation/FTIR method for
screening potential additive interactions; pictorial
representation.

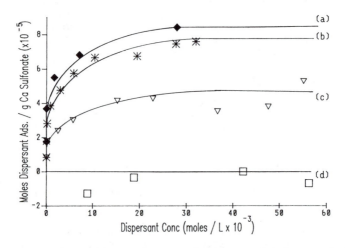

Figure 3. Adsorption of poly(isobutenyl)bissuccinimide
dispersants or polyisobutylene onto a basic calcium
sulfonate colloid from n-heptane solution: (a) H-100
PIB/PEHA/Bisimide; (b) H-1500 PIB/PEHA/Bisimide; (c)
H-1500 PIB/EDA/Bisimide; (d) H-1500 Polyisobutylene.

and

$$\frac{C}{n} = \frac{1}{n_m b} + \frac{1}{n_m} C \tag{2}$$

where C is the equilibrium dispersant concentration (in moles per liter) and n is the moles of succinimide dispersant adsorbed per gram of basic calcium sulfonate colloid (containing no carrier oil). As the dispersant concentration is increased, the amount of adsorbed succinimide dispersant will approach some limiting value, n_m. This value represents the maximum calcium sulfonate/succinimide interaction ratio. The value of the constant "b" is a measure of the adsorption strength. If true Langmuir adsorption behavior is observed, then a plot of C/n vs C should be linear, the slope is equal to $1/n_m$, and the y-intercept equal to $1/n_m b$. As shown in Figure 4, Langmuir adsorption behaviour was observed, and the various b and n values are tabulated in Table I.

Table I
Langmuir Constants for Adsorption of Succinimide
Dispersants onto the 20:1 Calcium Sulfonate Colloid

Dispersant (amine)	n_m, mol g^{-1}	b, M^{-1}	Disp./CaSf2 molar ratio
H-1500 bis(imides)			
a. EDA	4.5×10^{-5}	550	.135
b. PEHA	7.6×10^{-5}	1100	.227
H-100 bis(imides)			
a. PEHA	8.5×10^{-5}	1700	.254

The data in Figure 3 demonstrates that poly-(isobutenyl)succinimide dispersant molecules interact strongly with the calcium sulfonate colloidal particles. Evidence that this interaction occurs through the polar dispersant amine/succinimide headgroup and not the polyisobutylene polymer was obtained by substituting a sample of the nonfunctionalized polyisobutylene polymer (2060 av. M.W.) in place of the succinimide dispersant in the ultracentrifugation/FTIR procedure. The 1230 cm-1 polyisobutylene IR band (assigned to the gem-dimethyl structure on the PIB polymer) was used to follow the interaction. The polyisobutylene concentration was varied in mixtures with the basic calcium sulfonate colloid, and an adsorption "curve" was obtained. The results are compared with the adsorption curves obtained for the bissuccinimide dispersants. No adsorption of the polyisobutylene polymer was observed (curve "d", Figure 3) demonstrating that the polar dispersant amine/ succinimide headgroup is responsible for dispersant adsorption onto the calcium sulfonate colloid.

Polyisobutenyl Succinimide - Calcium Sulfonate Molar Interaction Ratios. The maximum molar succinimide/ sulfonate interaction ratios are shown in Table I. Since the basic calcium sulfonate itself is a sterically stabilized particle of calcium carbonate with an adsorbed layer of calcium sulfonate, the most plausible structure for the mixed micellar succinimide/sulfonate structure is that depicted pictorially in Figure 5. No clear relation between the molar interaction ratios and possible interaction ratios with either calcium cations or sulfonate anions is apparent, suggesting that these ratios may be determined primarily by steric packing constraints due to bulky PIB chains. No de-sorption of calcium sulfonate surfactant occurred, as determined by infrared spectroscopy.

Physical properties of the "mixed" sulfonate/ succinimide colloid are likely to be quite different from the original basic calcium sulfonate colloid. The succinimide dispersant has a surfactant "tail" that is approximately five times the length (extended conformation) of the sulfonate (for a 2060 average molecular weight polyisobutylene). In addition, the polyisobutylene chain structure is considerably more rigid than a simple hydrocarbon chain (due to steric repulsions from the gem di-methyl groups), and hence likely to adopt a more extended chain conformation. Surface properties of the mixed sulfonate-succinimide micelle are therefore likely to be dominated by the properties of the polyisobutylene polymer, and the overall size of the colloidal dispersion should increase. Light scattering studies demonstrate that succinimide dispersant adsorption results in measurable increases in colloidal partical sizes, providing additional evidence to support the pictorial representation of Figure 5. The basic calcium sulfonate colloid used in the light scattering studies had a base:sulfonate ratio of 12.5:1 and an average particle size of 107 Angstroms. When the H-100 PIB/PEHA/ bissuccinimide was added in a .16:1 molar ratio (roughly 40% of saturation), the measured particle size increased to 114 Angstroms. Adsorption of the larger H-1500 PIB/PEHA/ bissuccinimide in a 0.087:1 molar

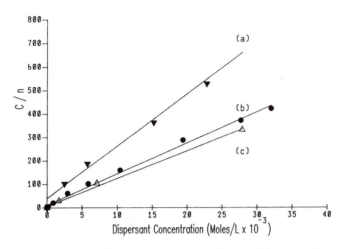

Figure 4. Adsorption data for poly(isobutenyl)-bissuccinimide dispersants onto a basic calcium sulfonate colloid from n-heptane solution plotted according to eqn. 2: (a) H-1500 PIB/EDA/Bisimide; (b) H-1500 PIB/PEHA/ Bisimide; (c) H-100 PIB/PEHA/Bisimide.

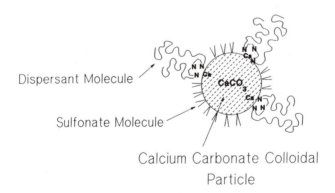

Figure 5. Pictorial representation of hypothesized interactions between a calcium sulfonate colloidal particle and adsorbed poly(isobutenyl)bissuccinimide dispersants.

ratio (also roughly 40% of saturation) resulted in an
additional increase in particle size (to 126 Angstroms).
Although these size increases are measurable and
significant, they are less than those expected for a
fully extended PIB chain conformation; 46 Å and 73 Å
maximum size increases are possible for the H-100 and the
H-1500 PIB dispersants respectively (based on an extended
chain conformation, dispersant adsorption onto the
colloid's inorganic core, and an adsorbed sulfonate
surfactant layer thickness of 20 Å). This data suggests
that at low (40%) coverage the PIB chains are in a more
collapsed conformation over the surface of the calcium
sulfonate colloidal particle. Higher levels of adsorbed
PIB succinimide dispersant should result in more extended
chain conformations and correspondingly larger size
increases.

Interaction Reversibility Studies. Although the
interaction between poly(isobutenyl)succinimides and
basic calcium sulfonate colloids follows typical
Langmuirian adsorption behavior, it can be demonstrated
that the adsorption interaction is irreversible in at
least certain cases. Adsorption reversibility was
evaluated through a three-way competitive interaction
experiment involving a "neutral" calcium sulfonate, a
basic calcium sulfonate colloid and a succinimide
dispersant, as described in the experimental section.
Since the "neutral" calcium sulfonate had a very slow
sedimentation rate, dispersants which interact
"irreversibly" with this sulfonate will not be removed
from solution during ultracentrifugation. Interaction
reversibility was measured using an order-of-addition
study; if the dispersant adsorption is reversible, then
the adsorption curves should be independent of the
addition order, provided sufficient time is allowed for
the mixture to reach equilibrium. A parallel control
experiment was also conducted, substituting a non-polar
mineral oil for the neutral calcium sulfonate.
    The results, shown in Figure 6, clearly demonstrate
that the order-of-addition is critical. Identical
adsorption curves are obtained when the "neutral"
sulfonate was added last or when a non-interacting
mineral oil was substituted for the "neutral" sulfonate
(curves (a) and (b) in Figure 6). However, when the
neutral sulfonate and dispersant were blended first, a
much weaker interaction between the dispersant and the
basic sulfonate was observed (curve (c), Figure 6). If
the dispersant adsorption were totally reversible, then
all three adsorption cureves should have been identical.
The fact that curves (a) and (b) are identical
demonstrates that succinimide dispersant adsorption onto
the basic calcium sulfonate colloid is essentially
irreversible under the conditions of this experiment.

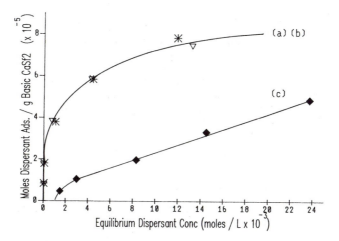

Figure 6. Competitive adsorption study: (a) "neutral"
Ca sulfonate added last; (b) no "neutral" Ca sulfonate;
(c) "neutral" Ca sulfonate added first.

Interaction between the succinimide dispersant and the
"neutral" sulfonate, in contrast, appears to show some
degree of reversibility. If the interaction were totally
irreversible, then curve (c) (Figure 6) should be a flat
line at first (until complete saturation of the "neutral"
sulfonate has occurred), followed by an adsorption curve
identical with curves (a) or (b) (but offset on the
dispersant concentration axis). The observed shape for
adsorption curve (c) suggests that a slow "release" of
dispersant from the neutral sulfonate may occur.

Irreversible adsorption that follows Langmuirian
behavior, as is observed for the interaction between the
polyisobutenyl succinimides and basic calcium sulfonate
colloids, presents an interesting paradox (18). If the
adsorption is irreversible (that is desorption cannot
occur), then why doesn't complete adsorption of the
succinimide dispersant continue until the maximum
saturation of the calcium sulfonate colloid is achieved?
The answer that has been proposed to this paradox is that
two adsorption interactions must occur - a labile
(reversible) initial interaction (resulting in the
observed Langmuirian behavior) and an irreversible
adsorption which occurs over a longer time period and
involves some type of "aging" transformation (18). In
the present study, one plausible model is that the
initial (reversible) interaction involves physi-
adsorption of the succinimide dispersant onto the
sulfonate colloid, followed by a slower irreversible
interaction (perhaps an amine/cation complexation
interaction). Additional studies are required to fully
explore the implications of the observed irreversibility
between the poly(isobutenyl)succinimide dispersants and
calcium sulfonate colloids.

CONCLUSIONS

Physical interactions between basic calcium
alkylarylsulfonate colloids and poly(isobutenyl)-
succinimide dispersants have been identified using an
ultracentrifugation/FTIR technique. Succinimide
dispersants are shown to adsorb onto calcium sulfonate
colloidal particles; the polar succinimide dispersant
headgroup is shown to be responsible for the observed
adsorption. Succinimide adsorption onto sulfonate
colloids is irreversible under certain conditions;
complexation between the polyamine and/or the succinimide
functionalities with the calcium cation may be the
driving force behind this interaction. Interactions
between succinimide dispersants and calcium sulfonate
colloids result in the formation of new mixed surfactant
colloids, as evidenced by measurable increases in
colloidal particle sizes.

ACKNOWLEDGEMENTS

The author thanks Texaco Inc. for permission to publish this work, and wishes to acknowledge the assistance of Dr. C. A. Migdal and Mr. J. F. Lucas in preparing the dispersant samples, and Ms. A. T. Arruza for conducting the adsorption and light scattering studies.

REFERENCES

(1) Singleterry, C. R. J. Am. Oil Chemists Soc. 1955, 32, 446.
(2) Stewart, W. T.; Stuart, F. A. In "Advances in Petroleum Chemistry and Refining", Vol. 7, Kob, K. A.; McKetta, J. J. Jr., Eds; Interscience: New York, 1963; pp 3-64.
(3) Marsh, J. F. Chem. and Indus. 1987, 20, 470.
(4) Markovic, I.; Ottewill, R. H.; Cebula, D. J.; Field, I.; Marsh, J. F. Colloid and Poly. Sci. 1984, 262, 648.
(5) Reading, K.; Dilks, A.; Graham, S. C. In "Petroanalysis '87"; Crump, G. B., Ed.; John Wiley and Sons Inc.: New York, 1988; pp 239-251.
(6) Tricaud, C.; Hipeaux, J. C.; Lemerle, J. Lub. Sci. 1989, 1, 207.
(7) Napper, D. H. In "Polymeric Stabilization of Colloidal Dispersions"; Academic Press: New York, 1983.
(8) Forbes, E. S.; Neustadter, E. L. Tribology 1972, 5, 72.
(9) Inoue, K.; Watanabe, H. ASLE Trans. 1983, 26, 189.
(10) Fontana, B. J. Macromol. 1968, 1, 139.
(11) Bradley, R. V.; Jaycock, M. J. ACS Div. Petrol. Chem. Preprints 1972, 17, G101.
(12) Spikes, H. A. Lub. Sci. 1990, 2, 3.
(13) Rounds, F. Lub. Sci. 1989, 1, 333.
(14) Jao, T.-C.; Joyce, W. S. Langmuir 1990, 6, 944.
(15) Nalesnik, T.E., U.S. Patent 4,636,322 (Jan. 13, 1987).
(16) Adamson, A. W. "The Physical Chemistry of Surfaces", 5th ed.; John Wiley and Sons, Inc.: New York, 1990, Chapter XI.
(17) Jao, T. C.; Kreuz, K. L. In "Phenomenon in Mixed Surfactant Systems", Scamehorn, J. F., Ed.; ACS Symposium Series 311; American Chemical Society: Washington, D.C., 1986; p 90.
(18) Zawadzki, M. E.; Harel, Y.; Adamson, A. W. Langmuir 1987, 3, 363.

RECEIVED January 6, 1992

PHASE BOUNDARIES AND SOLUBILIZATION IN MIXED SURFACTANT SYSTEMS

Chapter 27

Precipitation of Mixtures of Anionic Surfactants

John F. Scamehorn

Institute for Applied Surfactant Research, University of Oklahoma, Norman, OK 73019-0628

The tendency of an anionic surfactant to precipitate can be quantified in two different ways: (1) as a Krafft temperature or temperature below which surfactant precipitate appears, or (2) as a hardness or salinity tolerance - the minimum concentration of cation precipitating with the surfactant anion which is required to form any precipitate. Surfactants used in practical applications can contain many surfactant components, but these components are often similar enough so that they form nearly ideal mixed micelles. Yet precipitate initially formed from these mixtures contains only the one component most easily precipitated. As a result, the tendency of a multicomponent surfactant mixture to precipitate can be substantially less than that of a pure component typical of the mixture present at the same concentration.

Precipitation of surfactants from aqueous solution is one of their most important properties. In detergency, precipitation is undesirable and considerable effort has been expended to develop formulations with high hardness tolerance to permit washing in hard water (1). On the other hand, the tendency of surfactants to precipitate can be useful in such applications as recovery of surfactant from surfactant based separations (2), and surfactant-enhanced waterflooding of oil reservoirs (3).

Most systematic studies of surfactant precipitation have involved single surfactant components or well-defined binary surfactant mixtures. Yet, essentially all real applications of surfactants involve the use of extremely complex mixtures of individual surfactant components. For example, linear alkylbenzene sulfonate (LAS), often called the workhorse surfactant because of its dominance in practical applications, typically contains at least five alkyl chain lengths with a significant fraction of the surfactant (4), with the benzene sulfonate group attached to all the positions along each linear chain in significant amounts (5), generating at least 30 identifiable components.

The purpose of this short paper is to discuss the extension of single component surfactant precipitation theories to multicomponent systems and demonstrate important qualitative effects of the heterogeneous nature of practical anionic surfactant systems. Effects of added nonionic surfactant (6-8) or cationic surfactant (9,10) with anionic surfactants is addressed in other papers from our group - although

0097–6156/92/0501–0392$06.00/0

for highly purified co-surfactants. A recent review discusses published literature on precipitation in mixed surfactant systems in general (*11*).

Theory

Most practical detersive systems involve surfactant concentrations above the critical micelle concentration (CMC), so micelles are present. This equilibrium for anionic surfactants between monomer, micelle, and precipitate is shown for a ternary surfactant system in Figure 1. Whether or not precipitation of a surfactant occurs at equilibrium depends on the solution concentrations of the *monomeric* anionic surfactant and the *unbound* cationic counterion (e.g., Na+ or Ca²⁺) causing the precipitation. The monomeric surfactant concentration for each component depends on the monomer-micelle equilibrium since mixed micelles form between these similarly structured surfactants. In fact, we will assume here that the mixed micelles obey ideal solution theory as is commonly observed in systems containing similar surfactants (*12*).

On the other hand, the onset of precipitation as the counterion concentration is increased or the temperature is decreased corresponds to *precipitation of the single surfactant component whose monomeric solubility product is exceeded*. Precipitate containing multiple surfactant components in mixed surfactant systems when on a phase boundary (an infintesimal amount of precipitate present) is generally not observed (*6,13,14*). The fact that mixed micelles readily form, but mixed precipitate does not form is responsible for the remarkable precipitation properties we will discuss for these anionic surfactants.

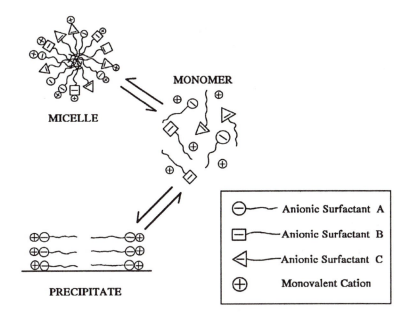

Figure 1.   **Schematic diagram of monomer-micelle-precipitate equilibrium in a ternary anionic surfactant system.**

Micelle Formation. For ideal mixed micelles, the mixed CMC and monomer concentration for each component can be predicted from (*15*):

$$y_i CMC_m = x_i CMC_i \tag{1}$$

$$\sum_i y_i = 1 \tag{2}$$

$$\sum_i x_i = 1 \tag{3}$$

$$C_i = y_i CMC_m \tag{4}$$

where y_i is the monomer mole fraction of surfactant component i, x_i is the micellar mole fraction of surfactant component i, CMC_m is the CMC of the mixture, CMC_i is the CMC of each component under the same conditions as the mixture, and C_i is the monomer concentration of surfactant component i. A surfactant-only basis is used for mole fractions y_i and x_i.

Since only the overall composition of a mixture is known, material balance equations must be used to account for the surfactant present as either micelles or as monomer (a negligible amount of surfactant is present as precipitate on a phase boundary):

$$z_i C_T = x_i C_{mic} + C_i \tag{5}$$

where z_i is the surfactant-only based mole fraction of the components in the feed (original solution), C_T is the total surfactant concentration in the feed, and C_{mic} is the total surfactant concentration present as micelles.

The CMC values of anionic surfactants are only mildly temperature dependent (*16,17*). Around room temperature, the CMC may increase or decrease slightly with increasing temperature (i.e., the heat of micellization is very small for these surfactants (*16*).) Therefore, we will assume in the modeling done in this paper that the values of CMC_i are independent of temperature.

Precipitation. Precipitation of surfactants at equilibrium is generally described quantitatively in two different ways: Krafft temperatures; or precipitation phase boundaries. A Krafft temperature is the maximum temperature at which precipitation occurs, while a phase boundary is generally isothermal and represents the minimum concentration of a counterion required to cause precipitation to occur. For example, a phase boundary can represent the hardness tolerance or minimum concentration of calcium required to precipitate an anionic surfactant (*7,8,11*). In either case, the Krafft temperature or phase boundary is considered attained when the first infintesimal amount of surfactant precipitate crystal appears; hence, all the surfactant is present as either micelles or monomer in the material balance.

For convenience in this paper, we will assume that sodium is the monovalent counterion and calcium is the divalent counterion in these derivations, although the theory is not restricted to this. We will also use concentration-based solubility products (K_{SP}), rather than activity based K_{SP} values because the former are more commonly reported in the literature. Inclusion of activity coefficients to convert a concentration-based value to an activity-based value would be straightforward.

Consider a counterion concentration on the phase boundary or at the Krafft temperature. For monovalent counterions:

$$K_{SPi} = [Na_u^+][C_i] \tag{6}$$

$$[Na_T^+] = [Na_u^+] + \beta_{Na}C_{mic} \tag{7}$$

where K_{SPi} is the solubility product of surfactant component i, $[Na_u^+]$ is the free (not bound to micelles) sodium concentration in solution, $[Na_T^+]$ is the total sodium concentration in solution (from surfactant and other added electrolytes), and β_{Na} is the fractional counterion binding of Na^+ on micelles (number of bound sodium ions divided by the number of charged surfactant molecules per micelle).

If a divalent counterion were causing precipitation:

$$K_{SPi} = [Ca_u^{2+}][C_i]^2 \tag{8}$$

$$[Ca_T^{2+}] = [Ca_u^{2+}] + 2\beta_{Ca}C_{mic} \tag{9}$$

where K_{SPi} is the solubility product of surfactant component i, $[Ca_u^{2+}]$ is the free calcium concentration in solution, $[Ca_T^{2+}]$ is the total calcium concentration in solution, and β_{Ca} is the fractional counterion binding of $[Ca^{2+}]$ on micelles (number of bound calcium ions divided by the number of charged surfactant molecules per micelle).

The heat of precipitation and the temperature dependence of the solubility product can be related by:

$$\frac{R\partial \ln (K_{SPi})}{\partial (1/T)} = \Delta H_{prei} \tag{10}$$

where R is the ideal gas constant, T is the absolute temperature, and ΔH_{prei} is the standard state heat or enthalpy of precipitation for component i (since a heat or enthalpy change is independent of standard state, we will dispense with the standard state designation on the heat term). Equation 10 is obtained from reference (*10*), but the negative signs in equations 11, 13, and 16 in that paper should not be there and are a typographical error. The numerical values and theoretical predictions in that paper are believed to be correct - there is just a sign error in the printed equations.

If the heat of precipitation is independent of temperature, integration of equation 10 yields:

$$\ln\left(\frac{[K_{SPi}]_{T2}}{[K_{SPi}]_{T1}}\right) = \frac{\Delta H_{prei}\left(\left[\frac{1}{T2}\right]-\left[\frac{1}{T1}\right]\right)}{R} \qquad (11)$$

where T2 is the higher temperature and T1 is the lower temperature.

Application of equations 1-11 permit calculation of Krafft temperatures or precipitation phase boundaries for mixtures of anionic surfactants being precipitated by either monovalent or divalent counterions. Several examples in the next section will illustrate application of the theory.

Results and Discussion

The effect of heterogeneity of the surfactant system will be illustrated in this paper by two examples: a Krafft temperature depression with a monovalent counterion, and a hardness tolerance enhancement in a binary system compared to the individual surfactant components involved.

Krafft Temperature.   Consider a binary mixture of two isomerically pure alkylbenzene sulfonates: a dodecylbenzene sulfonate with the benzene ring attached to the fourth carbon from the end of the alkyl chain (4-ϕ-C_{12}ABS), and a decylbenzene sulfonate with the benzene ring attached to the third carbon from the end of the alkyl chain (3-ϕ-C_{10}ABS). Sodium is the counterion from each surfactant and the mixture is present at a total surfactant concentration (C_T) of 0.01 M. There is 0.171 M added NaCl, so we will make the swamping electrolyte approximation and assume that

$[Na_u^+]$ is equal to 0.171 M. This implies that all of the sodium added to solution from dissolution of the surfactant is bound to micelles, which is not true, but this approximation is estimated to introduce an maximum error of 2.6 % in the value of

$[Na_u^+]$ and greatly simplifies the calculations.

Under these conditions, the CMC_i is 9 x 10^{-5} M and the Krafft temperature is 16°C for 4-ϕ-C_{12}ABS, and the CMC_i is 7.7 x 10^{-4} M and   the   Krafft temperature is 23°C for 3-ϕ-C_{10}ABS (18). In this system, the sodium salt of the alkylbenzene sulfonate is the precipitate (i.e., on the phase boundary, equation 6 must be satisfied). At the Krafft temperature, the monomer, micelles, and precipitate are all in simultaneous equilibrium (19), so for each pure component as a single surfactant system, the monomer concentration ($[C_i]$) is equal to the CMC_i. Substituting these values for $[C_i]$ and $[Na_u^+]$ into equation 6 yields $K_{SPi} = 1.54 \times 10^{-5}$ M^2 for 4-ϕ-C_{12}ABS and $K_{SPi} = 1.32 \times 10^{-4}$ M^2 for 3-ϕ-C_{10}ABS, these K_{SPi} values applying at the Krafft temperature of each component.   The only value not available to solve equations 1-6 and 11 simultaneously to predict the Krafft temperature of the mixed surfactant system at a specified feed composition is the heat of precipitation for each surfactant (ΔH_{prei}).   Therefore, we have assumed that ΔH_{prei} is the same for each component and used this as an adjustable parameter in the model to result in a one-parameter model to fit the Krafft temperature data shown in Figure 2 (*18*). The resulting best-fit value is ΔH_{prei} = -5.71 kcal/mole.

As seen in Figure 2, the theoretical values describe the data very well. A Krafft temperature satisfying equations 1-6 and 11 is obtained for each component; the component with the higher computed Krafft temperature will precipitate first and dictates the Krafft temperature of the mixture. This is the reason for the eutectic type behavior seen in Figure 2. The branch of the curve in the 4-ϕ-C_{12}ABS-rich portion of the diagram (right-hand side) corresponds to precipitation of that component and that in the 3-ϕ-C_{10}ABS-rich portion of the diagram corresponds to precipitation of this lower molecular weight surfactant component.

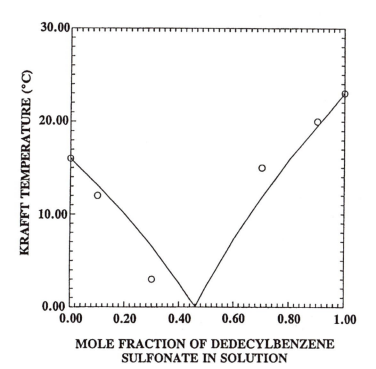

MOLE FRACTION OF DEDECYLBENZENE
SULFONATE IN SOLUTION

Figure 2.   Krafft temperature for a binary mixture of 4-ϕ-C_{12}ABS and 3-ϕ-C_{10}ABS at a total surfactant concentration of 0.01 M with 0.171 added NaCl. Data from reference 18.

For 4-ϕ-C_{12}ABS at 25°C, Noik et. al. (*20*) report K_{SPi} = 2.1 × 10⁻⁵ M². Using this value and the aforementioned K_{SPi} = 1.54 × 10⁻⁵ M² at 16°C in equation 11 yields a value of ΔH_{prei} = -5.90 kcal/mole, supporting the reasonableness of the value obtained by fitting the experimental data for the binary mixture. It might also be noted that for precipitation of Ca(2-ϕ-C_{12}ABS)$_2$ at 25°C, ΔH_{prei} = -12.0 kcal/mole

(21). One might expect that the heat of precipitation of this dialkylbenzene sulfonate might be approximately double that for the monoalkylbenzene sulfonate obtained in this calculation, as observed, providing more support for the validity of the heat of precipitation value calculated here.

It is remarkable that for only a binary mixture, such large Krafft point depressions are predicted and observed. From Figure 2, for an equimolar mixture of the two surfactants, the Krafft temperature is 3°C, compared to 16°C and 23°C for the single surfactant components.

As the number of similar surfactant components in the mixed systems increase, the Krafft point depression will become even greater. The eutectic-type of behavior seen in Figure 2 has been observed in previous experimental and theoretical work on binary mixtures of anionic surfactants *(13,14,22)*.

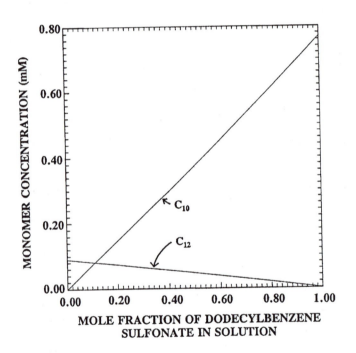

Figure 3. Monomer concentrations for $4\text{-}\phi\text{-}C_{12}ABS$ (C_{12}) and $3\text{-}\phi\text{-}C_{10}ABS$ (C_{10}) at a total surfactant concentration of 0.01 M with 0.171 added NaCl - system corresponds to that in Figure 2.

For this system, the monomer surfactant concentrations for both components are shown in Figure 3 as a function of the solution composition. The monomeric concentration of each component is less than that of that component in a single surfactant system due to formation of mixed micelles. The participation of both components in micelle formation and the participation of only one component in precipitate formation is responsible for this synergistic behavior, even though ideal micelles are formed.

Hardness Tolerance. Consider a binary mixture of sodium dodecyl sulfate (SDS) and 4-ϕ-C_{12}ABS at 25°C at a total surfactant concentration of 0.02 M with no added monovalent salt. The hardness tolerance is the minimum concentration of added calcium required to cause precipitation from this system (i.e., when equation 8 is satisfied for one of the components).

Under these conditions, the CMC_i is 1.62×10^{-3} M (18) and the K_{SPi} is 9×10^{-12} M^3 (20) for 4-ϕ-C_{12}ABS. Also, for SDS, the CMC_i is 6.65×10^{-3} M (7) and the K_{SPi} is 3.25×10^{-10} M^3. The K_{SPi} of SDS at 25°C was obtained from the K_{SPi} at 30°C of 5.02×10^{-10} (7) and a calorimetrically measured ΔH_{prei} of -15.6 kcal/mole (23) using equation 11. In the absence of added monovalent salt, it has been shown that the low level of calcium required to cause precipitation of SDS results in an *unbound* calcium concentration which does not affect the CMC_i of SDS significantly (7). Since 4-ϕ-C_{12}ABS has an even lower CMC_i and K_{SPi} than SDS, this same conclusion will apply for it also. Therefore, the aforementioned values of CMC_i will be used in this calculation. It has also been shown that β_{Ca} is 0.2 for SDS under the conditions used here (7); this value will be used here.

Using these parameters, simultaneous solution of equations 1-5, 8, and 9 permits prediction of the hardness tolerance of each of the two surfactant components. The hardness tolerance (total calcium concentration) of the mixture is dictated by the component with the lowest hardness tolerance since this component precipitates first. The results of this calculation are shown in Figure 4.

Under these conditions, the hardness tolerance varied monotonically between that of the individual surfactants. From these calculations, above a mole fraction of 4-ϕ-C_{12}ABS of 0.292, the precipitate contained that component; below this mole fraction, the precipitate contained SDS. The reason that almost no synergistic behavior is observed for hardness tolerance for these mixtures is that the vast majority of the calcium in solution is bound to micelles under these conditions. The total surfactant concentration is 3 times the CMC_i of SDS and 12.3 times that of the 4-ϕ-C_{12}ABS. The expected synergism was observed for the *unbound* calcium concentration corresponding to precipitation, as is also shown in Figure 4. At the maximum unbound calcium concentration calculated to be on a phase boundary (occurs where the mole fraction of 4-ϕ-C_{12}ABS is 0.292), the unbound calcium concentration is more than double that for the SDS-only case and more than 7 times that of the 4-ϕ-C_{12}ABS-only case. However, since the unbound calcium only contributes less than 1 % of the total   calcium in solution, this does not have significant influence on total calcium concentration in solution required to cause precipitation. Of course, if the total surfactant concentration were lower (lower concentration of micelles) or there was substantial added monovalent salt (to compete for binding sites on the micelles), the counterion binding of calcium to micelles would be less and synergisms of hetergeneity would be more substantial.

It is interesting to note that if there were no added electrolyte in the case of the system for which Krafft temperatures were measured and calculated (see Figure 2), instead of swamping electrolyte, at surfactant concentrations well above the CMC, a substantial fraction of the sodium in the system would have been bound on micelles and this would cause the synergisms in Krafft temperature depression due to heterogeneity to be reduced.

Therefore, other conditions, such as ionic strength of the solution, affect the importance of the distribution of surfactant components on the tendency of ionic surfactants to precipitate. For the arbitrary conditions selected here, the Krafft temperature was greatly influenced by heterogeneity, but the hardness tolerance was not. Future work will examine these factors systematically in more detail.

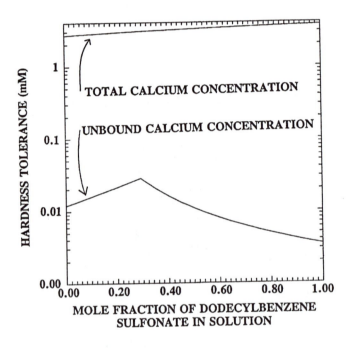

Figure 4.   Hardness tolerance (total calcium concentration required to cause precipitation) and unbound calcium concentration when precipitation occurs for a binary mixture of 4-ϕ-C$_{12}$ABS and SDS at a total surfactant concentration of 0.02 M at 25°C.

Acknowledgments

Financial support for this work was provided by the Department of Energy Office of Basic Energy Sciences Grant No. DE-FG05-84ER13678, National Science Foundation Grant No. CBT-8814147, U.S. Bureau of Mines Grant No. G1174140-4021, U.S. Environmental Protection Agency Grant No. R-817450-01-0, Oklahoma

Center for the Advancement of Science and Technology, the Oklahoma Mining and Minerals Resources Research Institute, the University of Oklahoma Energy Center, Aqualon Corp., Arco Oil and Gas Co., E.I. DuPont de Nemours & Co., Kerr-McGee Corp., Mobil Corp., Sandoz Chemical Co., Shell Development Co., Unilever Corp., and Union Carbide Corp. I would like to thank Sherril Christian, Ed Tucker, and Jeff Harwell for useful discussions, Cheryl Haskins for drawing Figure 1, and Mike Cox and Lee Matheson for supplying data on LAS component distribution.

Literature Cited

1. Coons, D.; Dankowski, M.; Diehl, M.; Jakobi, G.; Kuzel, P.; Sung, E.; Trabitzsch, U. In *Surfactants in Consumer Products, Theory, Technology, and Applications*; Falbe, J., Ed.; Springer-Verlag: Berlin, 1987; Ch. 5.
2. Brant, L. W.; Stellner, K. L.; Scamehorn, J. F. In *Surfactant-Based Separation Processes*; Scamehorn, J. F.; Harwell, J. H., Eds.; Marcel Dekker: New York, 1989; Ch. 12.
3. Arshad, S. A.; Harwell, J. H. SPE paper No. 14291, Presented at the 60th Annual Technical Conference of the Society of Petroleum Engineers, Las Vegas, September, 1985.
4. Matheson, K. L.; Cox, M. F.; Smith, D.L. *J. Am. Oil Chem. Soc.* 1985, **62**, 1391.
5. Matheson, K. L.; Cox, M. F., Vista Chemical Co., private communication, 1991.
6. Stellner, K. L.; Scamehorn, J. F. *J. Am. Oil Chem. Soc.* 1986, **63**, 566.
7. Stellner, K. L.; Scamehorn, J. F. *Langmuir* 1989, **5**, 70.
8. Stellner, K. L.; Scamehorn, J. F. *Langmuir* 1989, **5**, 77.
9. Stellner, K. L.; Amante, J. C.; Scamehorn, J. F.; Harwell, J. H. *J. Colloid Interface Sci.* 1988, **123**, 186.
10. Amante, J. C.; Scamehorn, J. F.; Harwell, J. H. *J. Colloid Interface Sci.* 1991, **144**, 243.
11. Scamehorn, J. F.; Harwell, J. H. In *Mixed Surfactant Systems*; Ogino, K.; Abe, M., Eds.; Marcel Dekker: New York, In Press.
12. Scamehorn, J. F. In *Phenomena in Mixed Surfactant Systems*; Scamehorn, J. F., Ed.; ACS Symp. Ser., Vol. 311, American Chemical Society, Washington, 1986, pp. 1.
13. Tsujii, K.; Saito, N.; Takeuchi, T. *J. Phys. Chem.* 1980, **84**, 2287.
14. Hato, M.; Shinoda, K. *J. Phys. Chem.* 1973, **77**, 378.
15. Scamehorn, J. F.; Schechter, R. S.; Wade, W. H. *J. Dispersion Sci. Technol.* 1982, **3**, 261.
16. Rosen, M. J. *Surfactants and Interfacial Phenomena*, 2nd Edition; Wiley: New York, 1989; Ch. 3.
17. Mukerjee, P.; Mysels, K. J. *Critical Micelle Concentrations of Aqueous Surfactant Systems*; National Bureau of Standards: Washington, 1971.
18. Scamehorn, J. F. PhD dissertation; University of Texas, 1980.
19. Shinoda, K. In *Colloidal Surfactants*; Shinoda, K.; Tamamushi, B.; Nakagawa, T.; Isemura, T., Eds.; Academic Press: New York, 1963; Ch. 1.
20. Noik, C.; Baviere, M.; Defives, D. *J. Colloid Interface Sci.* 1987, **36**, 115.
21. Peacock, J.M.; Matijevic, E. *J. Colloid Interface Sci.* 1977, **60**, 103.
23. Lowry, L. H.; Scamehorn, J. F.; Harwell, J. H. In Preparation.

RECEIVED March 23, 1992

Chapter 28

Effects of Structure on the Properties of Pseudononionic Complexes of Anionic and Cationic Surfactants

Ammanuel Mehreteab

Colgate-Palmolive Company, Corporate Technology Center, 909 River Road, Piscataway, NJ 08855–1343

Soluble anionic/cationic surfactant complexes can be formed from anionic and cationic surfactants either or both of which have hydrophilic group in addition to their charged heads. These complexes exhibit properties that are more similar to nonionic surfactants than to their ionic components. For example they exhibit cloud point phenomena and have low critical micelle concentration. In addition, the area of the hydrophilic group was much less than the sum of the areas of the anionic and cationic surfactant components.

Surfactants are unique as a class of compounds because they are soluble both in organic solvents and water. Their solubility in hydrocarbon solvents is due to their hydrophobic chain. Their solubility in water is due to the polarity and/or charge of their head group. The size of the hydrophilic group determines the degree of solubility of nonionic surfactants. Anionic and cationic surfactants are soluble due to their negative and positive charges respectively. When anionic and cationic surfactants are mixed the charges are neutralized and, consequently, the solubility is diminished. The resulting complex precipitates (Figure 1).

Because of insolubility there had not been many studies of mixtures of anionic and cationic surfactants. Recently, however, some anionic/cationic surfactant salts have been studied in detail. For example, the precipitation phase boundaries for sodium alkyl sufate/dodecylpyridium chloride were measured over a wide range of surfactant concentrations as a function of pH, temperature, and anionic surfactant alkyl chain length (1). The surface concentrations and molecule interactions in anionic-cationic mixed monolayers at various interfaces were studied (2). The surface activity and micellization were studied for systems of different hydrophobic chain length symmetry (3, 4).

Anionic/cationic surfactant complexes, though very surface active are rarely used as surfactants because of their low solubility. Recently, however, we have introduced soluble anionic/cationic surfactant complexes which are effective and efficient surfactant systems (5). We called these complexes pseudo-nonionic because they behaved more like nonionic surfactants than ionic surfactants. The criterion for

0097–6156/92/0501–0402$06.00/0

preparing these complexes was also given, which is, *"if either the anionic surfactant or the cationic surfactant or both have hydrophilic group in addition to their charged head groups, the resulting neutralized complex would be water soluble like nonionic surfactants if the additional hydrophilic group is large enough"* (Figure 2). The additional hydropilic group can be any charged group or noncharged polar group.

In this paper, previously reported results of surface and interfacial tensions and cloud point phenomena (5) and new investigations of additional properties of the complexes such as head group area and dynamic properties will be given.

Materials And Methods

Materials. The structure and abbreviations of the surfactants used in this study are shown in Table I. These surfactants can be classified as anionic and cationic surfactants. And within each class they can be categorized as ethoxylated and nonethoxylated.

Table I. Surfactant Structure and Abbreviation

Structure		Abbreviation
$CH_3(CH_2)_{13} SO_4 Na$		STS
$CH_3(CH_2)_m (OC_2H_4)_n SO_4 Na$		AEOS
$CH_3 (CH_2)_m \overset{+}{N} -(CH_3)_3 Br$	m=11	LTAB
	m=13	MTAB (TTAB)
	m=15	CTAB
$CH_3 (CH_2)_m \overset{+}{\underset{\underset{\bigcirc}{CH_2}}{N}} -(CH_3)_2 Cl$	m=11-15	VAR
x+1=coco $CH_3 (CH_2)_x \overset{(C_2H_4O)_m H}{\underset{(C_2H_4O)_n H}{\overset{+}{N}-CH_3}} Cl$	m+n=2	EQC(2EO)
	m+n=5	EQC(5EO)
	m+n=15	EQC(15EO)
	m+n=2	EQ18(2EO)
	m+n=5	EQ18(5EO)
x+1=18	m+n=15	EQ18(15EO)

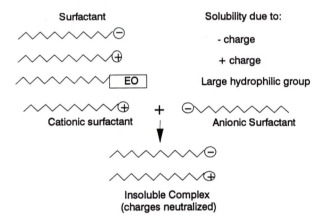

Figure 1. Surfactants and solubility.

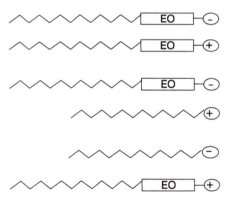

Figure 2. Pseudo-nonionic anionic/cationic complexes.

The anionic surfactants used were sodium tetradecyl sulfate (STS) from Eastman Kodak (Rochester, NY), alkylpolyethoxy(~9EO) sulfate (AEOS) i.e Alfonic 1214-65, with a carbon chain length of 12 to 14 and 65% degree of ethoxylation, from Vista Chemical Co., and an alkylphosphate ester (APE) i.e EMPHOS PS 236 from Witco Chemical Co. (Perth Amboy, NJ) which is a mixture of mono- and diester phosphate of hydroxy-terminated alkoxide condensate.

The nonethoxylated cationic surfactants used were lauryltrimethylammonium bromide (LTAB), myristoyltrimethylammonium bromide (MTAB also abbreviated TTAB for tetradecyltrimethylammonium bromide) from Sigma Chemical Co.(St. Louis, MO), cetyltrimethylammonium bromide (CTAB) from Aldrich Chemical Co. (Milwaukee, WI) and Variquat 50MC from Sherex Chemical Co. (Dublin, OH). Variquat 50MC is composed of 50% of alkyl(50% C_{14}, 40% C_{12}, 10% C_{16}) dimethylbenzylammonium chloride, 7.5% isopropyl alcohol and 42.5% H2O.

Ethoxylated cationic surfactants used were Ethoquad 18/12, Ethoquad 18/15, Ethoquad 18/20, Ethoquad 18/25, (methylbis(x-hydroxyethyl)octadecylammonium chloride where x = 2, 5, 10 and 15 respectively) and Ethoquad C/12, and Ethoquad C/25 (methylbis(x-hydroxyethyl)cocoammonium chloride where x = 2 and 15 respectively). These ethoxylated cationic surfactants were obtained from Akzo Chemie America (ARMAK Chemicals) as approximately 95% solutions.

All the above surfactants were used as supplied from the companies without further purification. It is to be understood, therefore, the quantitative results are subject to error based on the lack of purity. However, we do not believe the error substantially changes the interpretation of the results.

Method. Cloud point temperature and equilibrium and dynamic surface and interfacial tensions were measured as follows:

Cloud Point Temperature. Solutions containing both anionic and cationic surfactant were placed in vials and were heated slowly in a water bath while monitoring their temperature. The temperature at which the solutions turned cloudy were recorded as their cloud point temperatures.

Surface Tension. Several single and mixed surfactant solutions were prepared and their equilibrium and dynamic surface tensions and interfacial tensions were measured. Equilibrium surface tension was measured using Kruss Digital-Tensiometer K10T. Dynamic surface tension was measured using a modified SensaDyne Bubble Tensiometer using only the smaller orifice. The bubble rate was monitored using an oscilloscope connected to the pressure transducer of the instrument. Interfacial Tension was measured using the spinning drop tensiometer, hexadecane was the oil phase.

Results And Discussion

Formation of Soluble Complexes. When a neutral TTAB solution is added to an already acidic APE solution , the pH decreased (figure 3) indicating the replacement of the proton associated with the APE by TTAB resulting in the formation of APE/TTAB complex as shown in the equation below:

$$CH_3(CH_2)_m(OC_2H_4)_m\text{-O-}\overset{\overset{O}{\|}}{\underset{\underset{OH}{|}}{P}}\text{-OR} \quad + \quad CH_3(CH_2)_{13}\overset{+}{N}\text{-}(CH_3)_3$$

(APE)   (TTAB)

$$CH_3(CH_2)_m(OC_2H_4)_m\text{-O-}\overset{\overset{O}{\|}}{\underset{\underset{O^-}{|}}{P}}\text{-OR} \quad + \quad H^+$$
$$CH_3(CH_2)_{13}\overset{+}{N}\text{-}(CH_3)_3$$

(APE/TTAB)

If indeed a complex containing oxyethylene groups is formed, then we reasoned it should exhibit cloud point phenomena like ethoxylated nonionic surfactants. This is found to be so as shown in figure 4 (APE/TTAB at different pH's). pH affects the of dissociation of APE and, therefore, the neutraliztion point with TTAB.

The decrease in pH of an APE solution when TTAB is added to it and the exhibition of cloud point phenomena for mixtures of APE/TTAB is an indication that pseudo-nonionic complexes have been formed. Cloud point phenomena is exhibited only by the complexes and not the surfactant components.

Figures 5 and 6 show cloud point temperature vs. anionic mole fraction for two systems of anionic/cationic solutions where in one case the additional hydrophilic group is carried by the anionic surfactant and in the other case it is carried by the cationic surfactant respectively. The exhibition of cloud point phenomena of anionic/cationic surfactant solutions of compositions of around 1:1 mole ratio is another indication that a pseudo-nonionic complex is formed.

Any composition that deviates from the 1:1 mole ratio can be assumed to be a mixture of the pseudo-nonionic complex and the ionic surfactant in excess. The micelles of such mixtures are charged and result in higher cloud point temperature. This is similar to nonionic surfactants where addition of ionic surfactants raised their cloud point temperature (6).

The variables that affect cloud point values and other properties of the pseudo-nonionic complexes were studied with the results reported below.

Cloud Point Temperature. As with nonionic surfactants the cloud point temperature of the pseudo-nonionic surfactant complexes are dependent on total surfactant concentration and structure.

Effect of Surfactant Concentration. The cloud point temperature of the anionic/cationic surfactant mixtures depended on the total surfactant concentration and the relative concentrations of the anionic and cationic surfactants. Solutions with excess anionic surfactant showed one minimum in their cloud point temperature vs. total surfactant concentration (Figure 7). The cloud point temperature of the minimum increased with increase in the anionic surfactant mole fraction.

Solutions with excess cationic surfactant showed two minima (Figure 8). The cloud point temperature of the minima remained fairly constant. However, the maximum between the minima increased with increase of the cationic mole fraction.

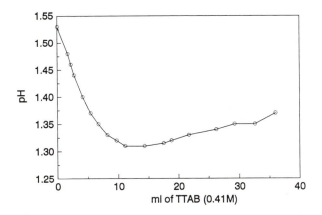

Figure 3. Titration of APE with TTAB. (Reproduced with permission from ref. 5. Copyright 1988 Academic Press, Inc.)

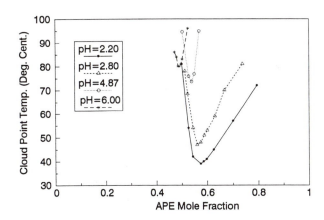

Figure 4. Effect of pH on cloud point temperature of APE/TTAB solutions. (Reproduced with permission from ref. 5. Copyright 1988 Academic Press, Inc.)

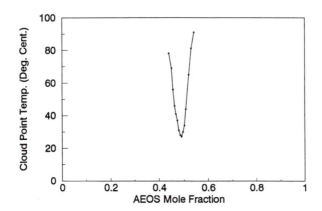

Figure 5. Cloud point temperature vs. AEOS mole fraction of AEOS/TTAB solutions. (Reproduced with permission from ref. 5. Copyright 1988 Academic Press, Inc.)

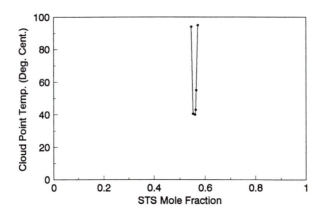

Figure 6. Cloud point temperature vs STS mole fraction of sodium tetradecyl sulfate/ethoxylated quat.

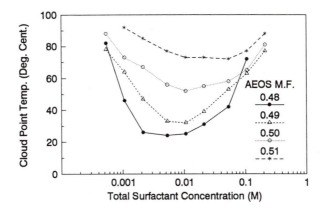

Figure 7. AEOS/TTAB + AEOS. Cloud point temperature vs. concentration.
(Reproduced with permission from ref. 5. Copyright 1988 Academic Press, Inc.)

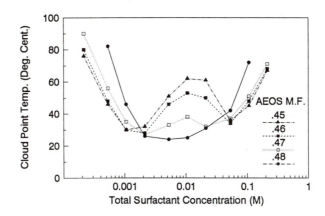

Figure 8. AEOS/TTAB + TTAB. Cloud point temperature vs. concentration.
(Reproduced with permission from ref. 5. Copyright 1988 Academic Press, Inc.)

Effect of Structure. As with nonionic surfactants, the cloud point temperature of the pseudo-nonionic complexes decreased with increase in the hydrophobicity of their surfactant components. The cloud point temperature of mixtures of AEOS and alkyltrimethylammonium bromides of different chain length is shown figure 9. The cloud point temperature decreased with approximately 10 degrees centigrade for every increase of methylene group.

The cloud point temperature of the anionic and cationic mixtures increased with increase in the hydrophilicity of the surfactant components. The cloud point temperature of mixtures of AEOS and two ethoxylated cationic surfactants with 3 and 5 ethylene groups is shown in figure 10. The cloud point temperature increased by more than 60 degrees Centigrade for an increase of two oxyethylene groups. These results are similar to those of ethoxylated nonionic surfactants whereby their cloud point decreases with increase in carbon chain length and their cloud point increases with increase in number of oxyethylene groups.

Equilibrium Surface and Interfacial Tensions. Equilibrium surface tension measurements showed that the pseudo-nonionic complexes are more efficient and effective than either of their ionic surfactant components, i.e. they have lower critical micelle concentration and lower attainable surface tension. The results are shown in figure 11.

The interfacial tension between hexadecane and solutions of AEOS/TTAB is shown in figure 12. It was observed that solutions with AEOS mole fractions of around 0.5 have interfacial tensions which are 10 to 30 fold smaller than those of their cationic (TTAB) and anionic (AEOS) surfactant components. There is no doubt that the pseudo-nonionic surfactant complexes, though soluble, are more surface active than their components.

Dynamic Surface Tension. Dynamic surface tension is found to be important to the flash foam and other properties of surfactants. Recently, a thorough theoretical investigation was made on the dynamic surface properties of mixed anionic-cationic surfactant solutions (7). The dynamic surface tension and the surface adsorption kinetics of the aqueous solutions of some anionic-cationic surfactant mixtures have been studied (8) using the oscillating jet method. In our study we used the maximum bubble pressure method to measure the dynamic surface tension of anionic and cationic surfactants and their mixtures at different bubble rates. As shown in figure 13, the dynamic surface tension of the pseudo-nonionic complex was found to be much lower than either of its surfactant components.

It is observed that dynamic surface tension depends on the bubble rate and the nature of the surfactant . The dependency of dynamic surface tension on the property of the surfactant, specifically static equilibrium surface tension and diffusion coeficient and the bubble rate is given by the equation below (9):

$$\gamma = \gamma_e + \frac{nRTT^2}{C}\left(\frac{1}{\pi Dt}\right)^{1/2}$$

where γ and γ_e are the dynamic and static surface tensions respectively, n=1 for nonionic surfactants and n=2 for ionic surfactants, R is gas constant, T is temperature,

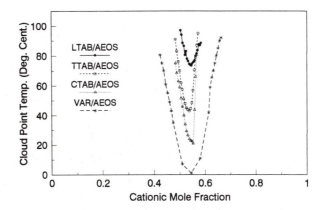

Figure 9. Effect of hydrophobic group on cloud point temperature.

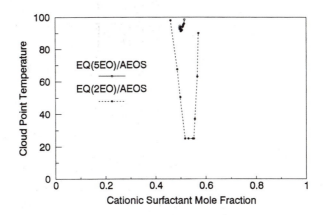

Figure 10. Effect of hyrophilic group on cloud point temperature.

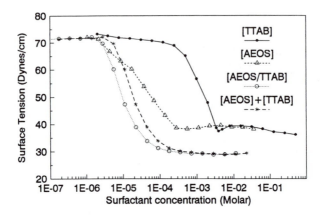

Figure 11. AEOS, TTAB and AEOS/TTAB solutions. Surface tension vs. concentration. (Reproduced with permission from ref. 5. Copyright 1988 Academic Press, Inc.)

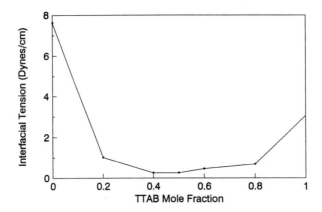

Figure 12. AEOS/TTAB. Interfacial Tension vs. TTAB mole fraction. (Reproduced with permission from ref. 5. Copyright 1988 Academic Press, Inc.)

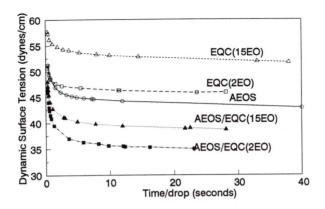

Figure 13. Dynamic surface tension of 0.02M total surfactant solutions.

Γ is adsorption, C is concentration, D is diffusion coefficient and t is time between bubbles.

A plot of γ vs $t^{-1/2}$ is linear as shown in figure 14 below.

Molecular area. The surface tensions, γ, of several cationic surfactants and their complexes with AEOS were measured as a function of their concentrations, C. Using Gibbs equation (where n=1 for nonionic and n=2 for ionic surfactants) the surface excess concentration, Γ, was calculated:

$$\Gamma = -\frac{1}{2.303nRT}\left(\frac{\partial \gamma}{\partial \log C}\right)_T$$

From the surface excess concentration, Γ, the area, a^s, of their hydrophilic group was obtained using the following equation:

$$a^s = \frac{10^{16}}{N\Gamma}$$

where N is Avogadro's number.

The pseudo-nonionic complexes were found to have a much lower area than the sum of the head areas of their anionic and cationic surfactant components. Figure 15 shows the head area for the surfactants studied. The reduction in areas per mole in mixed anionic/cationic surfactants at the air-solution interface was previously observed for nonethoxylated systems. For example, a decyltrimethylammonium/decyl sulphate complex gave an area which is about 70% of that corresponding to unmixed components (10, 11).

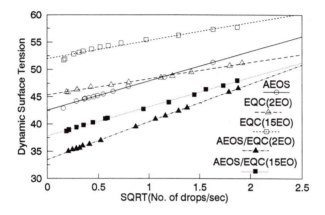

Figure 14. Dynamic surface tension of 0.02M total surfactant solutions.

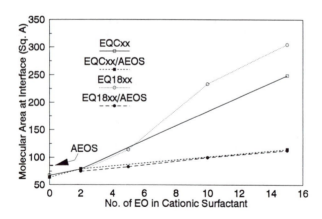

Figure 15. Effect of hydrophobe and hydrophile size on molecular area at interface.

Conclusion

Anionic-cationic surfactant complexes, often named catanionic, are usually too insoluble to be used as surfactants in aqueous solutions. Recently, however, we have introduced soluble anionic-cationic surfactant complexes, which in many ways behave like nonionic surfactants, thus named pseudo-nonionic surfactant complexes. Like nonionic surfactants they exhibit cloud point phenomena unlike their ionic surfactant components. Factors that affect the cloud point temperature of nonionic surfactants, such as structure and concentration, influence the cloud point temperature of pseudo-nonionic complexes. Pseudo-nonionic complexes are more effective and efficient than their ionic surfactant components as shown by their equilibrium and dynamic surface tensions and interfacial tensions. They pack at the interface more than their ionic components. Since, pseudo-nonionic complexes show their own characteristics, they can be treated as sepearate classes of surfactants distinct from ionic and nonionic surfactants.

Acknowledgment

I wish to thank Dr. F. J. Loprest for many helpful discussions.

Literature Cited

(1)   Amante, J. C.; Scamehorn, J. F.; Harwell, J. H. *J. Colloid Interface Sci.* **1991**, *144*, 243.
(2)   Gu, B.; Rosen M. J. *J. Colloid Interface Sci.* **1989**, *129*, 537.
(3)   Yu, Z.; Zhao, G. *J. Colloid Interface Sci.* **1989**, *130*, 414.
(4)   Yu, Z.; Zhao, G. *J. Colloid Interface Sci.* **1989**, *130*, 421.
(5)   Mehreteab, A.; Loprest, F.J. *J. Colloid Interface Sci.* **1988**, *125*, 602.
(5)   Mehreteab, A.; Loprest, F.J. *J. Colloid Interface Sci.* **1988**, *125*, 602.
(6)   Nakama, Y.; Harusawa, F.; Murotani, I. *J. Amer. Oil Chem.Soc.***1990**, *67*, 717.
(7)   Joos, P.; Hunsel, J.; Bleys, G. *J. Phys. Chem.* **1986**, *90*, 3386.
(8)   Zhang, L.; Zhao, G. *J. Colloid Interface Sci.* **1988**, *127*, 353.
(9))   Hansen, R. *J. Phys. Chem.* **1960**, *64*, 637.
(10)   Holland, P. M. In *Phenomena in Mixed Surfactant Systems;* Scamehorn J. F., Ed.; ACS Symposium 311; Amer. Chem. Soc.: Washington, DC, **1986**; p. 102.
(11)   Corkill, J. M.; Goodman, J. F.; Ogden, C. P.; Tate, J. R. *Proc. T. Soc. London, Ser. A* **1963**, *273*, 84.

RECEIVED March 9, 1992

Chapter 29

Magnetic Resonance of Phospholipids in Mixed Micelles and Membranes

Edward A. Dennis

Department of Chemistry, University of California—San Diego,
La Jolla, CA 92093

Magnetic resonance has provided an extremely powerful technique for the study of the phospholipid component of biological membranes and micelles. In many systems, the phospholipids behave as liquid crystals, such as the typical smectic mesophase of phosphatidylcholine bilayers (1,2). In such systems, order parameters derived from NMR studies on individual atoms in the phospholipid backbone have revealed important information about the packing and organization of phospholipids in membranes (3-6). In other systems, the phospholipids behave more as an isotropic liquid, such as phosphatidylcholine in mixed micelles with the nonionic surfactants. In the later cases, information about the structure and solution dynamics of the phospholipid can be obtained using high-resolution NMR techniques (7).

Although virtually all of the common nuclei have been useful in the study of phospholipids by NMR, ^{31}P-NMR studies have been particularly rewarding because of the single phosphorus atom in all common phospholipids. ^{31}P-NMR can readily differentiate between the bilayer phase of phosphatidylcholine and the hexagonal phase of phosphatidylethanolamine (8-10). In addition, ^{31}P-NMR is extremely useful in assessing the aggregation state of synthetic phospholipids containing short fatty acid chains and in determining the critical micelle concentration (CMC) of these phosphorus-containing surfactants. As such it can also be used to follow the micellization of phospholipid monomers as well as their co-micellization with other surfactants. Structural and chemical changes in phospholipids as well as reaction kinetics also can be advantageously followed. Of particular utility is the use of ^{31}P-NMR to follow the solubilization of membrane phospholipids.

^{1}H-NMR has the advantage of reflecting practically every part of the phospholipid molecule. Indeed, ^{1}H-NMR coupled with ^{2}H-NMR has provided a tremendous amount of information about the structure and ordering of phospholipids in both natural and synthetic membranes. For small sonicated phospholipid vesicles, high-resolution ^{1}H-NMR spectra have allowed the study of individual atoms as well as differences in the phospholipid on the inside and outside of the membrane (11,12). As with ^{31}P-NMR, synthetic phospholipids containing short fatty acid chains in mixed micelles with surfactants give rise to more isotropic structures in which all of

0097–6156/92/0501–0416$07.00/0
© 1992 American Chemical Society

the individual atoms can be resolved. Conformational analysis of phospholipids using ^{1}H-NMR has suggested that the phospholipid in aggregated structures is in a similar conformation to that found in the crystalline state *(13)*.

This overview of magnetic resonance studies of phospholipids in micelles and membranes will focus on phospholipids in isotropic structures utilizing high-resolution ^{1}H-NMR and ^{31}P-NMR approaches. It is in this area that NMR techniques are particularly useful in elucidating structural, conformational, and dynamic information. Although work in these areas has progressed in many laboratories around the world, we will draw on specific examples from work carried out in our laboratory over the past fifteen years. Parts of this overview are adapted with permission from a recent review on micellization and solubilization where more details on those aspects are provided *(14)*. In the last section, ^{1}H-NMR studies on a variety of phospholipids in membrane vesicles are summarized; more complete experimental details are provided elsewhere *(12)*.

Phospholipids

The most common phospholipids are derivatives of glycerol phosphorylcholine containing two long fatty acid chains. Both synthetic and natural phospholipids occur with a wide variation in chain composition and this can effect the solution properties of the phospholipid as shown in Figure 1. For normal phospholipids dispersed in water, many different structures are possible. They can form multilamellar vesicles with very large diameters or if treated in special ways they form large unilamellar vesicles (LUV) or small unilamellar vesicles (SMV) of about 250 Å diameter. In the presence of surfactants or detergents, much smaller mixed micelles are formed. Normal phospholipids do not exist as monomers in any appreciable concentration in water. However, synthetic phospholipids with short fatty acid chains are water-soluble forming monomers at low concentrations below the CMC and micelles at higher concentrations. Normal phospholipids form inverse micelles which are sometimes termed oil-water microemulsions in nonaqueous solvent combinations. Monolayers of phospholipids at the air-water-interface are also an attractive system for study.

The particular fatty acid chain composition can greatly affect the physical properties of the phospholipids in water as summarized in Table I. Thus if the fatty acids contain 2 to 4 carbons, the phospholipids exist as monomers in aqueous solution, with 6 to 8 carbons in the fatty acid chains at moderate concentrations, these phospholipids form micelles without surfactants added. With longer fatty acid chains, these phospholipids generally form bilayers. Furthermore, with long saturated fatty acids, a thermotropic phase transition occurs which can convert the phospholipid between a gel phase at lower temperature and a fluid phase at higher temperature. Even subtle changes in physical state including those indicated in the Table can effect the NMR spectrum of these phospholipids. With other polar groups than choline, the phospholipids exist in other phases as well.

^{31}P-NMR

Phospholipids give rise to relatively sharp resonances with narrow linewidths in mixed micelles with surfactants such as Triton X-100 *(16)* and cholate *(17)*. To ob-

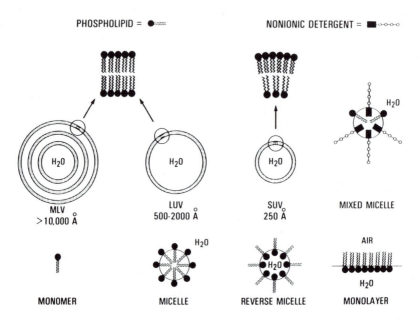

PHOSPHOLIPID = ●〰

NONIONIC DETERGENT = ■○-○-○-○

H₂0

MLV
>10,000 Å

LUV
500-2000 Å

SUV
250 Å

MIXED MICELLE

MONOMER

MICELLE

H₂0

REVERSE MICELLE

AIR

H₂0

MONOLAYER

Figure 1: Structures formed by phospholipids such as phosphatidylcholine are illustrated schematically. For vesicles, their diameters are also indicated. Multilamellar vesicles (MLV) and large unilamellar vesicles (LUV) have such large diameters that the outer surface, compared to the cross-section area of a phospholipid molecule, is relatively flat. For small unilamellar vesicles (SUV), however, the outer surface is highly curved as illustrated. In the presence of detergent, the phospholipids are solubilized into mixed micelles, which are illustrated here for the nonionic detergents. Synthetic phospholipids with short fatty acid chains can form monomers at low concentrations below the CMC and above they can form micelles in aqueous solution without the addition of detergents. In certain mixed solvent systems, natural phospholipids form reverse micelles or microemulsions with a small amount of water in the central core. Phospholipids also form monolayers at the air-water interface. Reproduced with permission from Ref. (15). Copyright 1983 Academic Press.

**TABLE I. SELECTED PHYSICAL PROPERTIES
OF DIACYL PHOSPHATIDYLCHOLINES[a]**

Acyl Chain Length	PC Name	State in Water	Thermotropic Phase Transition (°C)
2	diacetyl PC	monomer	-
3	dipropyl PC	monomer	-
4	dibutyroyl PC	monomer	-
6	dihexanoyl PC	micelle	-
7	diheptanoyl PC	micelle	-
8	dioctanoyl PC	micelle	-
12	dilauryl PC	bilayer	0
14	dimyristoyl PC	bilayer	23
16	dipalmitoyl PC	bilayer	41
18	distearoyl PC	bilayer	58
22	dibehenoyl PC	bilayer	75
18:1	dioleoyl PC	bilayer	-22
mixture	egg PC	bilayer	-11

[a]Although the most common state in water is indicated, the actual state depends on concentrations. Reproduced with permission from Ref. 1 where original references to literature values are given. Copyright 1983 Elsevier Science Publishers.

tain good resolution, it is essential that ^{1}H broad-band decoupling be employed with ^{31}P-NMR spectroscopy. ^{31}P-NMR chemical shifts for various phosphatidylcholine and phosphatidylethanolamine derivatives are summarized in Table II. For those phospholipids with very short fatty acid chains, their shift in both aqueous solution and in mixed micelles with Triton X-100 is given. Fatty acid differences and the degree of unsaturation have little effect on the chemical shift of the phosphorus atom. On the other hand, variations in the polar group and particularly substitution of the nitrogen atom is a sufficient change to differentiate the various phospholipid classes on the basis of the ^{31}P-NMR chemical shift. Lysophospholipids, which have only one fatty acid chain are themselves water soluble and form micelles at reasonable concentrations. They too can be differentiated by ^{31}P-NMR. Of particular interest is the observation that lysophospholipids can undergo migration reactions of either one of the remaining fatty acyl chains or of the phosphoryl group via an intramolecular arrangement. Our laboratory has utilized ^{31}P-NMR to follow these migration reactions as well as hydrolysis of phospholipids (18). This is reviewed in more detail elsewhere (7).

   ^{31}P-NMR chemical shifts are sensitive to minute structural changes in the phospholipid such as polar group substituents and varying pK's as well as temperature. However, of most importance for colloid chemists is the fact that the ^{31}P-NMR chemical shift is extremely sensitive to the physical state of the phospholipid. Hence changes of aggregation state can be readily followed by this technique. Thus increasing the concentration of a monomeric lipid through its CMC causes an upfield shift of the resonance line. Similarly, adding a surfactant to the monomeric phospholipid resulting in co-micellization also results in an upfield shift. For phospholipids in membranes (3,9), where very broad lines are observed, the addition of surfactant leads to the formation of mixed micelles with very sharp resonance lines characteristic of the particular phospholipid being examined. Prominent changes in relaxation times between monomers, micelles, and multibilayers or hexagonal phases have been reviewed in detail elsewhere (7). In the next section, we will concentrate on the use of ^{31}P-NMR chemical shift differences to follow the micellization and co-micellization of synthetic phospholipids. Solubilization issues will be discussed in terms of ^{1}H-NMR.

Micellization and Comicellization of Short Chain Phospholipids

^{31}P-NMR is an extremely useful tool for following the micellization of phospholipids (7,19). Dihexanoyl phosphatidylcholine is a phospholipid with six carbons in the fatty acid chains. Its ^{31}P-NMR chemical shift is shown is Figure 2. The shift of monomers and the change upon formation of micelles, which occurs at its CMC of about 11 mM is shown. The results agree with the CMC determined by other methods as well.

   In addition to determining the CMC, one can follow the co-micellization of monomeric phospholipids with surfactants. Figure 3 shows the shift for monomeric dihexanoyl phosphatidylcholine as it changes upon the addition of the surfactant Triton X-100. The chemical shift at the limit is similar to that for micelles of dihexanoyl phosphatidylcholine. From data of this sort, the partition coefficient for the phospholipid between aqueous solution and the mixed micelle can be calculated. For

TABLE II: ^{31}P-NMR SHIFTS OF PHOSPHOLIPIDS AND ANALOGUES IN
AQUEOUS SOLUTION WITH TRITON X-100[a,b]

	Triton X-100 (48 mM)	Chemical Shift (ppm)
dipalmitoyl PC	+	-0.86
1-palmitoyl lyso PC	-	-0.34
"	+	-0.38
2-palmitoyl lyso PC	-	-0.52
"	+	-0.55
dipalmitoyl β-PC	+	-1.45
palmitoyl lyso β-PC	-	-1.13
dibutyryl PC	-	-0.60
"	+	-0.61
1-butyryl lyso PC	-	-0.24
"	+	-0.25
2-butyryl lyso PC	-	-0.44
dipalmitoyl PE	+	-0.15
palmitoyl lyso PE	+	+0.26
dipalmitoyl β-PE	+	-0.71
palmitoyl lyso β-PE	+	-0.46
dipalmitoyl N-methyl PE	+	-0.30
palmitoyl lyso N-methyl PE	+	+0.10
dipalmitoyl N-methyl β-PE	+	-0.86
palmitoyl lyso N-methyl β-PE	+	-0.60
dipalmitoyl N,N-dimethyl PE	+	-0.42
palmitoyl lyso N,N-dimethyl PE	+	-0.02
dipalmitoyl N,N-dimethyl β-PE	+	-1.02
palmitoyl lyso N,N-dimethyl β-PE	+	-0.75
glycero-3-phosphorylcholine	-	-0.08
"	+	-0.09
glycero-2-phosphorylcholine	-	-0.79
glycero-3-phosphate[g]	-	+4.3
glycero-2-phosphate[g]	-	+3.9
dodecylphosphorylcholine	+	-0.39
bis(monoacylglyceryl)phosphate	+	-0.70

[a] Reproduced with permission from Ref. 7 where original references to literature values are given. Copyright 1984, Academic Press. Note that the signs of the chemical shifts have been switched from the original reports so that positive chemical shifts are now shown in the direction of decreasing field strength.

[b] $CaCl_2$ was generally present. When the compound readily dissolves in water in the absence of detergent, its chemical shift is also indicated. All compounds were analyzed at pH 8.0.

[c] Contained 10 mM EDTA.

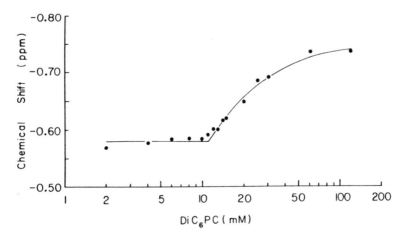

Figure 2: ^{31}P-NMR chemical shift of dihexanoyl/phosphatidylcholine (DiC$_6$PC) in D$_2$O as a function of concentration. The solution contained no further additives. The best-fit curve is calculated by assuming V$_{mono}$ = −0.58 ppm, V$_{mic}$ = −0.76 ppm, and CMC = 11 mM. Note that the original figure has been altered to now show positive chemical shifts in the direction of decreasing field strength. (Reproduced from ref. 19. Copyright 1981 American Chemical Society.)

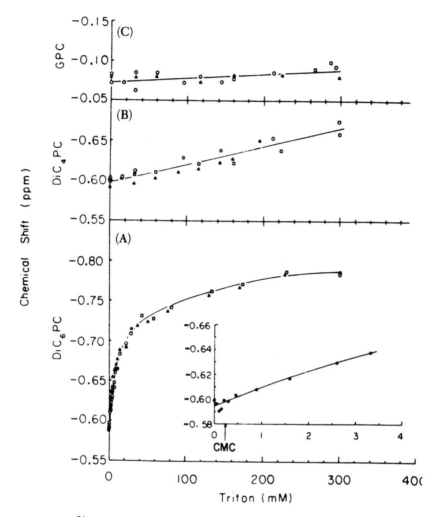

Figure 3: ^{31}P-NMR chemical shift of (A) dihexanoyl/phosphatidylcholine (DiC$_6$PC), (B) dibutyryl/phosphatidylcholine (DiC$_4$PC), and (C) glycerophosphorylcholine (GPC) as a function of the concentration of Triton X-100 employed. The initial concentration of phospholipid or GPC was 7.5 mM. The titration was carried out with 500 mM detergent except for the insert to (A) in which 7.5 mM Triton X-100 was employed. In each panel, the results of three experiments are plotted (circle, square, Δ). In (B), the titration was also carried out with mixed micelles (Δ) consisting of Triton X-100 (500 mM) and phosphatidylethanolamine (125 mM). Note that the original figure has been altered to now show positive chemical shifts in the direction of decreasing field strength. (Reproduced from ref. 19. Copyright 1981 American Chemical Society.)

similar phospholipids but with two less carbons in the fatty acid chains the titration is shown in Figure 3B. This phospholipid is monomeric to very high concentrations, probably about 80 mM because monomers of this lipid show very little change in shift with added Triton. Only the very beginning of this co-micellization curve is observed. The completely water-soluble analog, glycerol phosphorylcholine, with no fatty acid chains shows a slight shift with added Triton. This probably reflects the solvent dependence of these shifts in the presence of high concentrations of Triton. The partition coefficient for dihexanoyl phosphatidylcholine between the micellar and the aqueous phases is relatively constant over a large range of Triton. Of course, at extremely low Triton concentrations, some deviations are observed. The partition coefficients were found to be similar when determined by gel chromatography techniques for some of these Triton concentrations (19).

In summary, three methods have been employed to follow the micellization and co-micellization of synthetic short chain containing phospholipids either alone or with various surfactants: ^{31}P-NMR chemical shifts, ^{1}H-NMR spliting and the shift of the sn-2 alpha-methylene positions (as will be described in conjunction with the conformational studies in a latter section) and gel chromatography (19). All three lead to consistent results, suggesting the validity of the ^{31}P-NMR work.

^{1}H-NMR

As with ^{31}P-NMR, synthetic phospholipids containing short fatty acid chains which are completely soluble in aqueous solution give rise to high resolution ^{1}H-NMR spectra. Although significantly more complicated then ^{31}P-NMR spectra because of the numerous lines, the spectra of synthetic phosphatidylcholines (20) and phosphatidylethanolamines (21) have been assigned. ^{1}H-NMR can be used to follow micellization and co-micellization as well, but the later can be more complicated then with ^{31}P-NMR due to the resonance lines from the surfactant. In addition, the spectra of the various lysophospholipid isomers have been assigned (18). Utility of ^{1}H-NMR for following the solubilization of membrane phospholipids which themselves gives rise to very broad resonance lines was known quite early (22), but has only more recently been employed with a variety of surfactants. This use of ^{1}H-NMR will be considered in the next section.

Of greater utility has been the application of ^{1}H-NMR to conformational analysis of phospholipids. This has been possible for phospholipids in micellar and mixed micellar form. As discussed below, we have suggested that the phospholipid structure is similar to that observed in x-ray crystallography. We have more recently turned our attention to sonicated vesicles which are excellent membrane models. With the use of high-resolution NMR, it has been possible to obtain conformational information about these vesicles as well as differentiating the phospholipid on the inside and outside of the bilayer. These studies are also described below.

Solubilization of Membrane Phospholipids

The solubilization of phospholipids from lamellar structures is more complicated than the co-micellization experiments because phospholipid dispersed in water generally gives rise to very broad NMR resonances due to their packing, be they multibilayers, hexagonal phases or even sonicated vesicles. Although there are technical manipulations that can be performed, a dispersion of phosphatidylcholine in water does not give rise to a normal high-resolution NMR spectrum because of the very broad lines (22). However, if an excess of surfactant such as Triton X-100 is added to the phospholipid, a normal high-resolution spectrum is obtained in which most or all of the peaks can be assigned as shown in Figure 4. This change allows one to use NMR to follow the solubilization of phospholipids by Triton. Our studies have focused on "reporter groups" representing the hydrophilic and hydrophobic ends of both the Triton and phospholipid molecules. The appearance of the phospholipid peaks as Triton is added can be followed and changes in linewidths, intensities or relaxation times can be plotted.

The linewidth of the hydrophobic peak is not changed significantly between pure Triton micelles and mixed micelles in a 2:1 Triton to phospholipid mixed micelle. However, the hydrophobic region of the phospholipid is much narrower than in multibilayers or even sonicated vesicles. Solubilization of phospholipids can also be followed by other techniques such as gel chromatography (23) and this has given added to support the NMR conclusions.

From data obtained by NMR and gel chromatography, we (25) suggested a phase diagram for the Triton/dipalmitoyl phosphatidylcholine system. At lower temperatures, when the phospholipid is in the gel state, two phases coexist: a phospholipid bilayer phase and a mixed micelle phase. In one temperature range below the cloud point effects, one single homogenous phase of mixed micelles of Triton, phospholipid and water was found to occur. However, because of the much lower thermotropic phase transition, the egg phosphatidylcholine/Triton system is much easier to analyze (23) as shown schematically in Figure 5. If we start with a bilayer of egg phosphatidylcholine and add to it increasing amounts of Triton, the monomers of Triton will be intercalated into the phospholipid bilayer. This process will continue as more and more Triton is added until the phospholipid bilayers become saturated with Triton. For egg phosphatidylcholine, under the experimental conditions employed, this occurs at a mol ratio of Triton to phospholipid of about 1:1. The addition of more Triton generates Triton monomers and as the concentration of Triton is increased further, the free Triton concentration rises resulting in the solubilization of the phospholipid bilayers already saturated with Triton to make mixed micelles.

The minimum stoiciometry (23) for mixed micelles for Triton and egg phosphatidylcholine is about 2:1 Triton phospholipid. As more Triton is added, one is really in a two-phase region in which one just changes the proportion of phospholipid in bilayers and micelles until the point is reached in which all of the phospholipid is converted into mixed micelles. This is at an overall ratio of 2:1 Triton to phospholipid. Then, addition of Triton merely dilutes the phospholipid in the mixed micelles eventually approaching pure Triton micelles. In summary, one has initially a two-phase region consisting of bilayers of phospholipid and pure water and later consisting of two phases, the aqueous micellar phase and the bilayer phase, and finally a single phase region consisting only of mixed micelles. The structure of Triton micelles and these mixed micelles is considered in detail elsewhere (2,26,27).

This phase diagram was generated many years ago using the gel chromatography and NMR techniques described (22,23,25). Fortunately in generating such a diagram, we did not have to consider the CMCs of the phospholipid because the mono-

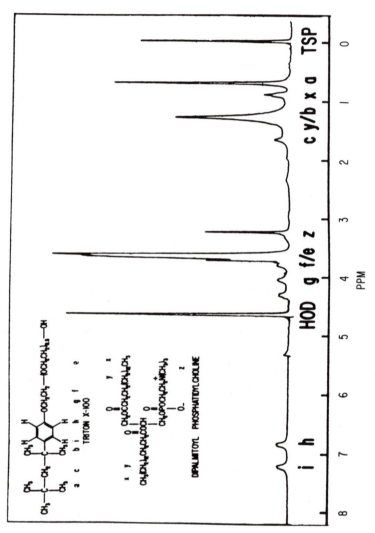

Figure 4: 220-MHz ^{1}H-NMR spectrum recorded at 37 °C of a mixture of 200 mM Triton X-100 and 100 mM dipalmitoyl phosphatidylcholine in D_2O and containing TSP. (Reproduced from ref. 24. Copyright 1976 American Chemical Society.)

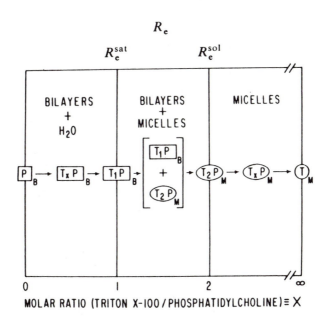

Figure 5: Schematic diagram of the average composition of the phases formed by Triton X-100 (T) and egg phosphatidylcholine (P) in the presence of an excess of water. This is shown as a function of the total molar ratio (X) of Triton/phospholipid. For simplicity, the stoichiometry of the phospholipid bilayers (B) in the presence of an excess of Triton is assumed to be 1:1 and the stoichiometry of Triton micelles (M) in the presence of an excess of phospholipid is assumed to be 2:1. The monomer concentration of phospholipid and Triton is negligible and is not indicated. Thus $X = R_e$, where R_e is the effective ratio of surfactant to phospholipid and the X or R_e corresponding to R_e^{sat} (R_e at saturation of the phospholipid bilayers with surfactant) and R_e^{sol} (R_e at complete solubilization of the phospholipid bilayers) are defined elsewhere (*1*). Adapted with permission from Ref. (*23*). Copyright 1974 Academic Press.

mer concentration of phospholipid is so low as to be negligible with natural phospholipids. The CMC of Triton is sufficiently low (0.3 mM) that at reasonable concentrations of Triton, the monomer contribution can also be ignored. Thus, these changes are expressed as a function of the total molar ratio of Triton/phospholipid. However, this is not generally the case for most common solubilizing surfactants have higher CMCs then Triton. We have more recently developed the equations for the more general case as well (1).

Conformation and Structure of Phospholipids in Micelles and Mixed Micelles

We (11,13,28) have tried to determine the conformation of phospholipids in micelles and mixed micelles. Even if the two fatty acids chains on the phospholipid are identical, the conformation details of each chain is different. Specifically, we have used ^{1}H-NMR to analyze the alpha-methylene group on each of the fatty acid chains, denoted as the sn-1 chain and the sn-2 chain as illustrated in Figure 6.

In all cases, the beta-methylene protons are decoupled to simplify the spectrum. The alpha-methylenes on the sn-1 chain appear as one large peak upfield from the sn-2 alpha-methylene which appears as an AB quartet. NMR experiments suggest that the sn-1 alpha-methylene is in a more hydrophobic environment then the sn-2 alpha-methylene and for the sn-2 chain, the two protons are different from one another and split each other giving rise to this AB quartet.

These differences can be illustrated for the synthetic phospholipid dihexanoyl phosphatidylcholine in which the fatty acid chains contain six carbons as shown in Figure 7. It is easiest to compare the beta-decoupled spectra; spectrum B is of the phospholipid in water below its CMC as monomers which gives rise to two peaks, the sn-1 and the sn-2 peak. Spectrum D is in the presence of Triton as mixed micelles and the shift between the sn-1 and sn-2 protons is dramatically increased. The AB quartet is seen for the sn-2 peak, but not for the sn-1 peak.

The NMR conclusions from these and other experiments lead to several generalizations (13): (i) The sn-1 and sn-2 alpha-methylenes have much greater shift differences in micelles above the CMC as well as in mixed micelles compared to monomers; (ii) The sn-2 alpha-methylene protons always give rise to the AB pattern and are nonequivalent in micelles; (iii) The sn-1 alpha-methylene protons do not show the AB quartet pattern in any system and are always upfield suggesting a more hydrophobic environment. Our NMR experiments can be most easily interpreted in terms of the x-ray crystal structure of a similar phospholipid (29). Here, the sn-2 chain is located near the polar portion and the two protons reflect different relative environments. On the other hand, this portion of the sn-1 chain is much more buried in the environment of the other hydrophobic groups.

In summary, (11,13) in micelles and mixed micelles, the phospholipid appears in a similar conformation to the x-ray and this is shown schematically relative to the interface in Figure 8. In other words, the sn-2 alpha-methylene is up at the interfacial part and the sn-1 alpha-methylene is much more buried. The sn-2 chain starts at the polar portion and takes a sharp bend at the alpha-methylene group. Our NMR experiments are most easily interpreted in terms of this conformation for phospholipids of various polar groups and in all kinds of surfactants from CTAB to Triton (11). NMR studies on the conformation of the surfactant portion are considered elsewhere

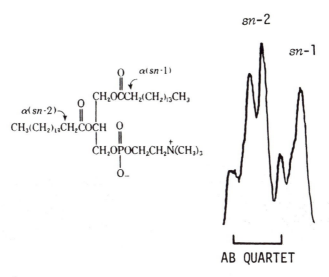

Figure 6: ^{1}H NMR spectrum in the 2.2-2.4-ppm region of Triton/phosphatidylcholine mixed micelles at a molar ratio of 4:1 Triton/phospholipid. The spectrum was taken at 360 MHz with the β-methylene protons decoupled and resolution enhancement. For more complete data, see Ref. (*11*).

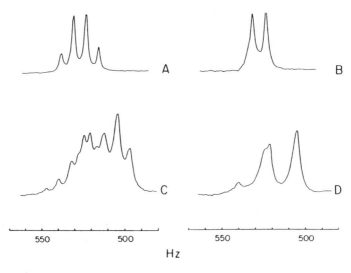

Figure 7: ^{1}H NMR spectra (220 MHz) of the α-methylene region of dihexanoyl phosphatidylcholine as monomers (A) without and (B) with the β-methylene groups decoupled and as mixed micelles at a molar ratio of Triton/phospholipid of 4:1 (C) without and (D) with the β-methylene groups decoupled. Reproduced from Ref. (*13*). Copyright 1978 American Chemical Society.

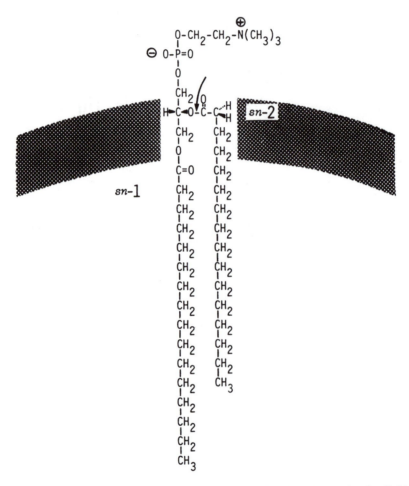

Figure 8: Schematic representation of a phospholipid molecule in the lipid-water interface emphasizing the difference in the carbonyl groups of the two fatty acyl chains. This picture is not intended to suggest a particular conformation of the phosphorylcholine portion. The *sn*-2 carbonyl is at the interface (arrow). Reproduced with permission from Ref. (*15*). Copyright 1983 Academic Press.

(24,30-32). The same arrangement of the phospholipid holds for micelles of short chain synthetic phospholipids alone above the CMC and also for sonicated vesicles of phospholipid (11,13). Seelig and coworkers (33) have extensively studied natural membranes and multibilayers using ^{2}H-NMR. Their results are interpreted in a similar manner. However, it appears that for monomeric phospholipids below the cmc, where they are not in aggregated states, the sn-2 chain is not so constrained and has free rotation (11,34).

Vesicles as Membrane Models

Small unilamellar vesicles prepared by sonication are among the most widely used membrane models. In the case of phosphatidylcholine, they have the advantage of occuring in a fairly homogenous population with an average diameter of about 250 A. They possess a much higher surface curvature then biological membranes, but they are more readily subject to study and do possess the two sides of a bilayer that are lacking in mixed micellar models of membranes. The resonance lines of the small sonicated vesicles are generally broader then those of mixed micelles; however, they are resolvable in contrast to those of multilamellar and large unilamellar vesicular preparations. Using resolution enhancement techniques, we (11) have found in general that sn-1 and sn-2 alpha-methylene protons can be resolved in sonicated vesicles. The chemical shift difference between these resonances is again about 0.1 ppm as is the case for mixed micelles. Furthermore, with decoupling of the beta-methylene protons and suitable resolution enhancement techniques, it was possible to identify the AB quartet pattern elucidated with micelles and mixed micelles as shown in Figure 9. In this way, we (11) generalize the results found with micelles and mixed micelles to sonicated vesicles and suggest that the arrangement of the fatty acid chains in the phospholipid in membrane bilayer structures is similar to that found in micelles and mixed micelles as discussed in the preceding section.

Our studies on sonicated vesicles focused on phospatidylethanolamine, phosphatidylserine, and phosphatidylglycerol rather then on phosphatidylcholine where the broadening of the resonance lines was greater. In the case of mixed micelles with surfactants, it was possible to obtain relatively narrow resonance lines with all of the above phospholipids including phosphatidylcholine. The ^{1}H-NMR spectrum of small unilamellar vesicles of phosphatidylcholine is shown in Figure 10. Here, even with resolution enhancement and beta-methylene decoupling, the pattern for the alpha-methylene protons is not as resolvable as for other phospholipid vesicles or mixed micelles.

Our current studies have focused on trying to explain the difference in the ^{1}H-NMR spectrum of phosphatidylcholine compared with other phospholipids. We (12) have now found that the reason for the more complex alpha-methylene region of phosphatidylcholine is that in those vesicles the alpha-methylene protons on the inside and the outside of the bilayer have different chemical shifts. In Figure 10, it is

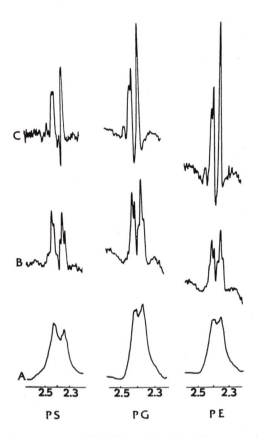

Figure 9: ^{1}H-NMR spectrum (360 MHz) of the 2.3–2.5 ppm region of sonicated vesicles of phosphatidylserine (PS), phosphatidylglycerol (PG), and phosphatidyl-ethanolamine (PE) at a concentration of 25 mM and pD 7.4, 7.4, and 9.1, respectively. Spectra A are without and spectra B are with resolution enhancement. Spectra C are with the β-methylene protons decoupled and resolution enhancement. (Reproduced from ref. 11. Copyright 1981 American Chemical Society.)

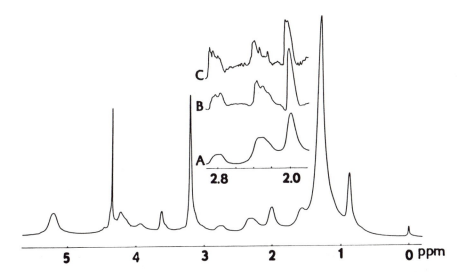

Figure 10: ^{1}H-NMR spectrum (360 MHz) of small unilamellar vesicles of phosphatidylcholine at a concentration of 25 mM and pD 6.9. Insert A is the expanded part of the spectrum between 2.0 and 2.8 ppm which shows the α-methylene protons. Insert B is the same as insert A with resolution enhancement. Insert C is the expanded part of the spectrum with the β-methylene protons decoupled and resolution enhancement. (Reproduced from ref. 11. Copyright 1981 American Chemical Society.)

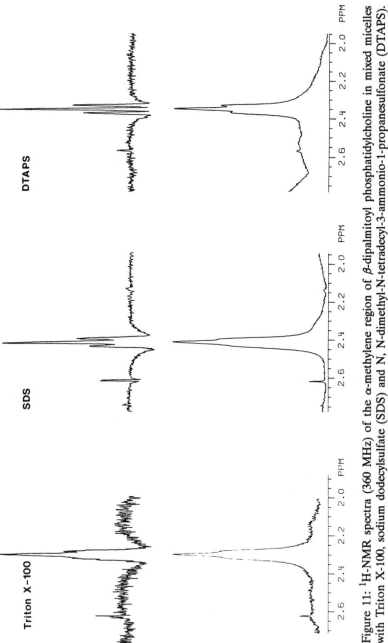

Figure 11: ^{1}H-NMR spectra (360 MHz) of the α-methylene region of β-dipalmitoyl phosphatidylcholine in mixed micelles with Triton X-100, sodium dodecylsulfate (SDS) and N, N-dimethyl-N-tetradecyl-3-ammonio-1-propanesulfonate (DTAPS). The detergent to phospholipid ratio was 8:1 in all cases. The spectra in the top row are with resolution enhancement, while the spectra in the bottom row are not. All spectra were obtained at 40 °C. Reproduced with permission from ref. 12. Copyright 1986 Elsevier Science Publishers.

apparent that there are three main resonances for the alpha-methylene region. These are due to the overlap of the *sn*-1 and *sn*-2 alpha-methylene protons in the inside of the vesicles with the *sn*-1 and *sn*-2 protons on the outside of the vesicles. This was shown by the synthesis of specifically deuterated phospholipids in which the alpha-methylene protons were replaced by deuterium atoms for either the *sn*-1 chain or the *sn*-2 chain (*12*). By comparison of these two deuterium substituted phosphatidylcholines, it was possible to unambiguously define the inside and outside resonances.

We (*12*) have now explored this phenomena more directly by studying a series of beta-phospholipids. These phospholipids are derivatives of 1,3-diglycerides and have the polar phosphoryl group on the *sn*-2 position of the phospholipid. This results in a symmetric compound as these phospholipids contain no asymmetric center. Thus there is no distinction between the alpha-methylene chain on the *sn*-1 and *sn*-3 positions and only one resonance line arises in the ^{1}H-NMR spectrum. This is shown in Figure 11 for beta- dipalmitoyl phosphatidylcholine in mixed micelles with various surfactants. Because beta-methylene decoupling was not applied to these spectra, sharp triplets are observed for the resolution enhanced alpha-methylene regions. With beta-methylene decoupling, sharp singlets would be observed.

Beta-phospholipids can be used advantageously to demonstrate the occurence of inside and outside alpha-methylene protons in sonicated vesicles as shown in Figure 12. Here we employ beta-decoupling along with resolution enhancement. This results in two sharp lines in a rough ratio of 2:1 in intensity reflecting the proportion of phospholipid molecules on the outside and inside of the vesicle bilayer, respectively. The finding of these two resonance lines for vesicles of beta-phosphatidylcholine confirms the assignment for alpha-phosphatidylcholines where the *sn*-1 and *sn*-2 alpha-methylene resonances are superimposed on the inside/outside differences giving rise to the complex spectrum shown in Figure 10.

The methylated derivatives of phosphatidylethanolamine are very important intermediates in the biosynthetic conversion of phosphatidylethanolamine to phosphatidylcholine. These intermediates have been suggested to play critical roles in membrane asymmetry and various membrane activation events. Figure 12 shows the ^{1}H-NMR spectrum of the alpha-methylene region of beta-phosphatidylethanolamine vesicles as well as its monomethylated, dimethylated, and trimethylated (phosphatidylcholine) derivatives. The vesicles formed in all four cases exhibit inside/outside differences as described above. Thus it appears to be a general phenomena that in sonicated vesicles, the alpha-methylene protons are good probes of the individual fatty acid chains as well as the sidedness of membrane vesicles. The conformational arrangement of the fatty acid chains in beta-phospholipids is of course different from that in alpha-phospholipids, but appears to be generally similar for all beta-phospholipids examined. Other studies are under way in our laboratory on the methylated derivatives of phosphatidylethanolamine as well as on the beta-phospholipids. These compounds have only recently been the subject of biophysical studies, yet they promise to provide important handles on questions of micelle and membrane structure.

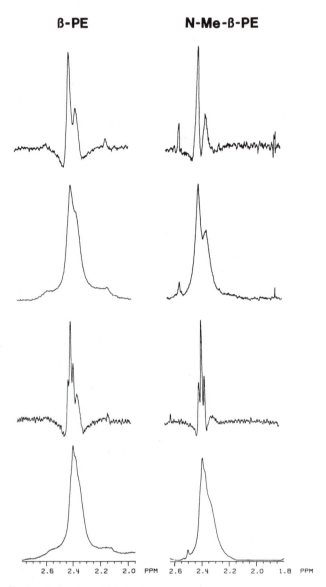

Figure 12: ^{1}H-NMR spectra (360 MHz) of the α-methylene region of sonicated vesicles of β-phosphatidylethanolamine, N-methyl-β-phosphatidylethanolamine, N,N-dimethyl-β-phosphatidylethanolamine, and β-phosphatidylcholine. Spectra with and without β-methylene decoupling as well as with and without resolution enhancement are shown, as indicated in the Figure. Reproduced with permission from Ref. (*12*). Copyright 1986 Elsevier Science Publishers.

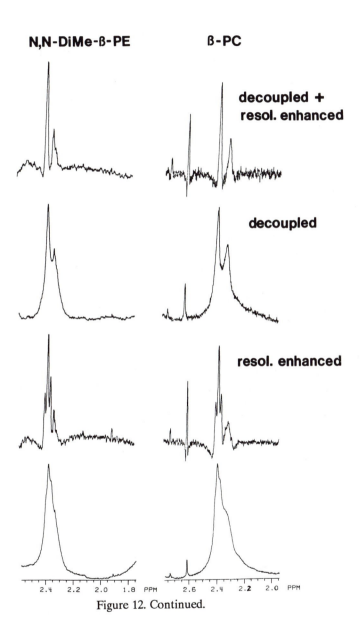

Figure 12. Continued.

Acknowledgment

This work was supported by grants from the National Science Foundation (DMB 88-17392) and the National Institutes of Health (GM-20,501). I wish to particularly thank the many graduate and postdoctoral students who contributed to this work over the years including especially Jacqueline DeBony, Tahsin Fanni, Jerry Owens, Andreas Plückthun, Anthony A. Ribeiro, and Mary F. Roberts.

Literature Cited

1. Lichtenberg, D.; Robson, R.J.; Dennis, E.A. *Biochim. Biophys. Acta Reviews on Biomembranes* **1983**, *737*, 285-304.
2. Robson, R.J.; Dennis, E.A. *Accounts of Chemical Research* **1983**, *16*, 251-258.
3. Smith, I.C.P.; Ekiel, I.H In *Phosphorus-31 NMR: Principles and Applications*; Gorentstein, D., Ed.; Academic Press: New York, NY, **1984**; pp 447-475.
4. Seelig, J. *Biochim. Biophys. Acta.* **1978**, *515*, 105-140.
5. Yeagle, P.L. *Acc. Chem. Res.* **1978**, *11*, 321-327.
6. Bocian, D.F.; Chan, S.I. *Annu. Rev. Phys. Chem.* **1978**, *29*, 307-335.
7. Dennis, E.A.; Plückthun, A. In *Phosphorus-31 NMR: Principles and Applications*; Gorenstein, D., Ed.; Academic Press, New York, NY, **1984**; pp 423-446.
8. de Kruyff, B.; Verkley, A.J.; Van Echteld, C.J.A.; Gerritsen, W.J.; Mombers, C.; Noordam, P.C.; de Gier, J. *Biochim. Biophys. Acta.* **1979**, *555*, 200-209.
9. Cullis, P.R.; de Kruyff, B. *Biochim. Biophys. Acta.* **1979**, *559*, 399-420.
10. Epand, R.M. *Biochemistry* **1985**, *24*, 7092-7095.
11. DeBony, J.; Dennis, E.A. *Biochemistry* **1981**, *20*, 5256-5260.
12. Plückthun, A.; DeBony, J.; Fanni, T.; Dennis, E.A. *Biochim. Biophys. Acta* **1986**, *856*, 144-154.
13. Roberts, M.F.; Bothner-By, A.A.; Dennis, E.A. *Biochemistry* **1978**, *17*, 935-942.
14. Dennis, E.A. *Advances in Colloid and Interface Science* **1986**, *26*, 155-175.
15. Dennis, E.A. In *The Enzymes*, 3rd ed.; Boyer, P., Ed.; Academic Press, New York, NY, **1983**; Vol. 16, pp 307-353.
16. Roberts, M.F.; Adamich, M.; Robson, R.J.; Dennis, E.A. *Biochemistry* **1979**, *15*, 3301-3308.
17. London, E.; Feigenson, G.W. *J. Lipid Res.* **1979**, *20*, 408-412.
18. Plückthun, A.; Dennis, E.A. *Biochemistry* **1982**, *21*, 1743-1750.
19. Plückthun, A.; Dennis, E.A. *J. Phys. Chem.* **1981**, *85*, 678-683.
20. Hershberg, R.D.; Reed, G.H.; Slotboom, A.J.; De Haas, G.H. *Biochim. Biophys. Acta.* **1976**, *424*, 73-81.
21. Plückthun, A.; Rohlfs, R.; Davidson, F.F.; Dennis, E.A. *Biochemistry* **1985**, *24*, 4201-4208.
22. Dennis, E.A.; Owens, J.M. *J. Supramol. Struc.* **1973**, *1*, 165-176.
23. Dennis, E.A. *Arch. Biochem. Biophys.* **1974**, *165*, 764-773.
24. Dennis, E.A.; Ribeiro, A.A. In *Magnetic Resonance in Colloid and Interface Science*; American Chemical Society Symposium Series, **1976**; Vol. 34, pp 453-466.

25. Ribeiro, A.A.; Dennis, E.A. *Biochim. Biophys. Acta.* **1974**, *332*, 26-35.
26. Robson, R.J.; Dennis, E.A. *J. Phys. Chem.* **1977**, *81*, 1075-1078.
27. Robson, R.J.; Dennis, E.A. *Biochim. Biophys. Acta.* **1978**, *508*, 513-524.
28. Roberts, M.F.; Dennis, E.A. *J. Am. Chem. Soc.* **1977**, *99*, 6142-6143.
29. Hitchcock, P.B.; Mason, R.; Thomas, K.M.; Shipley, G.G. *Proc. Natl. Acad. Sci. U.S.A.* **1974**, *72*, 3036-3040.
30. Ribeiro, A.A.; Dennis, E.A. *Biochemistry* **1975**, *14*, 3746-3755.
31. Ribeiro, A.A.; Dennis, E.A. *J. Phys. Chem.* **1976**, *80*, 1746-1753.
32. Dennis, E.A.; Ribeiro, A.A.; Roberts, M.F.; Robson, R.J. In *Solution Chemistry of Surfactants*; Mittal, K.L. and Kertes, A.S., Eds.; Plenum Press, New York, NY, **1979**; Vol. 1, pp 174-194.
33. Browning, J.L.; Seelig, J. *Biochemistry* **1980**, *19*, 1262-1270.
34. Burns, R.A.; Roberts, M.F. *Biochemistry* **1980**, *19*, 3100-3106.

RECEIVED February 26, 1992

Author Index

Affiliation Index

Subject Index

Production: Donna Lucas
Indexing: Deborah H. Steiner
Acquisition: Rhonda Bitterli, A. Maureen Rouhi, and Anne Wilson
Cover design: Alan Kahan

Printed and bound by Maple Press, York, PA

Other ACS Books

Chemical Structure Software for Personal Computers
Edited by Daniel E. Meyer, Wendy A. Warr, and Richard A. Love
ACS Professional Reference Book; 107 pp;
clothbound, ISBN 0–8412–1538–3; paperback, ISBN 0–8412–1539–1

Personal Computers for Scientists: A Byte at a Time
By Glenn I. Ouchi
276 pp; clothbound, ISBN 0–8412–1000–4; paperback, ISBN 0–8412–1001–2

Biotechnology and Materials Science: Chemistry for the Future
Edited by Mary L. Good
160 pp; clothbound, ISBN 0–8412–1472–7; paperback, ISBN 0–8412–1473–5

Polymeric Materials: Chemistry for the Future
By Joseph Alper and Gordon L. Nelson
110 pp; clothbound, ISBN 0–8412–1622–3; paperback, ISBN 0–8412–1613–4

The Language of Biotechnology: A Dictionary of Terms
By John M. Walker and Michael Cox
ACS Professional Reference Book; 256 pp;
clothbound, ISBN 0–8412–1489–1; paperback, ISBN 0–8412–1490–5

Cancer: The Outlaw Cell, Second Edition
Edited by Richard E. LaFond
274 pp; clothbound, ISBN 0–8412–1419–0; paperback, ISBN 0–8412–1420–4

Practical Statistics for the Physical Sciences
By Larry L. Havlicek
ACS Professional Reference Book; 198 pp; clothbound; ISBN 0–8412–1453–0

The Basics of Technical Communicating
By B. Edward Cain
ACS Professional Reference Book; 198 pp;
clothbound, ISBN 0–8412–1451–4; paperback, ISBN 0–8412–1452–2

The ACS Style Guide: A Manual for Authors and Editors
Edited by Janet S. Dodd
264 pp; clothbound, ISBN 0–8412–0917–0; paperback, ISBN 0–8412–0943–X

Chemistry and Crime: From Sherlock Holmes to Today's Courtroom
Edited by Samuel M. Gerber
135 pp; clothbound, ISBN 0–8412–0784–4; paperback, ISBN 0–8412–0785–2

For further information and a free catalog of ACS books, contact:
American Chemical Society
Distribution Office, Department 225
1155 16th Street, NW, Washington, DC 20036
Telephone 800–227–5558